Monte Proceedings of the Centro Stefano Franscini
Verità Ascona

Edited by K. Osterwalder, ETH Zürich

Muonic Atoms and Molecules

Edited by
L.A. Schaller
C. Petitjean

1993

Birkhäuser Verlag
Basel · Boston · Berlin

Editors' addresses:

Prof. Lukas A. Schaller
Institut de Physique
Université de Fribourg
CH-1700 Fribourg

Dr. Claude Petitjean
Paul Scherrer Institute
CH-5232 Villigen

A CIP catalogue record for this book is available from the Library of Congress Washington D.C., USA

Deutsche Bibliothek Cataloging-in-Publication Data
Muonic atoms and molecules / ed. by L.A. Schaller ; C. Petitjean. –
Basel ; Boston ; Berlin : Birkhäuser, 1993
   (Monte Verità)
   ISBN 3-7643-2851-7 (Basel ...)
   ISBN 0-8176-2851-7 (Boston)
NE: Schaller, Lukas A. [Hrsg.]

This work is subject to copyright. All rights are reserved, whether the whole or part of the material is concerned, specifically those of translation, reprinting, re-use of illustrations, broadcasting, reproduction by photocopying machine or similar means, and storage in data banks. Under § 54 of the German Copyright Law, where copies are made for other than private use a fee is payable to «Verwertungsgesellschaft Wort», Munich.

© 1993 Birkhäuser Verlag Basel, P.O. Box 133, CH-4010 Basel
Printed in Germany on acid-free paper, directly from the authors' camera-ready manuscripts
ISBN 3-7643-2851-7
ISBN 0-8176-2851-7

Contents

Preface IX

1. Nuclear Muon Capture, Fusion and Fission

Nuclear Muon Capture
J. Deutsch 3

Radiative Muon Capture on Hydrogen and the Induced
Pseudoscalar Coupling
M.D. Hasinoff 13

Muonic Atom Formation, Muon Transfer and Nuclear Fusion
in $D_2+{}^3He$ Mixture
G.G. Semenchuk 25

Recent Facets of Nuclear Fission Dynamics and Properties
of Heavy Muonic Atoms
P. David 35

2. Muonic Atoms Spectroscopy

Laser Spectroscopy of Muonic Hydrogen
E. Zavattini 53

Laser Induced $3D \rightarrow 3P$ and $4S \rightarrow 4P$ Transition in Muonic Hydrogen
P. Hauser 67

Muonium Spectroscopy
K.P. Jungmann 77

Relativistic Corrections to Particle Overlap in Atoms and Muon
Capture in Hydrogen
D. Bakalov 83

Muonic Atoms: Charge Radii and Nuclear Polarization
L.A. Schaller 89

Nuclear Polarization in Muonic Atoms
R. Rosenfelder 95

3. Muon Catalyzed Fusion and Cascade

Muon Catalyzed Fusion and Basic Reactions in Deuterium
and Hydrogen
P. Kammel 111

μ-Capture in the Mesic Molecule $pp\mu$
L.I. Ponomarev 129

Monte-Carlo Modeling of Ephitermal Effects in Muon-Catalyzed
dt Fusion
M. Jeitler 137

Direct Measurement of Sticking in Muon Catalyzed DT Fusion and
Physics of "Hot" μt Atoms
K. Lou 147

The Exotic-Atom Cascade
F.J. Hartmann 157

4. Muon Transfer

Muon Transfer Processes. Old and New Problems
S.S. Gershtein 169

Muon Transfer with a New High Pressure Gastarget
L. Schellenberg 187

On the Time Spectra of Muonic X-Rays in H_2+SO_2
F. Mulhauser 193

Muon Transfer from Excited Muonic Hydrogen to Helium Nuclei
A.V. Kravtsov 199

Muon Transfer to Elements with Z>1
H. Schneuwly 209

5. Hot Muonic Atoms

Kinetic Energies at the Formation and Cascade of μp-Atoms
F. Kottmann 219

Slowing Down of Negative Muons in Gaseous H_2 and
Determination of the Stopping Power
P. Hauser 235

Diffusion of Muonic Deuterium and Hydrogen Atoms
R.T. Siegel 243

Hot Muonic Deuterium and Tritium from Cold Targets
G.M. Marshall 251

Detection of Hot Muonic Hydrogen Atoms Emitted in Vacuum
Using X-Rays
R. Jacot-Guillarmod 261

Kinetics of Muon Catalyzed Fusion in H/D/T Mixtures
V.E. Markushin 267

6. Accelerator Plans and New Experimental Methods

Future Plans at TRIUMF
M.D. Hasinoff 277

Future Plans at PSI
H.K. Walter 289

Laser Polarized Muonic Helium
P.A. Souder 301

The Cyclotron Trap and Low Energy Muon Beams
L.M. Simons 307

Laser Spectroscopy of Muonic Hydrogen with a Phase Space
Compressed Muon Beam
D. Taqqu 313

A Device for Cooling Charged Particle Beams by Moderation
and Acceleration
M. Mühlbauer 325

Progress in Soft X-Ray Detection: The Case of Exotic Hydrogen
J.P. Egger 331

Measurement of the Stopping Power for μ^- and μ^+ at Energies between
3 keV and 50 keV
P. Wojciechowski 345

Public Lecture

Un Modo Insolito di Studiare le Proprietà Nucleari, Atomiche e Chimiche
G. Torelli 353

List of Conference Participants 369

PREFACE

From Sunday evening, April 5, until Thursday afternoon, April 9, 1992, 49 scientists from 10 countries met at the Centro Stefano Franscini on Monte Verità overlooking Ascona, in the state of Ticino in Switzerland, for an international workshop on Muonic Atoms and Molecules. More than two-thirds of the participants presented their results in talks of 20 to 40 minutes' duration. In addition, Prof. Gabriele Torelli gave, under the patronage of the Ministro del Ambiente of the state of Ticino, Dr. Mario Camani, a lecture in Italian entitled "Un modo insolito di studiare le proprietà nucleari, atomiche e chimiche".

The scientific program commenced on Monday morning with discussions centering on nuclear muon capture and nuclear fusion and fission, moving on to muonic atom spectroscopy in the afternoon. All of Tuesday was devoted to muon catalyzed fusion and muon transfer. On Wednesday morning, different aspects of hot muonic atoms were discussed, followed by informal gatherings in the afternoon and evening. On Thursday morning we took a look at the prospects for the TRIUMF and PSI meson factories, and new experimental methods. The conference was brought to a close in the afternoon with C.P. summarizing the events of the past days. The two organizers want to thank all participants for their contributions and for the lively discussions which often followed the different talks. One of the nicest outcomes of this workshop, apart from its having renewed the enthusiasm of scientists working in this exciting field of research, was the extent to which closer relationships between experimenters and theoreticians were developed.

We are indebted to the University of Fribourg, the Division F1 of PSI, the "3ème cycle de la physique en Suisse Romande" and especially to the Centro Stefano Franscini under the directorship of Prof. K. Osterwalder for financial support. We would like to emphasize the important role played by the secretary of this center, Ms. Katia Bastianelli, without whose skill and efficiency nothing would have worked as well as it did. Also, Mrs. Marie-Louise Raemy and Ms. Bernadette Picand of the Physics Institute of the University of Fribourg and Mrs. Marianne Wiedmer-Signer of the F1 division of PSI contributed greatly to the success of this conference. Last but not least, we acknowledge the tremendous "after-conference work" done by Mrs. Wiedmer in collecting, correcting and preparing camera-ready papers for print, and the cooperation of Dr. T. Hintermann of Birkhäuser Verlag in Basel in completing the present proceedings volume.

Lukas A. Schaller and Claude Petitjean

Nuclear Muon Capture, Fusion and Fission

Nuclear Muon Capture

J. DEUTSCH
Université Catholique de Louvain
Institut de Physique Nucléaire
Chemin du Cyclotron, 2, B-1348 Louvain-la-Neuve, Belgium

Abstract. After a brief reminder of the theoretical description of nuclear muon capture and of its main challenges, the information expected from precision-experiments on the $\mu^- + {}^3\text{He} \to \nu_\mu + {}^3\text{H}$ muon-capture reaction is discussed in some detail.

1. The theoretical scenary

In the following we briefly recall the description of nuclear muon-capture following it from the quark-level to the nucleon-level and from the nucleon-level to that of muon-capture in nuclei. For a detailed description cfr. e.g. ref. 1.

In the Standard Model nuclear muon-capture is attributed to the exchange of the charged electroweak gauge boson W between the (μ, ν_μ) and (u,d) pairs as illustrated in Fig. 1. On the hadron-vertex, the fundamental coupling-strength g is reduced by the V_{ud}-element of the Kobayashi-Maskawa-Cabibbo matrix. In the Standard Model the coupling is assumed to be of the purely left-handed Vector Axial-vector type ; the notations are obvious.

This description, simple on the _quark_-level, becomes more complicated on the _nucleon_-level. Due to the spatial extension and internal structure of the nucleons both the vector and axial-vector covariants of the hadron-current are modified : the coupling constants g_V and g_A change with the four-momentum transfer q^2 and additional terms (denoted as "weak magnetism" : g_M and "induced pseudo-scalar" : g_P) are introduced in the vector and axial-vector hadron currents as illustrated in Fig. 1. For the vector form-factors the Conserved Vector Current theory establishes the indicated links to electromagnetism ; as for the axial covariants similar predictions are provided by the

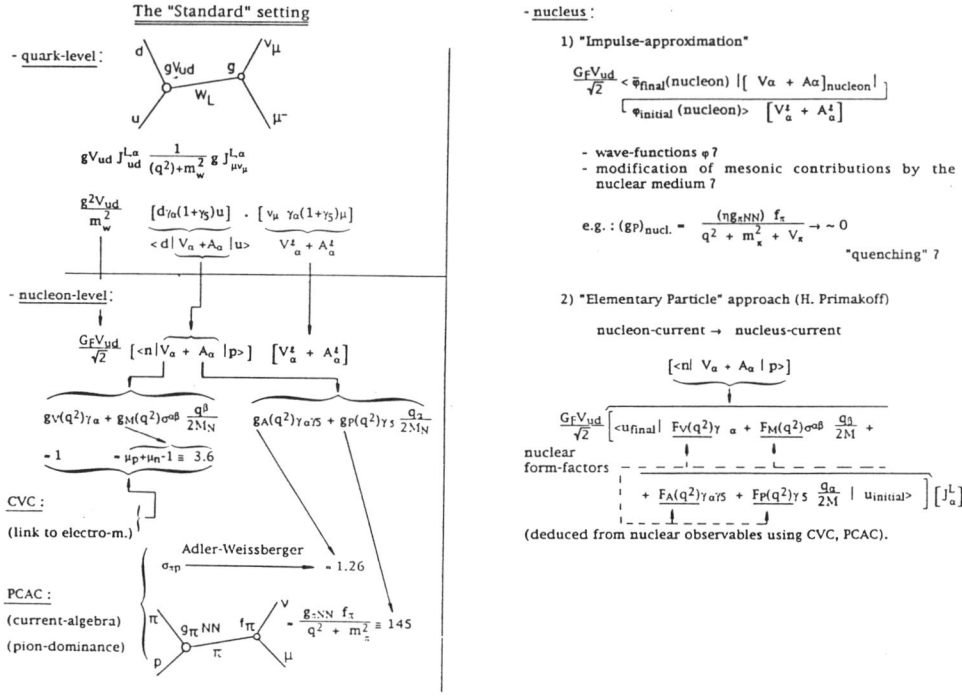

Figure 1: The description of muon-capture from the quark-level to the nucleus (cfr. text).

pion-pole dominance assumption attributing to the pion the non-conservation of the axial current.

The *nuclear* muon capture introduces additional complications which can be taken into account using two different recipes : the "impulse-approximation" and the so-called "elementary particle approach" pioneered by the late Henry Primakoff.

In the impulse-approximation approach the hadron covariant is constructed from that of the free nucleon taking into account the probability-distribution of these nucleons in the nucleus, i.e. their wave-function as indicated in Fig. 1. This approach requires the knowledge of a "reliable" wave-function ; moreover it does not take into account the mesonic exchange contributions influenced by the nuclear environment and possible modifications of the basic coupling-strength due to the nuclear medium. These hadronic effects are of-course related and their separation is not entirely warranted.

In the elementary-particle approach the nucleon form-factors are replaced by nuclear ones considered as unknown quantities to be deduced from other experiments on the same nuclei. This approach is free of the criticism one can address to the impulse-approximation but is not a genuine fundamental description as it merely relates the outcome of the investigations of the system performed with different probes.

2. Some challenges in nuclear muon-capture

The informations one may hope to extract from muon-capture are related to the electroweak (particle-physics) or hadronic (nuclear-physics) aspects of the capture process. For a discussion of these topics cfr. e.g. to refs. 2 and 3.

Amongst the particle-physics type of objectifs let us mention the exploration of the neutrino-sector searching for neutrino mass-eigenstates admixed into the muon-flavour and the exploration of the lepton-coupling testing the electron/muon universality. These two objectifs will be discussed below in conjunction with muon-capture in ^{3}He. The search for T-odd correlations in muon capture, i.e. searches for CP-violating scenarios beyond the classical one of the standard model was discussed at length in refs. 2 and 3 as well as in ref. 4.

Amongst the nuclear-physics aspects possibly the most interesting objectif is the exploration of the mesonic degrees of freedom and that of their alteration in nuclear matter. As illustrated in Fig. 1 the induced pseudoscalar form-factor probes directly the behaviour of the pion in nuclear matter and so is a sensitive probe of these effects. Actually various muon-capture experiments are planed or under execution at JINR (Dubna), TRIUMF and PSI with this aim. A discussion of the various experimental approaches is contained in refs. 1, 5 and 6. The topic will be taken up also by M. Hasinoff and P. Souder in their contribution to this workshop and by us in conjunction with muon capture in ^{3}He.

3. Muon capture in ^{3}He : precision-spectroscopy of the triton-recoil

As the $\mu + ^3\text{He} \to \nu_\mu + ^3\text{H}$ reaction has a two-body final state, the recoil energy of ^{3}H provides a measurement of the mass(es) of the emitted neutrino(s). In the standard model only one type of neutrino is emitted and it has zero mass. A precision-spectroscopy of the triton-recoil has then the potential to discover a deviation from this prediction if one of the emitted neutrinos has a sizeable mass above 20-30 MeV.

In the talk we described the ^{3}He/Xe gas-scintillation proportional chamber developed by the PSI/Louvain collaboration which achieved an energy-resolution in the %-range and a good background-rejection capacity. We do not duplicate this discussion in this written version and refer the interested reader to the description of our apparatus which appeared in Particle World [7].

4. Muon-capture in ^{3}He : the triton recoil asymmetry

The triton recoil presents a sizeable asymmetry with respect to the ^{3}He and muon-spins coupled to F = 1 [8] ; for the canonical value of g_P (cfr. Fig. 1), the asymmetry is expected to be 0.524 ± 0.006 and is a sensitive function of g_P : a 10 % variation of this form-factor induces a 3.8 % variation of the asymmetry [8].

Unfortunately only small polarizations of the F = 1 level could be obtained up to recently ; P. Souder will discuss in his talk the promises of "repolarization" offered by the use of polarized ^{3}He-targets.

5. Precision-measurement of the $\mu + ^3$He $\to \nu + ^3$H capture-rate : a test of the electron/muon universality

Our most precise test of the electron/muon universality, i.e. the equality of their coupling-strength to the W-boson is provided by the electron/muon branching-ratio in pion-decay [9]. Because of the well-known helicity-suppression of the electron-decay channel in the standard V-A interaction, the coupling-strength ratio deduced from the experiment is very sensitive to any hypothetical pseudoscalar coupling f_p as illustrated in Fig. 2.

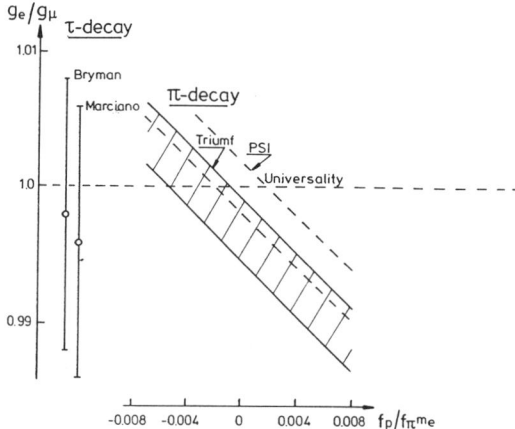

Figure 2: Ratio of the electron/muon coupling constant g_e/g_μ deduced from the experiments reported in ref. 9 as a function of a pseudoscalar coupling. Coupling-ratios deduced from τ-decay (cfr. text) are also shown.

We show also the strength-ratio deduced from τ-decay by D. Bryman [10] and W.J. Marciano [11] in good agreement with electron/muon universality ; it was stressed however by Marciano that even if the ratio of the electron/muon decay-branches

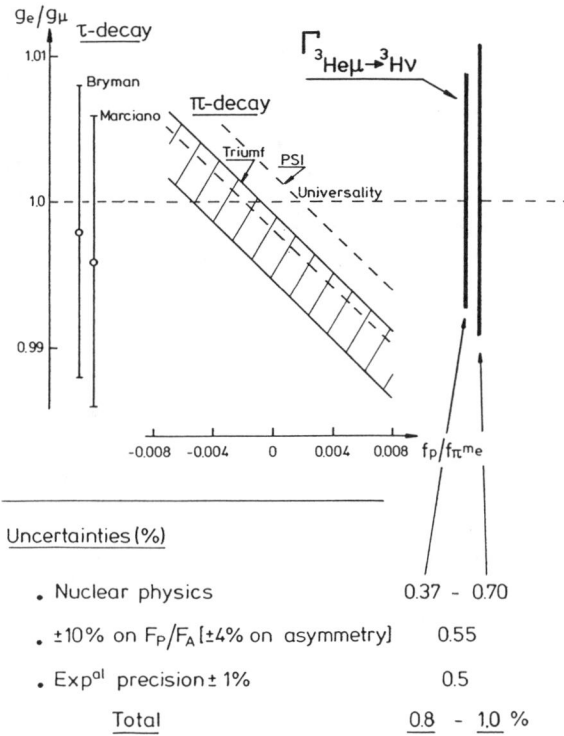

Figure 3: Muon-capture as universality-tests : comparaison of potentialities (cfr. text).

agrees with expectations, the absolute values of the partial decay-widths do not ... It seems worthwile to seek verifications of electron/muon universality from independent sources.

The potential use of muon capture to provide an independent verification of the electron/muon universality was first suggested by E. Zavattini [12]. It was noted however [13], that because of the high sensitivity of this test to g_p and the molecular problems involved in the analysis of such experiments (such as the ortho-to-para conversion in muonic hydrogen) the capture-reaction in ^{3}He may provide an interesting alternative. The interest of this approach is increased by the recent careful assessment of the potential errors introduced by nuclear-structure uncertainties of the A = 3 system [8] and ungoing independent experiments to measure g_p in this system (cfr. the preceeding section).

In Fig. 3 we compare the quality of the universality-test expected from a 1 % capture-rate measurement in this system with that of the particle-physics experiments reported in Fig. 2.

The comparaison of the electron/muon coupling-strength is based essentially on a comparaison of the ^{3}H beta-decay probability and that of the inverse muon-capture transition probability assumed to be measured to a precision of 1 %. On does _not_ assume the validity of PCAC to determine the induced pseudoscalar coupling (cfr. section and Fig. 1), but assumes conservatively that this quantity is determined by an independent experiment (cfr. section 4). The phenomenological inputs to the "elementary particle" evaluation contribute a "nuclear physics" error of 0.37 % ; this error-contribution increases to 0.7 % if one takes into account the reliability of the recipe in the extraction of the main axial form-factors q^2-dependence [8]. One notes that even with these generous error-assumptions the muon-capture approach to the electron/muon universality test is an interesting one.

The status of the capture-rate measurements is illustrated in Fig. 4. The experimental results (in increasing order of precision) are those of refs. 14, 15 and 16. As the experiment of ref. 16 was performed in pure ^{3}He (at 7 atm), a correction of 5 % was applied to take into account the eventuality of some of the captures occuring from the 2s-level[1][17].As this eventuality was not substantiated in other recent experiments [18], it may depend critically on the purity of the sample ; the reliability of the result described in ref. 16 correspondingly decreased, as suggested in Fig. 4. Fig. 4 indicates also the theoretical prediction of ref. 8 as discussed in section 1 : there is clearly need for an improvement of the experimental precision in a gas-mixture which avoids any potential capture from the 2s-level.

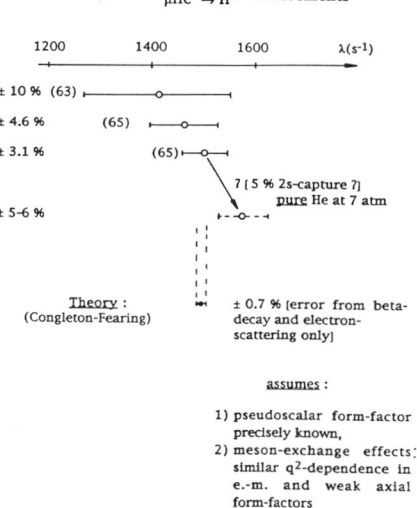

Figure 4: Status of the muon-capture rate determinations in ^{3}He : experiment and theory (cfr. text).

Such an experiment could make use of ^{3}He filled detectors as the one developed by the Louvain/PSI collaboration [7] or at LNPI[19]. In both detectors ^{3}He can be admixed to another gas[7,20] so as to certainly quench fast enough any 2s mesic-systems. Both detectors have a good localisation- and background-rejection capacity which should allow to determine with a %-precision or better the fraction of muons stopped in a given fiducial volume which gave rise to the emission of a triton. Muon-losses to the admixed gas can also be determined. The choice of a suitable detector is still under consideration.

We terminate this note discussing two effects which may influence a precision- measurement of the ^{3}He muon capture rate. The first one is a possible meta-stability of the 2s mesic level [17] which reduces the observed capture-rate. This possibility is controversial at a pressure of a couple of atmospheres even in pure He [18] ; it is expected to be completely negligible if one admixes some other gas into helium [21].

The other effect which can influence the observed capture rate is a transition from the upper (F = 1) hyperfine sublevel to the lower (F = 0) one during the lifetime of the muon. As the capture rate from the higher level is expected to be about 40 % lower than that from the lower level, an interdoublet transition probability higher than 5×10^4 s^{-1} will introduce a higher than 1 % correction to the observed capture rate. This can not be excluded purely phenomenologically on the basis of the time-evolution of the triton-appearance observed with limited precision by the authors of refs. 15 as can be seen in Fig. 5. Let us note that our results is in variance with that of ref. 22 who extracts a limit of R = 4×10^4 s^{-1} from the same data.

Fortunately there are good physics-reasons to reject any interdoublet conversion of significative probability. The system $(\mu ^3\text{He})^+$ has no electrons to absorb the M1-transition by Auger-effect. Even if $[\mu ^3\text{He}]^+e$ systems are formed, the interdoublet energy of 0.4 eV is negligible compared to the 13.6 eV binding of the electron and even in such systems no Auger-conversion can take place. Even if $[\text{He}_n(\mu ^3\text{He})^+]$ (n = 1, 2, ...) molecules are formed [23], the vibrational/rotational degrees of freedom of such a molecule cannot absorb the M1-transition [24]. There could be a potential danger to perform an Auger-conversion with one of the electrons of the ionisation-cloud produced by the stopping muon [24]. This will depend on the density of this cloud, i.e. on the drift-velocity in the ^{3}He gas. We know however that even in liquid helium (where the drift-velocity is smaller than in gas) the electron-cloud disperses fast enough so as not to allow a stopping μ^+ to pick up an electron and to produce muonium [25]. Consequently also this possibility to make an interdoublet transition can be safely excluded.

As a conclusion, it seems that a precision-determination of the ^{3}He + $\mu \to$ ^{3}H + ν

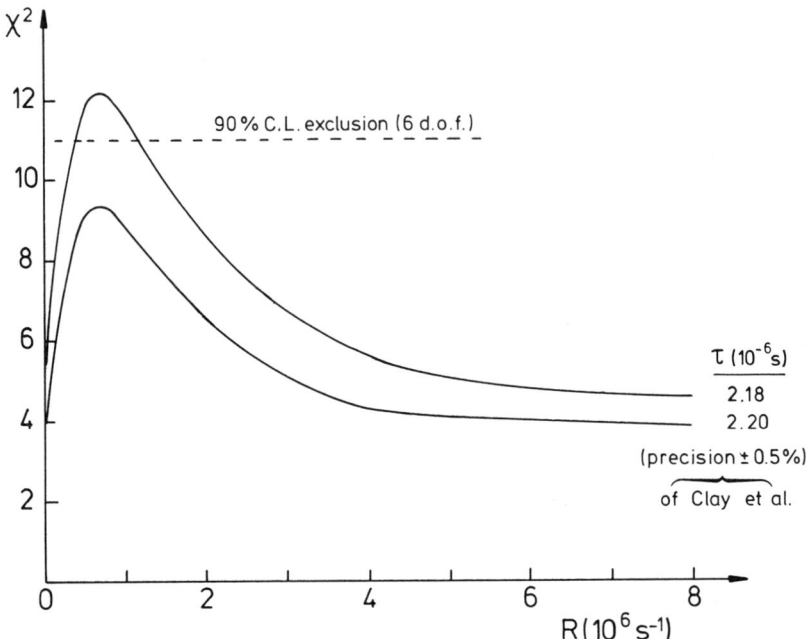

Figure 5: χ^2 exclusion-plot ; the interdoublet-conversion rate is R and τ is the observed disappearance-rate.

capture rate would be both interesting and possible

We wish to thank the organizers of the workshop, R. Prieels for help in the analysis, J. Egger and G. Semenchuk for many discussions on the potentialities of the detectors, J. Congleton and H. Fearing on the theoretical aspects and - finally - E. Zavattini and M. Leon for hints on the molecular question.

References

[1] N. Mukhopadhyay, Physics Reports **30C**, 1 (1977).

[2] Talks at LEMS-90, PSI, April 90, unpublished.

[3] Talks at the Workshop on the Future of Muon Physics, Heidelberg, May 1991, to be published.

[4] J. Deutsch : "A comment on T-violating triple correlations in muon capture", Workshop on Fundamental Muon Physics, Los Alamos Nat. Lab., Jan. 20-22, 1986.

[5] L. Grenacs, Ann. Rev. Nucl. Part. Sc. **35**, 455 (1985).

[6] J. Deutsch, in "Nuclear and Particle Physics at Intermediate Energies", Ed. J.B. Warren, Plenum Press, 1976.

[7] B. Tasiaux et al., Particle World **2**, 81 (1991).

[8] J.G. Congleton and H.W. Fearing, to be published.

[9] D.I. Britton et al., Phys. Rev. Letters **68**, 3000 (1992) ; G, Czapek et al., Univ. of Bern preprint BUHE-92-1 and Phys. Rev. Letters, to be published.

[10] D.A. Bryman, Triuf preprint TRI-PP-92-3, to be published, 1992.

[11] W.J. Marciano, Phys. Rev. **D45**, R721 (1992).

[12] E. Zavattini, CERN-EP/83-143, 1983, unpublished).

[13] J. Deutsch, inv. contrib., to the "Workshop on Fundamental Muon Physics", Los Alamos Nat. Lab., Jan. 20-22, 1986, unpublished.

[14] I.V. Falomkin et al., Phys. Letters **3**, 229 (1963).

[15] D.B. Clay et al., Phys. Rev. **B140**, 586 (1965).

[16] L.B. Auerbach, Phys. Rev. **B138**, 127 (1965).

[17] A. Placci et al., Nuovo Cim. **A1**, 445 (1971).

[18] H.P. von Arb, Phys. Letters **B136**, 232 (1984) ; M. Eckhause et al., Phys. Rev. **A33**, 1743 (1986).

[19] C. Petitjean et al., Muon Catalyzed Fusion 5/6, 1990/91, 261 ; the talk of K. Lou at this meeting.

[20] Talk of G. Semenchuk at this meeting.

[21] E. Zavattini, private communication.

[22] E. Zavattini in "Muon Physics", ed. V. Hughues.

[23] J.S. Cohen, Phys. Rev. **A25**, 1791 (1982).

[24] M. Leon, private communication.

[25] R.D. Stambaugh et al., Phys. Rev. Letters **33**, 568 (1974).

Radiative Muon Capture on Hydrogen and the Induced Pseudoscalar Coupling

RMC collaboration: M.D. HASINOFF[1], D.S. ARMSTRONG[2,+], G. AZUELOS[3,4], W. BERTL[5], M. BLECHER[2], C.Q. CHEN[1], P. DEPOMMIER[4], B. DOYLE[4], T. VON EGIDY[3,@], T.P. GORRINGE[6], P. GUMPLINGER[1], R. HENDERSON[3], G. JONKMANS[4], A.J. LARABEE[7], J.A. MACDONALD[3], S.C. MCDONALD[8], M. MUNRO[8], J.-M. POUTISSOU[3], R. POUTISSOU[3], B.C. ROBERTSON[9], D.G. SAMPLE[1], W. SCHOTT[1], G.N. TAYLOR[8], D.H. WRIGHT[3].

[1] University of British Columbia, Vancouver, B.C., Canada V6T 1Z1
[2] Virginia Polytechnic Institute & State University, Blacksburg, VA 24061 USA
[3] TRIUMF, Vancouver, B.C. Canada V6T 2A3
[4] Université de Montréal, Montréal PQ Canada H3C 3J7
[5] Paul Scherrer Institute, CH-5232 Villigen, Switzerland
[6] University of Kentucky, Lexington, KY 40506 USA
[7] Buena Vista College, Storm Lake, Iowa 50588 USA
[8] University of Melbourne, Parkville, Victoria, Australia 3001
[9] Queen's University, Kingston, Ontario, Canada K7L 3N6
[+] Present Address, University of California, Berkeley, CA 94720 USA
[@] Visitor from Technische Universität, Munich, Germany

Presented by M.D. Hasinoff

Abstract. Next to the hydrogen atom, the $\mu^- p$ system is the most fundamental lepton-hadron system; therefore it can be used to test and extend our basic knowledge of the weak semi-leptonic interaction between leptons and quarks. Radiative muon capture (RMC) is much more sensitive than ordinary muon capture (OMC) to the induced pseudoscalar coupling constant of the weak semi-leptonic current because of the much larger range of momentum transfers which can occur in RMC. A status report on TRIUMF E452, Radiative Muon Capture on Hydrogen, which is anticipating a 10% measurement of this coupling constant, is presented.

1. Introduction and Theoretical Background

According to the standard model the basic leptonic electro-weak interaction is mediated by the exchange of charged ($W^{\pm}$) or neutral (Z^0, γ) vector bosons. However, in the case of a semi-leptonic interaction between a lepton and a quark, or, in experimental terms, between a lepton and a hadron, the hadronic weak current is considerably more complicated because of the presence of the strongly interacting particles. In addition to the vector (V) and axial vector (A) currents of the purely leptonic weak interaction, the basic hadronic weak current may contain 4 additional "induced" terms – weak magnetism (M), pseudoscalar (P), scalar (S), and tensor (T) as shown below [1]

$$V_\mu = g_V \gamma_\mu + i\frac{g_M}{2M}\sigma_{\mu\nu}q^\nu + \frac{g_S}{m}q_\mu \qquad (1)$$

$$A_\mu = g_A \gamma_\mu \gamma_5 + i\frac{g_T}{2M}\sigma_{\mu\nu}q^\nu \gamma_5 + \frac{g_P}{m}q_\mu \gamma_5 \ . \qquad (2)$$

In general the g_i are complex coupling constants (related to the complex form factors, $F_i = g_i$ for $i = V, A$, $F_i = g_i/2M$ for $i = M, T$ and $F_i = g_i/m$ for $i = P, S$) which are functions of the momentum transferred, q^2; M is the nucleon mass and m is the lepton mass.[1] Assuming time reversal invariance, these complex coupling constants must be real numbers. The presence of g_S and g_T terms would require second-class currents which do not appear naturally in the minimal standard model [2] and which are also not observed experimentally [3,4,5]. The conserved vector current (CVC) hypothesis is a natural consequence of the standard model and it is well established experimentally. CVC can be used to relate the vector and weak magnetism form factors to the well known electromagnetic form factors. The pseudoscalar form factor can be related to the axial form factor (which is determined from the neutron lifetime) by use of the hypothesis of the partial conservation of the axial current (PCAC) assuming the dominance of the one pion exchange contribution at low q^2 and the Goldberger-Treiman (GT) relation for $g_A(0)$ [6,7]

$$g_A(0) = \frac{f_\pi g_{\pi NN}}{\sqrt{2}M} \qquad (3)$$

$$g_P(q^2) \simeq \frac{\sqrt{2}m_\mu f_\pi g_{\pi NN}(q^2)}{q^2 + m_\pi^2} \simeq \frac{2m_\mu M g_A(0)}{q^2 + m_\pi^2} \simeq 6.78 \qquad (4)$$

The GT prediction for g_A has been verified experimentally to better than 5% [8] but the PCAC prediction for g_P has thus far only been tested to 22%. There have been 5 OMC measurements of g_P/g_A in hydrogen over the past 30 years and the world average value is 7.0 ± 1.5. The most accurate measurement comes from the very

---
[1]The use of coupling constants rather than form factors is preferred, in this case, since the momentum dependence is small, except for g_P.

precise comparison of the $\mu^\pm$ lifetimes at Saclay [9] which itself gives an error of 42%. This is considerably worse than the few % error in the theoretical prediction for g_P and the 0.3% error achieved in measurements of g_A/g_V using free neutron decay [10]. It is therefore very timely to attempt a measurement of g_P/g_A to better than 10%.

Because of its variable momentum transfer (which brings it close to the pion pole) the RMC rate is considerably more sensitive to g_P. However, its extremely low branching ratio has thus far limited studies to nuclear targets where the nuclear structure adds considerable complications. Moreover, there is also the possibility that g_P is changed inside the nucleus due to a renormalization of the pion field within the nuclear medium [11,12]. The latest nuclear RMC measurements [13,14] seem to show a decrease of g_P with increasing Z but the data [15] for ^{16}O can be well fitted with both the PCAC value for g_P using the phenomenological model of Christillin and Gmitro [16] or a much larger value $g_P/g_A = 13.6 \pm 1.9$ using the microscopic model of Gmitro et al. [17]. This latter value is in good agreement with the value obtained from a comparison of OMC on ^{16}O to the $0^- \rightarrow 0^+$ β-decay of ^{16}N which gives $g_P/g_A \simeq 12.5$ [18]. Therefore, in addition to providing a test of PCAC and the dominance of the pion pole, the hydrogen measurement discussed below also serves as a benchmark for testing such basic ideas as renormalization of g_P/g_A in nuclei.

The first calculation of RMC on hydrogen was that of Huang et al. [19] although the effect of the μp relative spin orientation was not included until the work of Opat [20] which predicted the radiative rates for the spin-singlet and spin-triplet μp atoms, $\Gamma_s^\gamma = 4.96 \times 10^{-3} s^{-1}$, $\Gamma_t^\gamma = 9.00 \times 10^{-2} s^{-1}$, respectively. The prediction for the spin-doublet ortho [$p\mu p$] molecule was $\Gamma_{mol}^\gamma = 3.07 \times 10^{-2} s^{-1}$. The expected sensitivity to g_P was essentially equal to the accuracy in the rate determination. In contrast, for OMC a 4% rate measurement leads to a 42% error in g_P. However, the ortho-molecular state can also decay ($\lambda_{op} = 4.1 \pm 1.4 \times 10^4$) into the para-molecular state, which has a somewhat larger RMC rate. Figure 1 shows the relative contributions, $G_i(t)$, to the total RMC rate in liquid H_2 from the singlet μp atom and the two $p\mu p$ molecular states. The RMC rate is dominated by the ortho-molecular state only for the first few μ^- lifetimes so that the para-molecular contribution (about 20% of the total RMC rate) cannot be ignored. Figure 2 shows that the rate for RMC is rather insensitive to the exact value of the ortho-para conversion rate (λ_{op}). In OMC, on the other hand, the extracted value of g_P is reduced by nearly a factor of 2 by the inclusion of the para contribution [9]. The most recent calculation of Beder and Fearing [21] included the contribution of Δ excitation and concluded that the effect is at most 8% at the highest photon energies.

2. Experimental Difficulties

2.1 Liquid Hydrogen Target Chemistry. Figure 3 summarizes the 'fate' of the

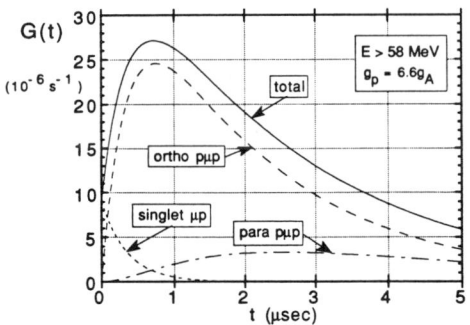

Fig. 1. RMC time distributions in liquid H_2. $G_i(t)$ is the product of the relative populations, $P_i(t)$, and the RMC rates, Γ_i, for each state [21].

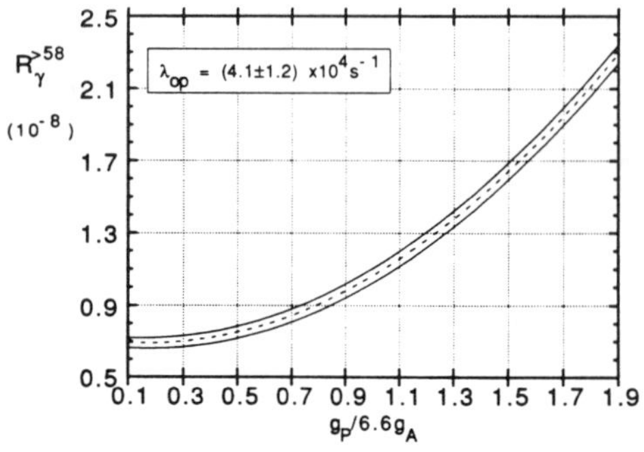

Fig. 2. The branching ratio ($E_\gamma > 58$ MeV) for RMC as a function of g_P/g_A for the measured value of the ortho-para conversion rate [21].

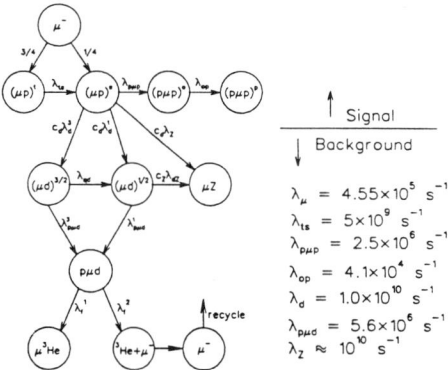

Fig. 3. Simplified diagram of μ^- reactions in LH_2 with small admixtures of deuterium (concentration c_d) and heavier impurities (concentration c_Z).

μ^- before capture. Extremely high purity of hydrogen in the liquid itself is critical. Otherwise the μ^- will be stolen from the proton before capture can take place. Impurity levels of $< 10^{-9}$ for all elements other than hydrogen are necessary since the capture probability increases as Z_{eff}^4 while the branching ratio for RMC/OMC remains nearly constant with increasing Z. Such purity is achieved by passing the hydrogen gas through a palladium filter each time it is transferred into the target vessel which itself is carefully outgassed for a period of at least 14 days prior to each run. An outgassing rate of $\approx 5 \times 10^{-11}$ l/min for the gold-walled cell has been achieved; this accumulates to $\approx 2 \times 10^{-6}$ liters in 30 days.[2] To reduce the formation of $p\mu d^+$ molecules and then ^3He by μ-induced fusion, the natural concentration

---

[2] The outgassing rate at LH_2 temperatures is several orders of magnitude lower than that at room temperature so this estimate is a conservative upper limit for the heavy impurities.

(100–150 ppm) of deuterium in hydrogen was reduced to <2.0 ppm by electrolyzing hydrogen from deuterium depleted water[3]. At this level the probability of RMC on ^{3}He is reduced to $\approx$ 2% of the RMC-H signal level. Based on extensive Monte Carlo simulations and detailed measurements of the stopping distribution of the beam, we chose Au for the target wall material and a cylindrical Au flask (16 cm $\phi \times$ 15 cm length $\times$ 250 μ thickness) was constructed. The RMC events from Au can be nearly eliminated by imposing a blanking time of $5\tau_\mu^{Au}=$ 365 ns following each μ-stop as well as by tracking cuts. Diffusion of $p\mu p$ atoms to the walls has been estimated and this is not expected to be a problem in such a large volume LH$_2$ target.

2.2 Other Backgrounds.

The unavoidable backgrounds of internal and external bremsstrahlung of ordinary muon decay electrons only occur below $E_\gamma=$ 53 MeV and they can be removed by energy cuts. As shown by Monte Carlo calculations, a comparison of the radiative muon decay (RMD) photon spectrum for both μ^+ and μ^- allows a check of the high-energy response function of the detector but only after a subtraction of the substantial contribution from the annihilation of the decay e^+ in the μ^+ case (see Fig. 4).

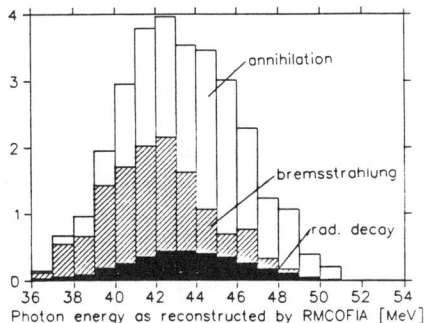

Fig. 4. Individual contributions to the low energy gamma spectrum for μ^+ decay indicating the very substantial contribution of the e^+ annihilation. All spectra have been folded with the RMC spectrometer acceptance and response functions to compare directly with the experimental data.

Photons from the decays of π^0's produced by charge exchange (B.R. $\simeq$ 0.6) from the few pions ($< 10^{-3}$) remaining in the muon beam after the RF separator occur exactly in the energy region of the RMC signal. However, these prompt γ's can be easily eliminated (below the 10^{-9} level) by rejecting any events which are in prompt time coincidence with a beam particle or where the pulse-height in any of the 4 beam counters is non-zero.

"SPURIOUS" photons, resulting from a pile-up of a decay e^- with an e^+ from a very asymmetric photon conversion in the Pb converter during the drift time of the tracks in the drift chamber, must also be accurately identified and rejected. This is easily done in software using the multi-hit timing information available from the trigger scintillators.

---

[3]Manufactured by AECL-Canada and sold by ISOTEC Inc., Miamisburg, Ohio, USA

Even cosmic ray induced photons constitute a troublesome background. The entire detector is therefore shielded on the top and sides with both scintillators and drift chambers to reject such events. In addition, the use of a tracking detector with multi-hit capabilities allows rejection of any γ event in which there is any additional track indicating the presence of an initiating cosmic ray shower. We have accumulated substantial data without beam in order to characterize and measure this background. These data agree with the energy spectrum measured above the endpoint of the RMC signal in the μ^- running period.

3. TRIUMF RMC Spectrometer

Since the RMC process on hydrogen is extremely rare (branching ratio for $E_\gamma > 58$ MeV $\simeq 1.6 \times 10^{-8}$), it must be separated from the extremely large number of low energy γ's from RMD as mentioned above. A detector with a large solid angle, good detection efficiency and moderate energy resolution is essential. The choice of a cylindrical ($\Omega \approx 3\pi$) pair spectrometer for a tracking detector provides a unique photon signature and therefore completely eliminates the neutron problem which has plagued the earlier nuclear RMC measurements with NaI detectors. A schematic view of the TRIUMF RMC facility is given in Fig. 5. The LH$_2$ target is surrounded by 5 cylindrical assemblies of concentric segmented scintillators (A,A',B,C,D) which are used to form the fast first level trigger and also to reject the single charged tracks from ordinary muon decay. A thin (1.08 mm) Pb converter is placed between the 12-fold segmented B/C scintillators. A proportional wire chamber (IWC) is inserted just inside the inner wall of the main drift chamber which has 4 superlayers (3 axial and 1 stereo). The stereo layer provides one z point per track,

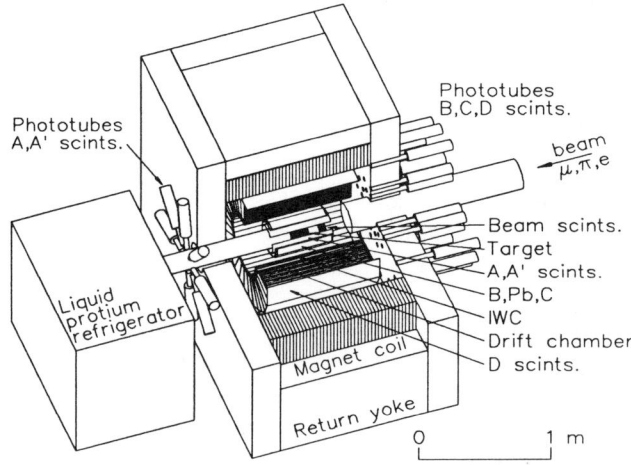

Fig. 5. Schematic view of the TRIUMF RMC spectrometer facility (see text for details).

while the second z point is provided by the IWC which has 2 layers of spiral cathodes. Each of the 6 sense wires in each of the 272 drift cells is equipped with a multi-hit LeCroy 1879 FastbusTDC to measure the drift times in 2 nsec buckets. A C·D trigger pattern check is made for the proper curvature of both $e^{\pm}$ tracks and then an adequate number of active drift cells in each of the independent layers of the drift chamber is required in the 2nd level trigger within 1 μs. This second level hardware trigger reduces the trigger rate from 1500 Hz down to about 300 Hz. A 3rd trigger decision using software cuts in the Fastbus μprocessor reduces the taping rate to less than 100 Hz with a typical event size of 2 Kbytes.

4. Preliminary Results

The detector acceptance and line shape are obtained by tuning the beam for π^- and using the well known reactions $\pi^- p \to \gamma n$ (40%) and $\pi^- p \to \pi^0 n \to \gamma\gamma n$ (60%). A typical event where both of the $\pi^0 \gamma$'s have converted in the Pb sheet (10% conversion probability) is shown in Fig. 6. The overall spatial resolution for the $e^{\pm}$ tracks in the DC is $\approx$ 140 μ and the spectrometer energy resolution at B = 0.24 T is $\approx$ 10% (FWHM) between 50 and 130 MeV. Figure 7 shows the measured photon energy spectrum and the simulated Monte Carlo distribution. The agreement is excellent except in the region of the 129 MeV γ where the $e^{\pm}$ tracks can be very close together. Such $\pi^- p$ runs are taken at regular intervals during the course of the $\mu^- p$ data collection to monitor the acceptance of the spectrometer.

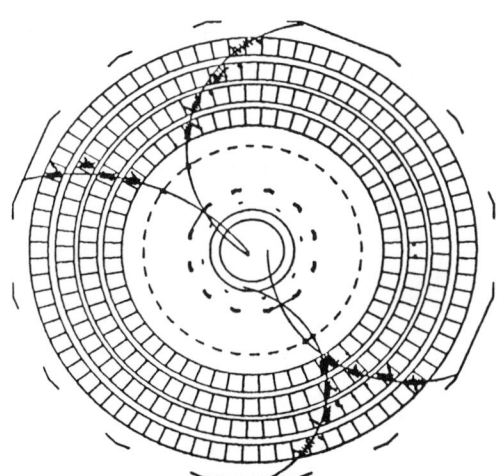

Fig. 6. End view of a gamma event from the $\pi^0 \to \gamma\gamma$ decay in which each of the γ's has converted in the Pb converter. The hit scintillators are shown as solid lines. The hits in the 3rd (stereo) superlayer are displaced from the circular tracks until the z displacement of the tracks from the central plane of the spectrometer is included.

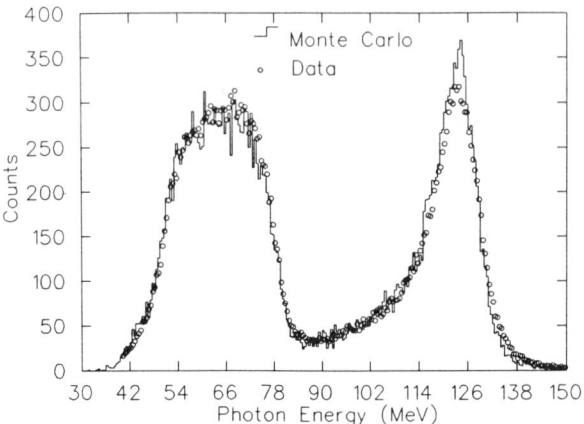

Fig. 7. Photon energy spectrum from the reactions $\pi^- p \to \gamma n$ and $\pi^- p \to \pi^0 n \to \gamma\gamma n$ compared with Monte Carlo simulation (solid histogram).

Figure 8 shows the RMC-H data before the application of any cuts. The spectrum is dominated by the RMD bremsstrahlung events below 55 MeV and the $\pi^- p$ events above 55 MeV. Figure 9 shows the same spectrum after the application of all cuts. The small insert shows the region above 58 MeV where we can extract the RMC-H signal from the RMD and bremsstrahlung gammas. The π^- induced gammas are totally removed and the few events observed in the spectrum above 100 MeV are consistent with measurements of the cosmic-ray background spectrum when the cyclotron is operating and delivering beam to the nearby channels. Thus we believe

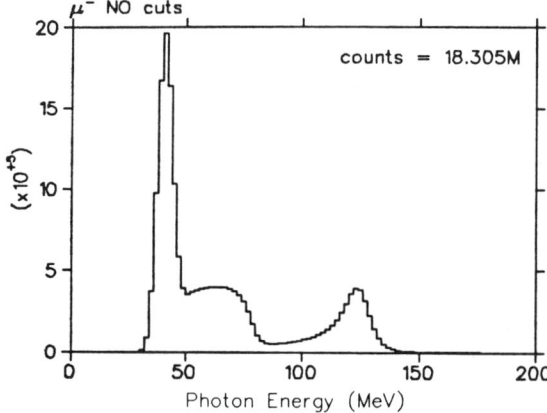

Fig. 8. Reconstructed RMC photon spectrum with no cuts applied except very general vertex requirements.

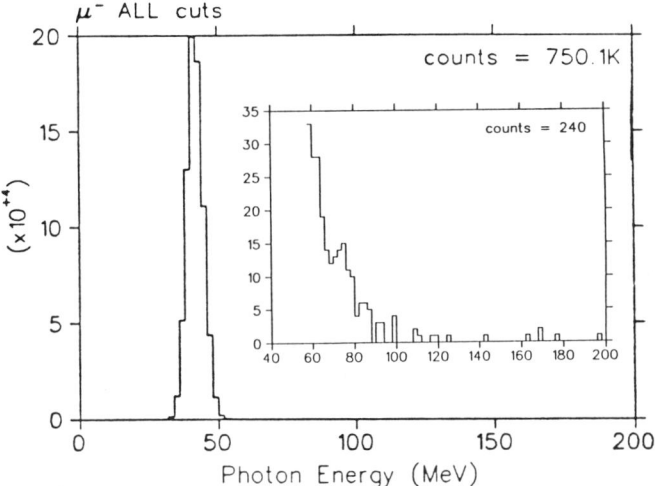

Fig. 9. Reconstructed RMC photon spectrum with all cuts applied (including the blanking time cut). The insert shows the region of the RMC-H signal above 58 MeV.

that $\approx 50\%$ of our high energy background is due to neutrons produced at the production target or beam dump which enter the spectrometer and produce γ's by (n,γ) or (n,π^0) reactions in the yoke or coil. Further data are now being obtained to reduce the error on this background subtraction.

The time distribution of these remaining γ events is shown in Fig. 10. The solid curve is a fit with 2 exponentials ($\tau_\mu^H, \tau_\mu^{Au}$) plus a constant background to account for the pileup of 2 or more muons. From such a fit we can obtain the relative numbers of μ's stopping in LH$_2$ and Au and then extrapolate beyond $5\tau_\mu^{Au} = 365$ ns to obtain an estimate of the number of Au RMC events in our final RMC-H spectrum (12 events).

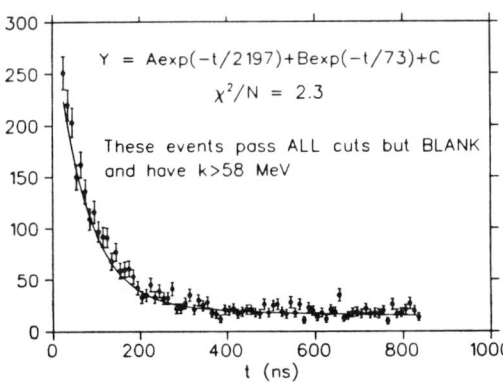

Fig. 10. Time distribution of all photon events with $E_\gamma > 58$ MeV which pass all cuts except the blanking time cut. The solid line is a fit to the data with 2 different lifetimes for Au and H plus a constant background.

5. Summary and Outlook

The final RMC-H spectrum contains 240 events with $E_\gamma > 58$ MeV. This represents all the data collected in 4 running periods from August '90–June '91 ($\approx 2 \times 10^{12}$ μ^- stops). There are 228 events in the RMC region of the spectrum (58–100 MeV). After subtraction of the backgrounds from cosmic rays (9%), the Au wall stops (5%), the high energy tail of the bremsstrahlung γ's (3%), and $p\mu d$ fusion (2%), we obtain about 200 events with an overall statistical error of $\approx 9\%$. Additional data from December '91 and January '92 are currently being analyzed. Runs with ordinary hydrogen gas with ≈ 100 ppm deuterium contamination have also been performed ($\approx 3.9 \times 10^{11} \mu$ stops) to check the calculated number of RMC photons expected from ^{3}He following $p\mu d$ fusion. The extraction of g_P/g_A from the measured RMC rate requires a careful determination of all of our systematic errors and verification of our Monte Carlo simulation (because of the finite energy resolution of the spectrometer). We hope to complete this analysis and publish our first results for g_P/g_A by the end of the year. The expected goal of ≈ 400 RMC events from hydrogen should be obtained by April '93. The combined systematic errors are expected to be $\approx 4\%$ excluding the 4% contribution due to the uncertainty in λ_{op}. When all the errors, including the λ_{op} uncertainty are combined with the final statistical error, we expect an 8% determination of the capture rate and therefore a 10% value for g_P.

Acknowledgements.

This work is supported by the Natural Sciences and Engineering Research Council (NSERC) and the National Research Council (NRC) of Canada, the U.S. National Science Foundation (NSF), the Paul Scherrer Institute (Switzerland) and the Australia Research Council.

References

[1] J.R. Luyten, H.P. Rood and H.A. Tolhoek, Nucl. Phys. **41**, 236 (1963).

[2] P. Langacker, Phys. Rev. **D14**, 2340 (1976).

[3] A.I. Boothroyd, J. Markey and P. Vogel, Phys. Rev. **C29**, 603 (1984).

[4] P. Lebrun *et al.*, Phys. Rev. Lett. **40**, 302 (1978).

[5] T. Minamisono *et al.*, in *Proc. XXIIIth Yamada Conf. on Nuclear Weak Processes and Nuclear Structure, Osaka, June, 1989* (World Scientific, 1989) p.58.

[6] M.L. Goldberger, and S.B. Treiman, Phys. Rev. **110**, 1178, 1478 (1958).

[7] L. Wolfenstein, in *High Energy Physics and Nuclear Structure*, ed. S. Devons (Plenum, New York, 1970) p.661.

[8] C. Dominguez, Phys. Rev. **D25**, 1937 (1982).

[9] G. Bardin *et al.*, Nucl. Phys. **A352**, 365 (1981), Phys. Lett. **B104**, 320 (1981), Riv. Nuovo Cim. **8** 1 (1985).

[10] D. Dubbers *et al.*, Europhys. Lett. **11**, (3), 195 (1990).

[11] J. Delorme, M. Ericson, A. Figureau, and C. Thevenet. Ann. of Phys. (N.Y.), **102**, 273 (1976).

[12] M. Ericson, *Progress in Nuclear and Particle Physics* (Pergamon, Oxford, 1978) p.67.

[13] M. Döbeli *et al.*, Phys. Rev. **C37**, 1633 (1988).

[14] D. A. Armstrong *et al.*, accepted by Phys. Rev. May 1992.

[15] D.S. Armstrong *et al.*, Phys. Rev. **C40**, R1100 (1989).

[16] P. Christillin and M. Gmitro, Phys. Lett. **B150**, 50 (1985).

[17] M. Gmitro, A.A. Ovchinnikova, and T.V. Tetereva, Nucl. Phys. **A453**, 685 (1986).

[18] A.R. Heath and G.T. Garvey, Phys. Rev. **C31**, 2190 (1985).

[19] K. Huang, C.N. Yang, and T.D. Lee, Phys. Rev. **108**, 1348 (1957).

[20] G.I. Opat, Phys. Rev. **B134**, 428 (1964).

[21] D.S. Beder and H. Fearing, Phys. Rev. **D35**, 2130 (1987).

Muonic Atom Formation, Muon Transfer and Nuclear Fusion in a $D_2 + {}^3\text{He}$ Gas Mixture

D.V. BALIN, V.N. BATURIN, Yu.A. CHESTNOV, E.M.MAEV, G.E. PETROV,
G.G. SEMENCHUK, Yu.V. SMIRENIN, A.A.VOROBYOV, N.I.VOROPAEV
St.Petersburg Nuclear Physics Institute
Gatchina 188350, Russia

V. CZAPLINSKI, M. FILIPOWICZ, A. GULA
Institute of Physics and Nuclear Techniques
Cracow 30-059, Poland

E. GULA
Institute of Nuclear Physics,
Cracow 31-342, Poland

presented by G.G. Semenchuk

Abstract. In the gas mixture $D_2 + {}^3\text{He}$ the probability of $d\mu$-atom formation in the ground state was determined, W=0.535±0.017, and also the rate of molecular muon transfer $\lambda_{d^3 He} = (1.24 \pm 0.05) \cdot 10^8 \text{s}^{-1}$ and the upper limit of the nuclear fusion rate in the ${}^3\text{He}\mu d$ molecule, $\lambda_f < 7 \cdot 10^7 \text{s}^{-1}$, by means of registration of charged fusion products of $d\mu d$ catalysis and muon capture in the ionization chamber. The experiment was done at the muon channel of PNPI at a gas pressure of 87 atm and an atomic concentration of ${}^3\text{He}$ of 11%.

1. Introduction

The muon catalyzed fusion investigations in the last years [1-3] lead to the conclusion that a more detailed knowledge of the primary stage of muonic hydrogen atom formation is required, including deexcitation, thermalization and muon exchange. The complicated kinetics of these competing processes determine the energy spectra of muonic atoms in the ground and in excited states. The rates of subsequent molecular reactions and hence all the kinetics of catalysis depend on the shapes of these spectra.

Because of this, the fast muon exchange between the hydrogen isotopes $p\mu \to d\mu$, $p\mu \to t\mu$, $d\mu \to t\mu$ became a field of experimental research with the aim to determine the probability q_{1S} for muonic atoms to reach the ground state in muonic cascade deexcitation [4]. The fast exchange of muons from excited hydrogen isotopes to helium is also interesting, because helium is always present in the d-t mixture due to tritium decay and fusion, and it truncates the catalytic cycles. It also has an influence on the muonic atom formation, because of different cross-sections of muon capture by hydrogen and helium isotopes ($\lambda^a_{H\mu}$ and $\lambda^a_{He\mu}$). This problem is important for the understanding of muon deceleration and capture in matter and was considered in various theoretical approaches [5,6].

In this paper we describe the attempt to evaluate the contribution of fast muon exchange from excited states of $d\mu$ atoms to helium (1-W). Here W is the probability of the muon to form a $d\mu$ atom and to reach its ground state. It is related to q_{1S}, the ground state population, as follows:

$$W = q_{1S}/(1 + AC) \qquad (1)$$

where $A = \lambda^a_{He\mu}/\lambda^a_{d\mu}$ is the ratio of muonic helium to muonic deuterium formation rates, $C = C_{He}/C_d$ is the ratio of atomic concentrations of helium and deuterium.

Earlier we have obtained the first data [7] on muon exchange from the ground state of $d\mu$ atoms to ^{3}He and ^{4}He with the ionization chamber (IC) technique [8], that used registration of consecutive muon catalyzed $d\mu d$ fusions. That experiment has confirmed the molecular mechanism of the transfer [9]. It was done with a small admixture of helium, less than 2%, to ensure the exchange from the ground state, and to reduce the probability of direct Heμ formation. In the new experiment the helium concentration was chosen to be 11%, in order to increase the direct Heμ formation and the muon transfer from excited states. A new IC with larger sensitive volume was used [10], because the yield of dd fusions at this helium concentration was much lower.

2. The kinetics of muonic processes in the D_2+^3He mixture

We have treated the energy and time spectra of the IC pulses according to the kinetic diagram of Fig.1. In the present experiment we have specially used the ^{3}He isotope to be able to mesure the peculiar branch in kinetics: the nuclear muon capture in ^{3}He,

$$^3He + \mu \to {}^3H + \nu_\mu ,$$

observing the tritium particles with 1.9 MeV energy.

The quantity W was determined in two ways. In the first one, the ratio of the yields of consecutive first and second dd-fusions was used,

$$N_{F2}/N_{F1} = W\lambda_{dd\mu}(1 - w_{dd})/(\lambda_{dd\mu} + \lambda_{d^3He} + \lambda_0) \qquad (2)$$

where $\lambda_{dd\mu}$ is the $dd\mu$ molecule formation rate, λ_{d^3He} is the molecular muon transfer rate from $d\mu$ to ^{3}He, λ_0 is the muon decay rate, and w_{dd} is the muon sticking probability in $d\mu d$ fusion. All these kinetic parameters have been measured in our previous experiments [7,8]. As one can see in Fig. 1, the yield of dd fusions depends on the number of $d\mu$ atoms in the ground state; so a shortage of second fusions could be due to the presence of ^{3}He, through atomic capture of muons and fast transfer of muons from excited $d\mu$ atoms.

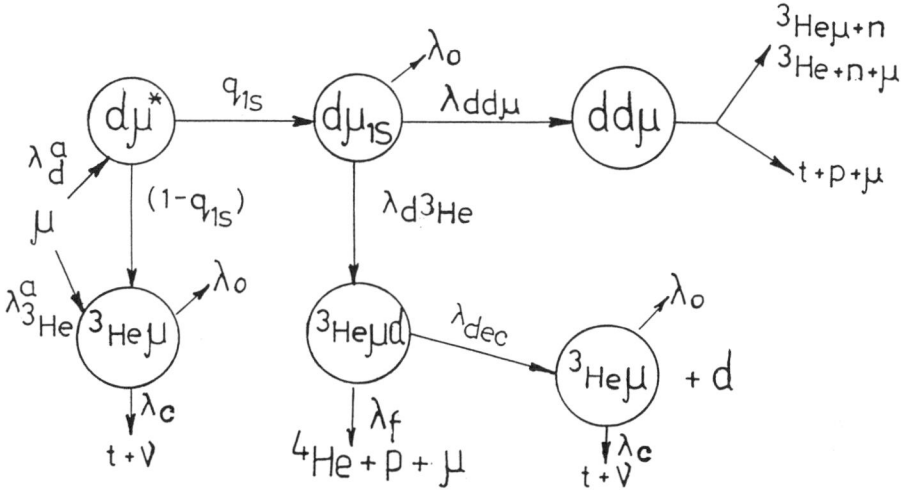

Figure 1: Kinetics of the μCF processes in D$_2$+^{3}He mixture.

In the second method the new feature of the experiment was used for the first time to control the muon's fate: the nuclear capture of muons by ^{3}He. The number of tritons, normalized to the number of first dd fusions, is equal to

$$N_t/N_{F1} = \frac{\lambda_c}{\lambda_0 + \lambda_{tot}}(1 - W + \frac{W\lambda_{d^3He}}{\lambda_{dd\mu} + \lambda_{d^3He} + \lambda_0})/(\frac{W\lambda_{dd\mu}}{\lambda_{dd\mu} + \lambda_{d^3He} + \lambda_0}). \qquad (3)$$

Here λ_c=1.55 10^3 s^{-1} is the t+ν fraction of the nuclear muon capture rate, λ_{tot}=2.05 10^3s^{-1} is the total muon capture rate [11].

It should be noted that the first two terms, 1-W, describe the fast muon transfer, and the third term the transfer from the ground state of $d\mu$ atoms. The relative contributions of these channels depend on the helium concentration and were in our experimental conditions approximately equal.

We would like to mention specifically the possible observation of the nuclear fusion in the intermediate molecular ^{3}Heμd system, with the rate λ_f predicted in [12,13] to be in the range of 10^8 to 10^{11}s^{-1}. The charged products of the fusion reaction

$$d + {}^3\text{He} \rightarrow {}^4\text{He} (3.66 \text{ MeV}) + p(14.64 \text{ MeV})$$

are especially appropriate for observation by our IC method.

3. Data analysis and results

The experimental conditions and the setup with a multianode IC were the same as described in [10]. The signals of stopping muons, subsequent fusions and capture products in the sensitive IC volume were registered. The higher concentration of ^{3}He in this experiment has lead to a special new kind of background - the charged particles emerging from ^{3}He capture of thermal neutrons dissipated in the experimental hall:

$$^3\text{He} + n \rightarrow t + p + 0.774 \text{ MeV} . \tag{4}$$

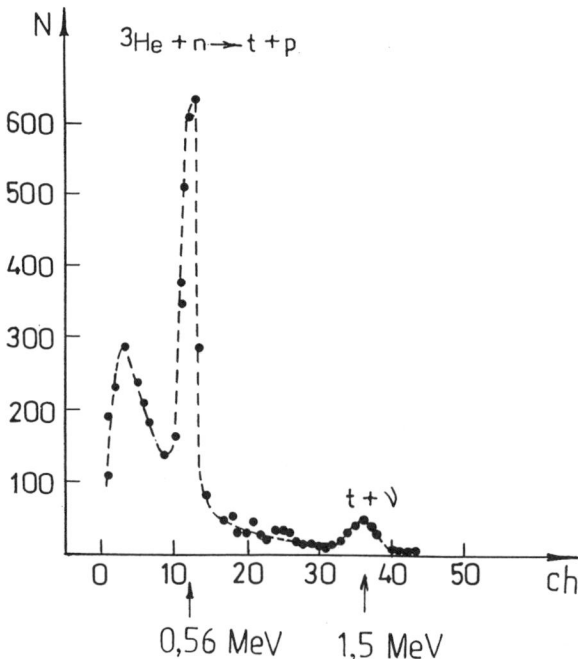

Figure 2: The amplitude spectrum of background events registered by the IC in the time interval $3.5 < T < 8.5$ μs after a muon stop.

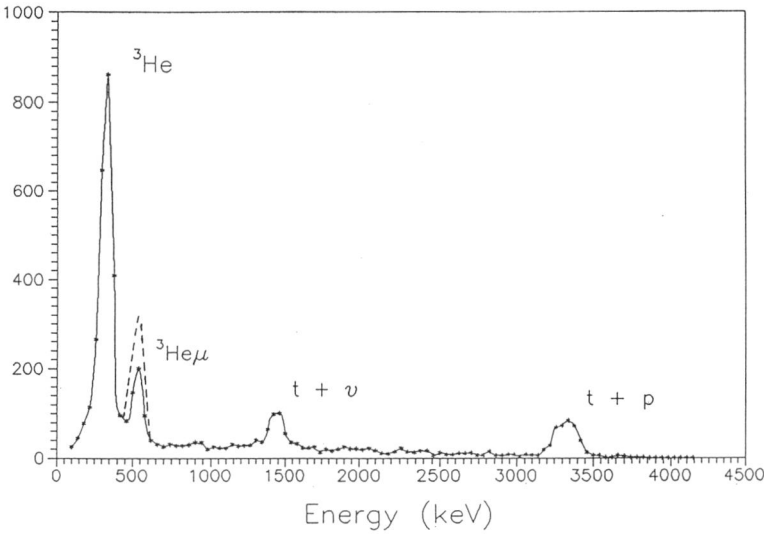

Figure 3: The energy spectrum of charged particles. The dashed line shows the ^{3}Heμ peak region of the spectrum prior to the background subtraction.

The energy of these particles is in the same range as the ^{3}He+n and ^{3}Heμ+n peaks of $d\mu d$ fusions, and their contribution should be estimated. Fig. 2 demonstrates the raw spectrum of ^{3}He+n events in the time interval 3.5 to 8.5 μs after a muon stop. Besides the usual selection of fusion events by the criteria described in [8,10], we have used the property of uniform time distribution of the background events to get rid of them. Comparing the amplitude spectrum of all events in the full time interval (0.5-8.5 μs) with the spectrum in the interval (3.5-8.5 μs), where almost all $d\mu d$ fusions are over, one can subtract the background after appropriate normalization. The spectrum of $d\mu d$ fusion and muon capture after such cleaning is shown in Fig. 3. The registered energies are less than the initial ones because of charge recombination due to the high pressure (87 atm) of the gas mixture in the IC. The dashed line shows the ^{3}Heμ peak region of the spectrum prior to the background subtraction. It was taken into account by analysis of the time distribution of fusions and captures, shown in Fig. 4. This spectrum was approximated by the sum of 3 exponentials that describe the kinetics processes of Fig. 1 - the $dd\mu$ chain, the ^{3}Heμd molecule formation and the fast formation and decay of ^{3}Heμ atoms:

$$dN/dt = A exp(-\lambda_\Sigma t) + B[exp(-\lambda_\Sigma t) - exp(-(\lambda_0+\lambda_c)t)] + C exp(-(\lambda_0+\lambda_c)t) + D \quad (5)$$

where $\lambda_\Sigma = \lambda_{dd\mu} + \lambda_{d^3He} + \lambda_0 + \lambda_{dz}$, $D \sim 1/\text{bin}$ is the constant background of neutron capture, λ_{dz} is the transfer rate of muons to impurities (mostly nitrogen). Their concentration was subject to chromatography checks before and after the experiment and the value of this rate was about 10% of the measured rate λ_{d^3He}.

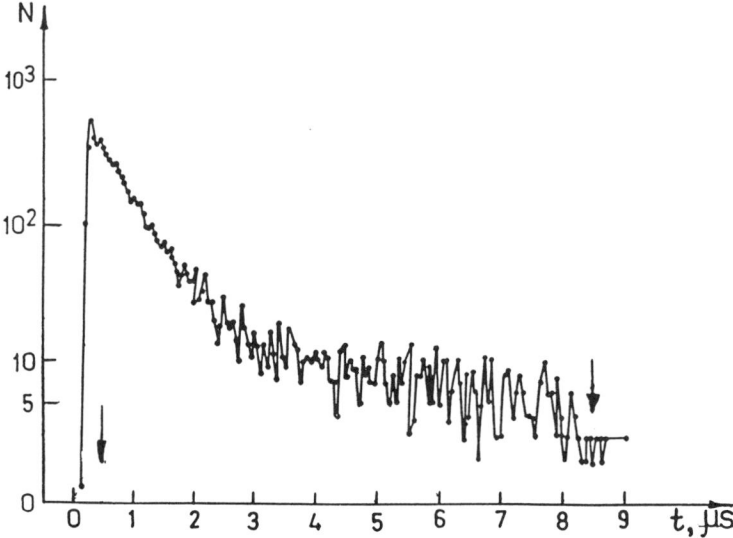

Figure 4: Time distribution of all registered charged particles. The arrows indicate the interval chosen for the fit.

Table 1 contains the statistics of $d\mu d$ fusion and muon capture events, selected for the data analysis.

Table 1:

Total number of registered events	19000
Number of events selected for analysis	5153
Number of ^{3}He+n background events	1008
First dd fusions N_{F1}	3610
Number of tritons N_t	535
Second dd fusions N_{F2}	56

This data has been corrected for the dead time of the IC, taking into account the kinetics, in a way similar to that described in [8]. From the analysis of the ratio N_{F2}/N_{F1} (the first method) we obtain W=0.53±0.08. The second method provided much better accuracy, W=0.535±0.017. This is due to the rise in the number of ^{3}Heμ atoms with increasing ^{3}He concentration C_{He}, because of direct atomic capture and fast exchange, decreasing the number of first fusions N_{F1} and consequently, even more rapidly, of N_{F2}.

To estimate the ground state population q_{1S} at our experimental conditions, we use the data of [14], i.e. measurements of $p\mu$ and ^{4}Heμ formation in their ground states by observation of muonic X-rays at 6 torr pressure and helium concentration of 11%, the same as in our experiment. According to the experimental method, we get

$$W = N_{p\mu 1s}/(N_{p\mu 1s} + N_{He\mu 1s}) = 0.82 \pm 0.02 , \qquad (6)$$

and can neglect at such a low pressure the fast muon transfer in the muonic cascade and assume $q_{1S}=1$. Then, according to definition (1) for W, we have A=1.77±0.04. If we neglect the small difference in relative atomic muon capture rates (5-7%) in H_2+He and D_2+He mixtures [5], we can use this value of A in the analysis of our data. So, for W=0.535 and A=1.77 we get q_{1S}=0.65±0.06. It is interesting to compare this result with the calculations of q_{1S} in [15]. Fig. 5 shows the calculated dependences of q_{1S} on helium concentration for various densities. At C_{He}=11% one can see a nice agreement of our result with the calculations, which we may consider as an experimental evidence of intense exchange of muons from excited muonic atoms to helium. But it should be treated as qualitative, because of a number of assumptions taken by this comparison.

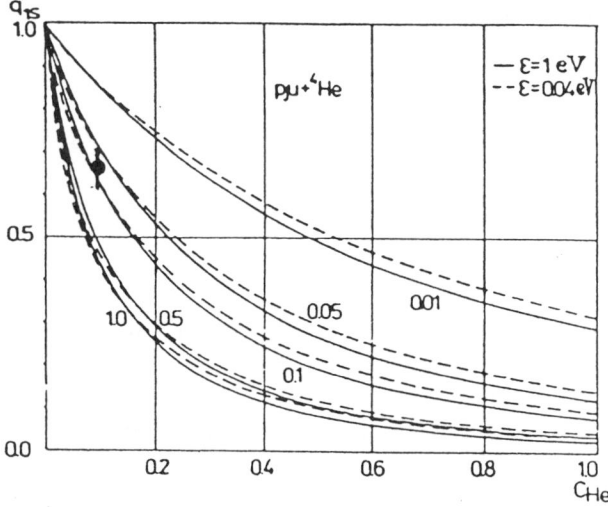

Figure 5: The ground state population q_{1S} vs. concentration of He [15]. The point is our experimental result.

If we assume, as a "zero approximation", q_{1S}=1, then we get A=7.0±0.5, larger than all the modern theoretical estimates [5,6] that give A~(1.5-2). This also indicates the significance of fast transfer. We would like to note that in 1984 we have made the first attempt to measure the relative rates of muonic deuterium and muonic helium formation by counting $d\mu d$ fusions by the first method [8] and obtained A=7±6, with the error so large due to low statistics of the second fusions.

It seems important that the value A=1.77 is very close to the result of [16], A=1.84±0.09, obtained for the transfer of pions from excited pionic hydrogen. This could be evidence for a similar probability of muon and pion capture by atoms.

The analysis of the time distribution of $d\mu d$ fusion provided the rate of molecular muon exchange from $d\mu$ to ^{3}Heμ,

$$\lambda_{d^3He} = (1.24 \pm 0.05) \cdot 10^8 \text{s}^{-1}$$

with twice improved precision compared to our first measurement [8], $(1.27\pm0.11)\cdot 10^8$ s^{-1}. Both values rather confirm the calculated rate $1.3\cdot 10^8$ s^{-1} that was obtained in the model of statistical reconstruction of the "unfrozen core", than that of $1.43\cdot 10^8$ s^{-1} in the model of "frozen core" [17]. More information about the influence of the external electronic shell on the molecular muon exchange could be obtained from further experiments with other isotopes of hydrogen and helium.

The data of this experiment concerning the search of nuclear fusion in the ^{3}Heμd system was described in our report [18]. Here we present only the upper limit of the fusion rate: $\lambda_f < 7 \cdot 10^7$ s^{-1}, assuming the deexcitation rate of the complex [^{3}Heμd] is $\lambda_{dec}=1.55\cdot 10^{11}$s^{-1} [19] by X-ray emission (85%) and Auger process (15%). But if we consider the proposition of Kamimura [20] about an additional channel of two-particle decay of the complex with the rate $\sim 8\cdot 10^{11}$s^{-1},

$$[^3\text{He}\mu d] \rightarrow [^3\text{He}\mu]_{1s} + d, \qquad (7)$$

then the estimate becomes $\lambda_f < 4 \cdot 10^8$s^{-1}.

Conclusion

The ionization chamber method proved to be very fruitful in investigating fast processes of muon exchange, especially in the D_2+^3He mixture, due to registration of the new kinetics channel of muon capture that controls the fate of an individual muon. We are going to perform measurements at various pressures and concentrations in order to measure separately the parameters q_{1S} and A.

An experiment to measure the nuclear muon capture rate in ^{3}He with high precision also seems possible.

Acknowledgements

One of us (G.G.S.) is especially grateful to Dr. C. Petitjean and to Prof. L. Schaller for invitation and for hospitality during this meeting in Ascona.

References

[1] L.I. Menshikov, L.I. Ponomarev, Pisma ZhETF, **39**, 542 (1984), JETP Lett. **39**, 663 (1984).

[2] V.M. Bystritsky, A.V. Kravtsov, N.P. Popov, ZhETF **97**, 73 (1990).

[3] A.V. Kravtsov et al, ZhETF **96**, 439 (1989).

[4] W.H. Breunlich et al, Ann. Rev. Nucl. Part. Sci. **39**, 311 (1989).
P. Ackerbauer, W.H. Breunlich et al, Addendum and beam request for PSI proposal R-81-05.1, 1991.

[5] J.S.Cohen et al, Phys. Rev. **A27**, 1891 (1983).

[6] V.N. Dolinov, G.Ya. Korenman et al, Muon Cat. Fusion **4**, 169 (1989), Muon Cat. Fusion **7**, 182 (1992).

[7] D.V. Balin et al, Pisma ZhETF, **42**, 236 (1985).

[8] D.V. Balin et al, Phys. Lett. **B141**, 173 (1984), Muon Cat. Fusion **7**, 1-36 (1992) (translation of the preprint LNPI-964,1984).

[9] Yu.A. Aristov et al, Yad. Fiz. 33(1981)1066.

[10] D.V. Balin et al, Preprint LNPI-1630, 1990.

[11] L.B. Auerbach et al, Phys. Rev. **B138**, 127 (1965).

[12] M. Kamimura, quoted in the talk of K. Nagamine at AIP Conf. Proc. **181**, 23 (1989).

[13] D. Harley, B. Muller and J. Rafelski, J. Phys. **G16**, 281 (1990).

[14] F. Kottmann, in Proc. II Int. Symp. on Muon and Pion Inter. with Matter, Dubna, 268 (1987).

[15] A. Gula, A.V.Kravtsov, N.P.Popov, Proc. Int. Symp. MCF, Oxford, 55, (1989).

[16] V.I. Petrukhin, V.M. Suvorov, ZhETF **70**, 1145 (1976).

[17] V.K. Ivanov, A.V. Kravtsov, A.I. Mikhailov, ZhETF **91**, 358 (1986).

[18] D.V. Balin et al, Report INT-250/PS, Krakow 1991, Muon Cat. Fusion **7**, 301 (1992).

[19] T. Ishihara, S. Hara, Phys. Rev. **A39**, 5633 (1989).

[20] M. Kamimura, private comm., quoted in the talk of K.Nagamine at the 3rd Int. Conf. on Particle Nuclear Physics, Intersee (1991).

Recent Facets of Nuclear Fission Dynamics and Properties of Heavy Muonic Atoms

P. DAVID[1], Ch. RÖSEL[1], F. F. KARPESHIN[1,2], B. SABIROV[1,3]
presented by P. David[†]

[1)] Institut f. Strahlen- u. Kernphysik, Universität Bonn, D-5300 Bonn 1, F. R. Germany
[2)] Institute of Physics, St. Petersburg State University, Petrodvorets, 198904 Russia
[3)] Laboratory of Nuclear Problems, JINR, Dubna, 141980 Russia

Abstract: The results from quite different and independent experimental approaches of measuring the fission times of hot and cold heavy nuclei are described. The interpretations of the respective experimental observables in the framework of models, the derived corresponding fission times, and a viscosity constant are presented.

1. Introduction

The dynamical behaviour of large amplitude collective nuclear motion, as occuring in heavy ion reactions and in nuclear fission is difficult to observe. This is so, since it is strongly determined as well by the rate energy is dissipated from the collective nuclear motion into the internal single particle excitation energy, as also by the mechanism the dissipation propagates.

In the picture of a potential energy surface of a heavy fissioning nucleus especially the transition from saddle to scission is of interest. In the interaction of the collective nuclear motion with the single particle degrees of freedom by non-conservative forces, i.e. forces which may be treated in the picture of the nuclear viscous fluid by a phenomenological viscosity constant, energy and/or angular momentum may be transferred and the characteristic time is determined by the energy dissipation i.e. the friction process. The theoretical work to describe momentum transfer, particle multiplicities, mass distributions etc. of energetic heavy ion reactions in the intermediate

---

[†]Invited Talk Presented at the Workshop on Muonic Atoms and Molecules, Monte Verità, Ascona. April 5-9, 1992

energy range between low incident energies $\leq$ 5-10 MeV/u, where time dependent Hartree-Fock calculations are successful, and high energies $\geq$ 100 MeV/u, where the intranuclear cascade model is used, must account for continuous exchange of energy between the collective motion of the mean field and the individual nucleonic excitations in a manner consistent with usual conservation laws. From introducing the concept of nuclear dissipation by describing the process of induced fission as a diffusion process in terms of the Fokker-Planck equation (Weidenmüller et al. [1]) and from TDHF calculations (Negele et al. [2]) the time scale involved for this process is determined to be several 10^{-21} s. Dissipation, as the irreversable flow of energy between various degrees of freedom of the system, is an intrinsic nuclear phenomenon, since nuclear reaction times are short ($\leq 10^{-21}$s) and the nuclear system can be regarded as being isolated, i.e. not in contact with an external heat bath [1, 3, 4, 5].

The existence of two widely different time scales is an important property of dissipative processes. In this sense compound nucleus formation and decay may be considered as a dissipative process if the internal equilibration time of the compound nucleus (the heat bath) is short in comparison with the decay time. The fulfillment of this condition has to be proven for the particular compound processes under investigation, eventually precompound processes have to be included.

There is need and interest in understanding nuclear friction as the phenomenological model of conceiving dissipation not only at the nuclear surface but all over the deep nuclear interior. Phenomenologically two quite distinct mechanisms of dissipation have been derived [6]:

a) dissipation that is built on the one-body theory. In the theory of infinte systems this would be called Landau-damping.

b) dissipation due to two-body collisions.

The damping called Landau-damping follows from the fact that the particles carrying the dirturbance travel at different velocities, causing eventual incoherence. The time scale for Landau-damping is different from that of viscosity.

Evidence for dissipation in fission is indirectly derived mainly from the multiplicity of neutrons emitted prior to fission. In heavy-ion fusion-fission and fast fission processes, where neutrons are measured as originating from different reaction phases, therefore, it has to be duly investigated that the time preceeding the transient state is short. The transient state follows from describing the fission process in terms of a diffusive probability current over the fission barrier, which is calculated by solving a Fokker-Planck equation [1]. At time zero, when a nuclear reaction induces fission, the fission degree of freedom is in its ground state, located near the minimum of the fission potential. If, in comparison with the characteristic time scales of the fission process, the nucleonic degrees of freedom excited by the nuclear reaction equilibrate quickly, the fission degree of freedom may be viewed as being in contact with a heat bath of temperature $T = (E_x/a)^{1/2}$, E_x being the nuclear excitation energy, a the level density parameter. The diffusive probability current over the fission barrier rises smoothly from zero (at time zero) to a quasistationary value. The transient time τ_t is defined as the time required for this current to reach 90 % of its quasistationary value [3, 4].

The timescale of 10^{-21} s is one reason why up to now it has hardly been possible to devise experiments, such that strength and mechanism of nuclear dissipation could clearly be derived, despite their strong influence on the dynamic motion. Another reason is the disentangling of influences of dissipation on measurable quantities from effects of collective degrees of freedom of non-fission type.

It is the aim of this contribution to compare different experimental accesses to the time scales of dissipative processes in heavy ion fusion-fission and fast fission reactions, in antiproton and in prompt muon induced fission in considering neutrons, the Λ-baryon, and the muon as chronometers. A schematic overview is displayed in fig. 1.

2. Heavy ion induced fission reactions - Neutrons as a chronometer

The large amplitude collective nuclear motion of binary fission of an excited nuclear system involves collective degrees of freedom and for this reason may be inherently slower than decays based on the coupling of only nucleon degrees of freedom. The collective relaxation times (10^{-21} - 10^{-20} s) are up to three orders of magnitude larger than the typical relaxation time of single particle degrees of freedom.

In the picture of the potential energy surface of a liquid drop the parameters for deformation, elongation, and separation into two fragments define the static properties of the nuclear shape and its transitions as a function of the excitation energy. For the cold nucleus and for low excitation energies the level density is rather low and is increasing with growing excitation energy. The time within which the system proceeds towards scission is rather large for the less excited nuclei and decreases exponentially with increasing temperature. For a sufficiently high nuclear temperature it will be smaller than the transition time τ_{ss} determined by the dynamics between saddle and scission points.

In order to get information on the dynamical behaviour of fissioning composite systems the prescission and postscission neutron emission in coincidence with fusion-fission events has been investigated [7, 8, 9, 10, 11, 12]. Different mass asymmetries in the entrance channel correlate with different formation times of the composite systems. The number of the emitted neutrons originating from the fusion phase of the reaction allows to deduce these times. The comparison of the measured prescission neutron multiplicities with those resulting from statistical model calculations indicates that there is no considerable neutron emission during the formation. This may be interpreted for the formation times to be short as compared to the decay times and, moreover, all observed prescission neutrons to be emitted by the composite system after relaxation of the mass assymmetry [10]. For heavy ion reactions covering the excitation energy range from about 100 to 800 MeV with respective fusion-fission and fast fission character, the duration of the scission time τ_{sc} is considered to be composed of three subsequent time intervals, the equilibration time for the compound nucleus formation τ_e, the transient time τ_t to reach a quasistationary probability flow across the saddle point irreversibly developing towards scission, and the saddle-to-scission time τ up to the scission configuration.

The dynamics of fusion is such that in comparison to the other times the fusion time may become negligibly small by a proper choice of the projectile-target combination. Principally the transient (fission) time is then determined from comparing it with known times of other nuclear decay types, e.g. the number of neutrons evaporated by deexciting heavy nuclei. The calibration of the basic time interval is defined by the average time for evaporating one neutron. This may be calculated in the frame of the statistical model from the level densities. To be able to measure the scission time τ it is necessary to determine those neutrons and their multiplicity M_{pre} evaporated by the nucleus before reaching the scission point and to separate them from the postscission neutrons, since during all the stages of fusion-fission reactions neutrons are emitted. The concept is as follows:

For *spontaneously fissioning* nuclei the characteristic time can be defined as the decay lifetime τ. With decreasing fissility parameter, this time increases (to infinity). The time to reach the scission point τ_{ss} from the saddle poinmt configuration is comparably short.

For higher excitation energies E_x, when E_x exceeds the fission barrier, the fission time gets shorter.

In *fusion-fission* reactions, with a fission barrier the total fission time scale ranging from compound nucleus formation to scission is determined by the sum of several time intervals:

i) The thermal equilibration time τ_e, which typically is several 10^{-22} s. It defines the starting time for statistical equilibrium evaporation of particles from the compound nucleus, but preequilibrium emission may be present. τ_e = equilibrium time for compound nucleus formation.

ii) The statistical model mean life time τ_{sm}. The statistical model mean life time τ_{sm} of the compound nucleus until the first decay takes place is given by $\tau_{sm,1} = \hbar/\Gamma_{tot}$ with $\Gamma_{tot} = \sum_i \Gamma_i$. It starts after the equilibration phase of duration τ_e has reached the equilibrium ditribution with E_x of the compound nucleus, spin I and a shape corresponding to the potential energy ground state minimum. The widths Γ_i depend on E_x and infer lifetimes of 10^{-15} to 10^{-22} s with increasing E_x. For the neutron decay width holds with transmission coefficients $T_\ell(\varepsilon_n)$

$$\Gamma_n(E_x, I) = \frac{2(s_n+1)}{2\pi\rho(E_x, I)} \sum_{\ell=0}^{\infty} \sum_{J=I-\ell}^{I+\ell} \int_0^{E_x-s_n} \rho(E_x - B_n - \varepsilon_n, J) T_\ell(\varepsilon_n) d\varepsilon_n,$$

with similar expressions for charged particle emission. The level densities are calculated by

$$\rho(E_x, J) \sim \frac{2J+1}{E_x - E_{rot}} \exp\left[2\sqrt{a_n(E_x - E_{rot})}\right]$$

E_x compound nucleus excitation energy
B_n neutron separation energy
s_n neutron spin
ε_n neutron kinetic energy
I compound nucleus spin
J final state spin
T_ℓ transmission coefficient
a_n level density parameter ($\sim A/10$)
E_{rot} rotational energy

Since excitation energies $E_x > 100$ MeV in heavy nuclei lead to emission times $\tau_n > 10^{-20}$ s prescission neutron multiplicities M_{pre} can be converted into scission time scales from thermal equilibration to scission (τ_f) using the relation:

$$\tau_{sm}^{(n)} = \sum_{i=1}^{M_{pre}} \tau_{sm,i}^{(n)} = \sum_{i=1}^{M_{pre}} \frac{\hbar}{\Gamma_{n,i}} \left[\frac{\Gamma_{n,i}}{\Gamma_{n,i} + \Gamma_{CN,i}} \right]$$

(for $T = \sqrt{E_x/a_n} < 5$ MeV).

To this time $\tau_{sm}^{(n)}$ the last few neutrons contribute significantly, emitted from a system cooled down to $E_x < 100$ MeV.

iii) The transient delay time τ_t. During this time the quasi-stationary population distribution in the potential well is established including the saddle point configuration by the coupling of the collective to the intrinsic single particle degrees of freedom (by friction). Once after the time τ_t the escape rate has become constant, a quasistationary probability flow across the saddle point is built up for an irreversible development towards scission. Before this delay time the full fission width Γ_f is not developped.

iv) The saddle to scission time τ_{ss} is the time the system takes to proceed towards the scission configuration.
The sum of these times may range from 10^{-15} s for low values of E_x to 10^{-21} s for high excitation energy. In the latter case τ_{sm} is negligible.

To calculate the prescission time the effective fusion-fission lifetime for first chance fission is used [13]:

$$\tau_f = \tau_{sm} + 0.5\,\tau_t + \tau_{ss}$$

Fusion-fission reactions as the most central heavy ion collisions are the slowest process ($> 10^{-21}$ s). The identity of the projectile and target nuclei are completely lost.

Collisions with intermediate impact parameters lead to *quasi-fission* ($\sim (5\text{-}10) \times 10^{-21}$ s). Full momentum transfer to the compound system is observed. Mainly symmetric fission occurs with a slight shoulder in the mass distribution near the projectile mass at forward angles.

In *fast fission*, following the most rapid and peripheral deep inelastic collisions ($\sim 10^{-21}$ s), the mass symmetric fission barrier of the compound nucleus is reduced to zero due to the high angular momentum. The nucleus is trapped behind the mass asymmetric fission barrier at the entrance channel mass asymmetry but it can escape freely after the mass equilibration. In this case of a missing fission barrier the statistical-model pre-fission neutron multiplicity does not exist and the time scales can be deduced from the neutron decay times alone.

Fission time scales have been and are being investigated with neutrons, charged particles and photons emitted before fission. The majority of published work deals with the dominant neutron decay channel.

Within this concept the fission times calculated in the frame of the statistical model decrease exponentially with increasing nuclear temperature. The experimentally determined scission times in fusion-fission and fast fission reactions decrease less steeply (see figs. 9 and 12 of ref. [7]). The discrepancy between the calculated and the measured fission times for temperatures T between 1.5 MeV and 2.5 MeV may only be explained by a dynamical slackening of the separation process due to a strong damping of the collective motion. Similar heavy ion reactions, leading to temperatures up to T=5 MeV in the fused system, reveal also these nuclei to fission slowly within 1 to $3 \cdot 10^{-20}$ s.

Measuring the neutrons with their multiplicities M_{post} evaporated from the separated fission fragments gives the temperature of the nascent fragments at the scission point. This temperature is found to be independent of the initial excitation energy and smaller than T=1.5 to 2.5 MeV. This means that, the later fission occurs, the more efficiently the nuclear system gets cooled by neutron evaporation until the evaporation time, exponentially increasing with decreasing temperature, becomes comparable to the fission time. So, independently of the primary excitation energy in heavy ion fusion-fission and fast fission reactions, fission as the last step in the decay chain occurs for a relatively cold nucleus and is slow, taking 10^{-20} to 10^{-19}s. Fission is the slowest decay process due to a strongly damped collective motion. This is expressed in figs. 9 and 12 of ref. [7].

It has been shown in calculations with the Boltzmann master equation (BME) [14] that the equilibration phase for reactions with ^{27}Al projectiles of energies 10 - 100 MeV/u extends to typically $5 \cdot 10^{-22}$ s depending only little on the injection mechanism [15]. Simulations in the Landau-Vlassov model give the linear momentum transfer to be completed within this time [16]. The fission degrees of freedom must be coupled by a nuclear dissipation mechanism to the intrinsically equilibrated reaction system having a fission barrier. The rate of this dissipation corresponds to a rather large viscosity with the consequence of an overdamped motion towards scission.

The times τ_t and τ_{ss} have been derived in the reaction ^{16}O + ^{142}Nd at 207 MeV in refs. [3] and [17] to be of the order of $6 \cdot 10^{-21}$ and $2\text{-}4 \cdot 10^{-21}$, respectively. It is this sum τ_t + τ_{ss} that is determined in neutron multiplicity experiments, which allow to separate the prescission multiplicity M_{pre} from the multiplicity M_{post} of the fragments by means of the different kinematical focussing of the respective moving sources.

To be able to use the equation for $\tau_{sm}^{(n)}$ as a neutron clock, the number of neutrons M_{pre} minus the statistical model contribution originating during τ_{sm} must be known. In the case of the reaction systems ^{32}S + ^{197}Au, ^{232}Th at 838 MeV [11] τ_{sm} is much smaller than τ_f [16] and for this reason τ_{sm} with the full value M_{pre} can be taken to define the dynamical time scale of the transient phenomenon referred to $0.5\tau_t$ + τ_{ss}. The fission time $\tau_f = \tau_e + \tau_{sm} + 0.5\tau_t + \tau_{ss}$ increases exponentially with the number of prescission neutrons, i.e. the absolute times are essentially determined by the last two neutrons. For symmetric fragmentations τ_f is about 2×10^{-20}s, independent of the linear momentum transfer (LMT) and is systematically higher than for the corresponding asymmetric fragmentations. The fission time τ_f is also increasing with the fissility parameter.

Within the outlined model the fission study of the reaction systems ^{32}S + ^{197}Au and ^{32}S + ^{232}Th with 26.2 MeV/u has been performed and yield the following results [11]: The LMT has been obtained from the fragment velocity vectors and converted into the excitation energies E_x (LMT). The neutron energy spectra indicate four moving sources (preequilibrium, prescission, two fragments) with isotropic Maxwellian emission characteristics. For both reactions and in all LMT bins about 4 preequilibrium neutrons are emitted per fission, the multiplicities and energies being well reproduced by the BME model calculation. The prescission neutron multiplicities increase with LMT according to E_x(LMT). The postscission multiplicities remain constant at about 6 - 8 neutrons (two neutrons more for ^{232}Th due to the higher Q_{eff}-value). The postscission multiplicities and the temperature parameters obtained for the fragment sources imply fission to take place for an elongated system cooled down to 50 - 60 MeV of intrinsic excitation. The accumulation of the evaporation times of all prescission neutrons shows fission of the studied systems to be an inherently slow process taking 5×10^{-21} to 3×10^{-20} s rather independently of E_x(LMT). Fission proceeds faster for asymmetric than for symmetric mass splits. The fission time scale increases with the fissility κ of the reaction system, indicating an increasing distance between saddle and scission point.

The evaluations sketched above depend strongly on the particle level densities used. The fission times derived for the given heavy-ion fusion-fission and fast fission reactions may come out shorter with more realistic level densities [18]. Presently measurements of the charged particles emitted before fission are performed and will give insight into the phenomena of fission dynamics in particular due to their higher sensitivity to the nuclear shape configurations.

2. Antiproton Induced Fission
2.1. Neutrons as a Chronometer
In contrast to these observations in heavy ion induced fusion-fission reactions antiproton induced fission fragment spectroscopy in the experiment PS203 at LEAR [19] indicates a different result for the post-scission neutron multiplicities as a function of the excitation energy of the fissioning nucleus.

A nucleus is excited by annihilation of stopped antiprotons with a small angular momentum transfer. If no direct nucleon measurements are performed the emitted particles cannot easily by ascribed to the emitting phase. This is so for the case of particles emitted during the cascade and in the evaporation phase, because the difference between these two processes has no effect on the fission fragment characteristics. Differently, particle emission before and after scission do have a different effect: the velocity of a fragment will be increased as a result of neutron emission before scission M_{pre}, whereas that after scission M_{post} will not affect the velocity on the average. This allows to estimate how many particles are emitted before and after scission $M_{total} =$

$M_{pre} + M_{post}$.
From the total mass distribution follows, that the $\bar{p}$-annihilation leads on the average to the removal of 26 ± 2 nucleons from ^{238}U, corresponding to an average thermal excitation energy of 220 ± 20 MeV. Basing the calculation of the average particle number emitted before scission on the correlation between the total mass and the total kinetic energy and on the TKE-systematics of fission fragments by Viola [20] gives the ratios M_{pre}/M_{total} as a function of the mass loss. The value of this ratio is relatively constant with the mass loss with an average of 0.44 ± 0.06. This value is definitely smaller than the one derived from the heavy-ion reactions and may indicate for $\bar{p}$ induced fission the emission of relatively more post fission neutrons from hot fragments. This in turn is the indication of a fast fission process leaving fragments behind that evaporate a different number of neutrons at different excitation energies.

2.2. The Λ Hyperon as a Chronometer

The Λ hyperon lifetime is longer than the fission time, and once deposited within a nucleus the strange baryon survives all stages of nuclear fission finally sticking to one of the fragments.

By antiproton annihilation nuclei may be excited to a rather high energy and, due to meson and nucleon emission, may get momenta up to several hundreds of MeV/c. The heavy Λ-hypernucleus is created in antiproton annihilation at rest in about 2 % of all cases with an excitation energy of about 160 MeV (1) $\bar{p}N \to K\bar{K}$ + pions, 2) $\bar{K}N \to \Lambda\pi$), equilibrating thermally to $E_x \approx$ 50 to 100 MeV and T ≈ 1.5 to 2 MeV in about 10^{-22} s with the possibility for the hypernucleus to fission. Whereas heavy hypernuclei decay predominantly via weak nonmesonic interaction $\Lambda + N \to N + N$, light hypernuclei suffer mesonic decay $\Lambda \to N + \pi$.

The spectroscopy of hypernuclei of fission fragments provides the possibility to investigate the dynamics of highly excited fissioning nuclei [21]. In particular the knowledge of the lambda sticking probability as a function of the fragment mass seems valuable for this purpose. Two extreme views are possible:

a) If the formation of fission fragments occurs within a short time after the Λ production, the probability for the hyperon to remain in either fragment of different mass should be proportional to the mass number.

b) If the scission process is slow this probability of Λ caging in one fragment should be influenced by e.g. the Λ binding energies.

In the experiment PS177 at LEAR [21, 22, 23] using the recoil distance technique together with detecting the secondary electron emission from the target by a microchannel plate, prompt and delayed fission events induced by antiprotons stopped in the uranium target could be distinguished. The fission fragment masses were determined by the double velocity method. Fission fragments originating from three different reactions were separated in the experiment:

i) Prompt fission of an excited nucleus induced by antiproton annihilation occuring in the target and giving a large electron signal in the microchannel plate (MCP). The fission fragments are recorded in the unshadowed region only.

ii) Delayed fission of a hypernucleus recoiling from the target with a small MCP signal. The fission fragments are recorded in the shadowed and unshadowed regions.

iii) Prompt fission of a hypernucleus after its formation due to the high excitation energy deposited in the annihilation process. A high MCP signal is registered. The lambda particle is caged in one of the fission fragments, escaping from the target. During the flight of the fragment and within $\tau(\Lambda_{free}) = 10^{-10}$ s, the lambda decays. The released energy gives rise to particle emission changing by this the momentum of the fragment, which may then be deflected into the shadowed region. These are the events searched for.

Events as described under iii) allow to study hypernuclei of fission fragments, and by this fission dynamics, since the lambda particle must have been attached to the fragment detected in the shadowed region from the moment of its binary fission origin. The mass of this fragment can be determined by means of the double velocity technique. ($A = m_1 + m_2$; $m_1 v_1 = m_2 v_2 \rightarrow m_1 v_1 + m_1 v_2 = m_2 v_2 + m_1 v_2 \rightarrow R_1 = m_1/(m_1 + m_2)$ $= v_2/(v_1 + v_2)$; neutrons evaporated isotropically).

In figs. 2 a and b, 3 a and b, and 4 of ref. [23] the experimental results are plotted showing clearly that heavier mass fragments are more often carrying a Λ hyperon than lighter mass fragments. In delayed hypernuclear fission no such correlation is observed (3 b).

The lambda attachment probability may be described in the region of non-vanishing mass yields R_1 by:

$$A_\Lambda(R_1) = 0.5 + \alpha(R_1 - 0.5) \text{ with } \alpha \approx 1.7$$

$\alpha \approx 1.7$ means that the dependence is stronger than the limiting case $\alpha=1$, i.e. proportional to the fragment mass. Regarding the lambda baryon as a probe of the dynamics of neck formation, Nifenecker and Malek [24] in the framework of a schematic scission model based on the experimental data of refs. [21, 22, 23] show the attachment function to be sensitive to the scission configuration, dynamics of neck formation (scission time) and nuclear temperature at scission.

The slope parameter α of the attachment probability is investigated in its relation to the fission dynamics [24]:

To estimate the state the lambda occupies in the nucleus at the moment of scission, the values of the lifetimes of the hyperonic states have to be compared with the fission time. From the values of the transient delay times for fission as given in refs. [7, 8, 9, 10, 11] the Λ decay width of $\Gamma_c = 6.5$ keV determined for the shortest time gives the limit that lambda states with larger widths will have decayed to their thermal equilibrium population before fission. But due to its coupling to higher lying levels by the lambda-nucleon collisions even the lowest energy level of the lambda gets a width. Therefore, the lambda degree of freedom may be considered as thermally equilibrated at the time of fission. Accordingly the occupation of excited states is governed by the Boltzmann factor $\exp(-\varepsilon_{\Lambda i}/T)$ ($\varepsilon_{\Lambda i}$ = excitation energy of the Λ state).

Schematizing the exit configuration for the phase of the neck formation as a cylindrical shape

a) in the sudden approximation the slope of the lambda attachment function is obtained to be much smaller than in the experiment. This condition seems not to match with nature.

b) in the adiabatic approximation, where the fission process is slow enough, the lambda may be expected to be in a statistical equilibrium with the nuclear system until the final state of two separated fragments is formed. The slope of the attachment probability comes out to be much larger than the experimental value indicating that thermal equilibrium in the final fragments is not reached.

Intermediate velocities of the neck formation are considered in a master equation approach for the lambda localization [24].
$P(R_1,t)$ is the probability to observe the lambda in a prefragment with mass ratio R_1. Prefragments are assumed to exist at each step of the transition in the exit scission configurations and would become fragments in a sudden scission geometry at the time t. The change of $P(R_1,t)$ in time Δt is

$$\Delta P(R_1,t) = (-W(1 \to 2) \cdot P(R_1,t) + W(2 \to 1) \cdot (1 - P(R_1,t))) \cdot \Delta t$$

where $W(1 \to 2)$ is the transition of the lambda from fragment 1, with mass ratio R_1, to fragment 2, with the mass ratio $1 - R_1$; $W(2 \to 1)$ is the reverse process.
For almost separated fragments with a small time-independent coupling the equilibrium solution is known and corresponds to the adiabatic case. At equlibrium $dP(R_1,t)/dt=0$ and

$$\frac{W(1 \to 2)}{W(2 \to 1)} = \frac{A(1-R_1,T)}{A(R_1,T)} = \frac{\rho_2}{\rho_1},$$

ρ_i (i=1,2) are the level densities of the lambda baryon in the prefragment i.
For a cylindrical fragment containing the lambda hyperon with the transmission coefficient Θ of the Λ-barrier between the fragments, the transition probabilities are given as

$$W(1 \to 2) = \frac{\Theta T}{2\pi\hbar} \exp(-\{\varepsilon\}/T) / \sum_{n=1}^{\infty} \exp(-n^2 \cdot \varepsilon_{1\|}/T)$$

$$W(2 \to 1) = \frac{\Theta T}{2\pi\hbar} \exp(-\{\varepsilon\}/T) / \sum_{n=1}^{\infty} \exp(-n^2 \cdot \varepsilon_{2\|}/T)$$

Herein is
$\{\varepsilon\}$ $\max(\varepsilon_{1\|}\varepsilon_{2\|})$,
$\varepsilon_{i\|}$ (i=1,2)   longitudinal energy of the lambda in the ith fragment
$\frac{n\varepsilon_{i\|}}{\pi\hbar}$ frequency of the lambda motion along the fragment axis in the nth mode
T   nuclear temperature

The solution of the master equation for T = 1.5 MeV near $R_1 = 0.5$ yields the experimental slope of the attachment probability $\alpha = 1.7$ for a time of neck formation of 2×10^{-21} s. Fig. 3b in ref. [24] displays the comparison between the experimental data and the result from the master equation solution.

Recently Krappe and Pashkevich [25] using the raw data of ref. [23] have presented a preliminary estimate of the lambda attachment rate in solving the time dependent Schrödinger equation assuming adiabaticity for the lambda motion in the ground state of a Saxon-Woods potential with the primary nucleus. This decays along a symmetric fission path α ($\alpha=0$: sphere, $\alpha=0.7$: saddle, $\alpha=1$: scission point). Two velocities are considered, moderately damped and rather undamped motion. Preliminary results of saddle to scission times are of the order of several 10^{-21} s, indicating the strong sensitivity of the attachment probability to the fission time scale also in this approach not including the influence of the collective motion during the phase of neck formation.

3. Muon attachment to fission fragments

Differently to heavy ion and antiproton induced fission reactions, where the fissioning nuclei are highly excited, nuclei in prompt muon induced fission have excitation energies between about 6 and 10 MeV only.

In a number of theoretical works [26, 27, 28, 30, 31, 32, 33, 34, 35] the possibility has been discussed to get information on the dynamics in prompt nuclear fission from the value of the probability P_L with which the muon is attached to the light fission fragment. The results are controversial. The necessity to measure this probability with high statistical accuracy has become obvious. In particular the measurement of the dependence of this quantity on mass number A and total kinetic energy TKE(A) would give a much more solid base for further discussions.

Belovitskij et al. [36] have performed very illustrative work with emulsions and have obtained a value for P_L and results on the muon conversion probability as well. Here we discuss the results of a first electronical measurement of P_L [37].

In the prompt muon induced fission the muon interacts via the electroweak force only. After the fission process it is attached to either fission fragment with a probability of more than 0.98 [38]. The ratio P_L/P_H, of the attachment probabilities of the light and heavy fragment, respectively, depends on the fission time. Because of the higher binding energy of the muon P_H is bigger than P_L and it increases with the fission time. The muon bound to a fission fragment decays into an electron, its antineutrino, and muonic neutrino with a probability of 0.05. Because the lifetime of the bound muon is large as compared to the time of flight elapsing for the fragments to be spatially separated, the corresponding fission fragment can be determined by detecting the fission fragment and the track of the decay electron.

The fission fragments are registered in large area semiconductor detectors. The decay electron is detected with the magnetic spectrometer SINDRUM [39], which consists of 5 multiwire proportional chambers (MWPC) and a hodoscope (fig. 1 of ref. [37]).

In the data analysis electron tracks were accepted if they intersect only one detector area in order to assign the electrons to the fission fragments unambiguously. By applying windows on light and heavy fragments and requiring prompt fission, the attachment probabilities are obtained from the corresponding time spectra $t(\mu^-,e^-)$ of light and heavy fragments shown in figure 2a and b of ref. [37], respectively. Lifetimes were fitted to the data of this figure.

The results listed in table 1 show a decreasing attachment probability with increasing TKE for all three mass windows. The errors of the values are large, however.

The theoretical predictions concerning the possibility to draw conclusions on the dynamics of the fission process from the attachment probabilities as calculated in the respective models are controversial. Since the calculations have been either performed for the nuclei ^{238}U or ^{242}Pu, the result obtained experimentally for ^{237}Np has been transformed to these nuclei. For this purpose the attachment probability has been approximated by a Fermi-distribution. The attachment probability $\bar{P}_L$ averaged over the light fragment mass includes, within the error bars, the value of Karpeshin et al. [35] but also that of Ma et al. [30] with P_L = 4.9 % and 5.7 %, respectively. The calculations of Bracci et al. [33, 34] and Maruhn et al. [31, 32] have been performed for the fragment mass ratio $A_H/A_L = 1.2$ for ^{238}U and ^{242}Pu, respectively. A direct comparison with the corresponding experimental value of the mass window m_L = (104.5 - 111.5) u and correcting for the experimental mass resolution shows agreement with the value 0.10 of Bracci et al. [33, 34]. The model of Maruhn et al. [31, 32] gives the dependence of the value of the attachment probability on the friction strength. The experimental result corresponds to a viscosity constant of $\lambda \approx$ 1900 MeVfm/c.

Bracci et al. [33] and Karpeshin et al. [35] predict an increase of P_L for the TKE increasing within the mass window. The results given in table 1 have the opposite trend.

Karpeshin [40] has recently described the muon distribution in the adiabatical base of two quasimolecular muonic states $1s\sigma$ and $2p\sigma$ for the Np-data given in tab. 1. The result is shown in fig. 2. The mass dependence of the attachment probability has been transformed into the charge dependence. Whilst a good agreement between the theoretical and the experimental mean value of the attachment probability averaged over all the fragment masses and kinetic energies is obtained, the slopes of the curves in fig. 2 deviate from each other. It has been shown that this difference cannot be attributed to the difference between fissioning nuclei, but might rather be due to a dynamical origin enlarging $P_L(Z)$ for small Z and (perhaps) diminishing it for the light fragments close to the symmetric yields.

In extension of the previous work [40] the following dynamical effects have been considered by Karpeshin: Scission of the neck occurs at $R_{sc} \approx$ 20 fm and the preceeding stage of the nuclear deformation from zero to the scission value takes place within a time interval (fission time) of several 10^{-21} s. The nascent spherical fragments were considered as continously and uniformly accelerated. Thus including the fission stage really influences the P_L-values and restricts the fission time to t$\leq 10^{-21}$s. The P_L-value is diminished for more symmetric fragments, but increased for more asymmetric fragments as compared to the case of omitting the fissioning stage. This diminishes the deviation between the slopes of the curves in fig. 2.

Furthermore, the calculated dependence of P_L on TKE gives P_L = 4.4% for TKE = 160 MeV and t = $2 \cdot 10^{-21}$ s. For larger TKE but three times shorter fission time t = $0.7 \cdot 10^{-21}$ s one obtains P_L = 1.9%, i.e. a 2.3 times smaller value. Thus the attachment probability to the light fragment happens to be approximately proportional to the scission time, despite different TKE. It is commonly believed that higher TKE correspond to a shorter prescission time. Hence, if the fragments with TKE > 170 MeV originate in an approximately 20 to 30 % smaller fission time than the fragments with TKE < 170 MeV this gives P_L-values for the faster fragments about 20 % smaller in agreement with the experimental values.

The result may be interpreted as indication of a short fission time t $\sim 10^{-21}$ s in prompt muon induced fission. It is obvious that improving the spectroscopic data of tab. 1 will provide a firm basis for safe conclusions.

Principally, strong anomalous E1 internal conversion may occur as a result of the superposition of nuclear odd-even surface vibrations. Considering anomalous E1 conversion in nuclei with octupole and simultaneously quadrupole deformations and muon shake-off in prompt fission Karpeshin [41] has suggested two phenomena that may be used to understand the fission dynamics. Such deformations give rise to strong penetration effects in internal conversion for E1 transitions. Applying this idea to evaluate the muon shake-off probability P_{sh} gives a value of $P_{sh} \approx 0.5$ % per prompt fission, in good agreement with the experimental result [36]. The shake-off probability is proportional to the dissipative time t_{diss} of the oscillating fragments. Therefore, the experimental proof of converted muons in connection with fission fragment spectroscopy would give direct information concerning dissipation in nuclei. For cold fission when fragments with small deformation and high final kinetic energy are formed, the shake-off contribution must be small.

4. Conclusion

In summarizing, results from quite different and independent experimental approaches have been described tackling the problem of measuring the fission times of hot and cold heavy nuclei. The experimental data deserve refinement, in particular with respect to the fission fragment spectroscopy, but also with respect to the statistical accuracy. The experimental methods described all rely on interpretations within the framework of theoretical models. Improving eihter part will mutually stimulate refinements or new ways in the other field. The challenge to understand more microscopically the collective phenomenon of nuclear fission requires the knowledge of the fission times, of the friction process and their propagation, and of the least known vibrations and oscillations together with their dissipation times during the stage of neck formation.

References

[1] H. A. Weidenmüller, Zhang Jing-Shang, Phys. Rev. **C29**, 879, (1984)

[2] J. W. Negele, S. E. Koonin, P. Möller, J. R. Nix, A. J. Sierk, Phys. Rev. **C17**, 1098 (1978)
J. W. Negele, Nucl. Phys. **A502**, 371c (1989)

[3] P. Grangé, S. Hassani, H. A. Weidenmüller, A. Gavron, J. R. Nix, A. J. Sierk, Phys. Rev. **C34**, 209 (1986)

[4] K. H. Bhatt, P. Grangé, B. Hiller, Phys. Rev. **C33**, 954 (1986)

[5] H. A. Weidenmüller, Nucl. Phys. **A502**, 387, (1989)

[6] G. F. Bertsch, Les Houches, Session XXX, Nuclear Physics with Heavy Ions and Mesons, Vol. 1, 1978, p. 175, North Holland

[7] D. J. Hinde, D. Hilscher, H. Rossner, Nucl. Phys. **A502**, 497c (1989)

[8] D. Hilscher, Phys. Blätter DPG **47**, 1073 (1991), and refs. therein

[9] D. J. Hinde, D. Hilscher, H. Rossner, Nucl. Phys. **A538**, 243c (1992)

[10] W. P. Zank, Hilscher, G. Ingold, U. Jahnke, M. Lehmann, H. Rossner, Phys. Rev. **C33**, 519 (1986)

[11] E. Mordhorst, M. Strecker, H. Frobeen, M. Gasthuber, W. Scobel, B. Gebauer, D. Hilscher, M. Lehmann, H. Rossner, Th. Wilpert, Phys. Rev. **C43**, 716 (1991) and refs. therein

[12] H. Rossner, D. J. Hinde, J. R. Leigh, J. P. Lestone, J. O. Newton, J. X. Wei, S. Elfström, Phys. Rev. **C45**, 719 (1992)

[13] K. H. Bhatt, P. Grangé, B. Hiller, Phys. Rev. **C33**, 954, (1986)
D. J. Hinde, A. Ogata, M. Tanaka, T. Shimoda, N. Takahashi, A. Shinohara, S. Wakamatsu, K. Katori, H. Okamura, Phys. Rev. **C39**, 2268 (1989)

[14] M. Blann, Phys. Rev. **C31**, 1245 (1985)

[15] W. Scobel, E. Mordhorst, M. Strecker, R. Caplar, Proc. 6th Int. Adriatic Conf. Nucl. Phys. Dubrovnik 1987, Ed. N. Cindio, W. Greiner, R. Caplar, World Scientific, Singapore 1987 p. 267

[16] J. Galin, Nucl. Phys. **A488**, 297c (1988)
D. Jaquet, G. F. Peaslee, J. M. Alexander, R. Borderine, E. Duek, J. Galin, D. Gardes, C. Grégoire, D. Guerreau, H. Fuchs, M. Lefort, M. F. Rivet, X. Tarrago, Nucl. Phys. **A511**, 195 (1990)
D. Jaquet, J. Galin, R. Borderine, D. Gardes, D. Guerreau, M. Lefort, F. Monnet, M. F. Rivet, X. Tarrago, E. Duek, J. M. Alexander, Phys. Rev. **C32**, 1594 (1985)

[17] A. Gavron, Nucl. Phys. **A502**, 515c (1989)

[18] M. G., Mustafa, M. Blann, A. V. Ignatyuk, S. M. Grimes, Phys. Rev. **C45**, 1078 (1992)

[19] T. von Egidy, P. Baumann. H. Daniel, T. Haninger, F. J. Hartmann, P. Hofmann, Y. S. Kim, M. S. Lotfranaei, W. Schmid, A. S. Botvina, Ye. S. Golubeva, A. S. Iljinov, V. G. Nedorezov, A. S. Sudov, J. Jatrzebski, W. Kurcewicz, P. Lubinski, A. Grabowska, A. Stolarz, D. Hilscher, D. Polster, H. Rossner, H. Machner, G. Riepe, P. David, H. S. Plendl, J. Lieb, B. Wright, K. Ziock, Int. Nucl. Phys. Conf. Wiesbaden, July 26 - August 1, 1992

[20] V. E. Viola, Nucl. Data Section **A1**, 391 (1966)

[21] S. Polikanov, Nucl. Phys. **A502**, 195c (1989)

[22] J. P. Bocquet, M. Rey-Campagnolle, G. Ericson, T. Johansson, J. Konijn, T. Krogulski, M. Maurel, E. Monnand, J. Mougey, H. Nifenecker, P. Perrin, S. Polikanov, C. Ristori, G. Tibell, Phys. Lett. **B182**, 146 (1986); Phys. Lett. **B192**, 312 (1987)

[23] M. Rey-Campagnolle et al., in First Biennal Conf. on Low Energy Antiproton Physics, Eds. P. Carlson, A. Kerek, S. Szilagyi, World Scientific 1991, p. 385
F. Malek, et al., ibid. p. 391

[24] H. Nifenecker, F. Malek, Nucl. Phys. **A531**, 539 (1991)

[25] H. J. Krappe, V. V. Pashkevich, Verhdlg. der DPG **1**, 94 (1992)
H. J. Krappe, private communication

[26] V. A. Karnaukhov, Sov. J. Nucl. Phys. **28**, 621 (1978)

[27] Y. N. Demkov, D. F. Zaretski, F. F. Karpeshin, M. A. Listengarten, JETP Letters **28**, 287 (1978)

[28] D. F. Zaretski, F. F. Karpeshin, M. A. Listengarten, V. N. Ostrovski, Sov. J. Nucl. Phys. **31**, 24 (1980)

[29] P. Olanders, S. G. Nilsson, P. Möller, Phys. Lett. **B90**, 193 (1980)

[30] Z. Y. Ma, X. Y. Wu, G. S. Zhang, Y. C. Cho, Y. S. Wang, J. H. Chiou, S. T. Sen, F. C. Yang, J. O. Rasmussen, Nucl. Phys. **A348**, 446 (1980), Phys. Lett. **B106**, 159 (1981)

[31] J. A. Maruhn, V. E. Oberacker, C. Maruhn-Rezwani, Phys. Rev. Lett. **44**, 1576 (1980)

[32] V. E. Oberacker, J. A. Maruhn, LAMPF II, Workshop Los Alamos 1982, **LA-9798-P**, UC-28

[33] L. Bracci, G. Fiorettini, Nucl. Phys. **A423**, 429 (1984)

[34] P. Boi, L. Bracci, G. Fiorettini, P. Quarati, Phys. Lett. **B132**, 39 (1983)

[35] F. F. Karpeshin, M. S. Kashiev, V. A. Kashieva, Sov. J. Nucl. Phys. **45**, 965 (1987)

[36] G. E. Belovitskij V. N. Baranov, C. Petitjean, XVIIIth International Symposium on "Nuclear Physics and Chemistry of Fission" in Gaussig/Dresden 1988, Report Zentralinstitut für Kernforschung Rossendorf, **ZfK-732** (1991) 249, Ed. H. Märten, D. Seeliger

[37] F. Risse, W. Bertl, P. David, R. Engfer, H. Hänscheid, E. Hermes, J. Konijn, C. T. A. M. de Laat, H. Pruys, Ch. Rösel, W. Schrieder, A. Taal, and D. Vermeulen, Z. Phys. A - Hadrons and Nuclei **339** (1991) 427

[38] G. E. Belovitskij, L. V. Suhov, C. Petitjean, Lecture notes in Physics Vol. **158**, 71 (1982)

[39] SINDRUM, W. Bertl. et al., SIN-Proposal R-80-06.2, 1981

[40] F. F. Karpeshin, Yaden. Fizika, **55,1** 1992 (Engl. transl. Sov. J. Nucl. Phys.)

[41] F. F. Karpeshin, Yaden. Fizika, in press (Engl. transl. Sov. J. Nucl. Phys.)

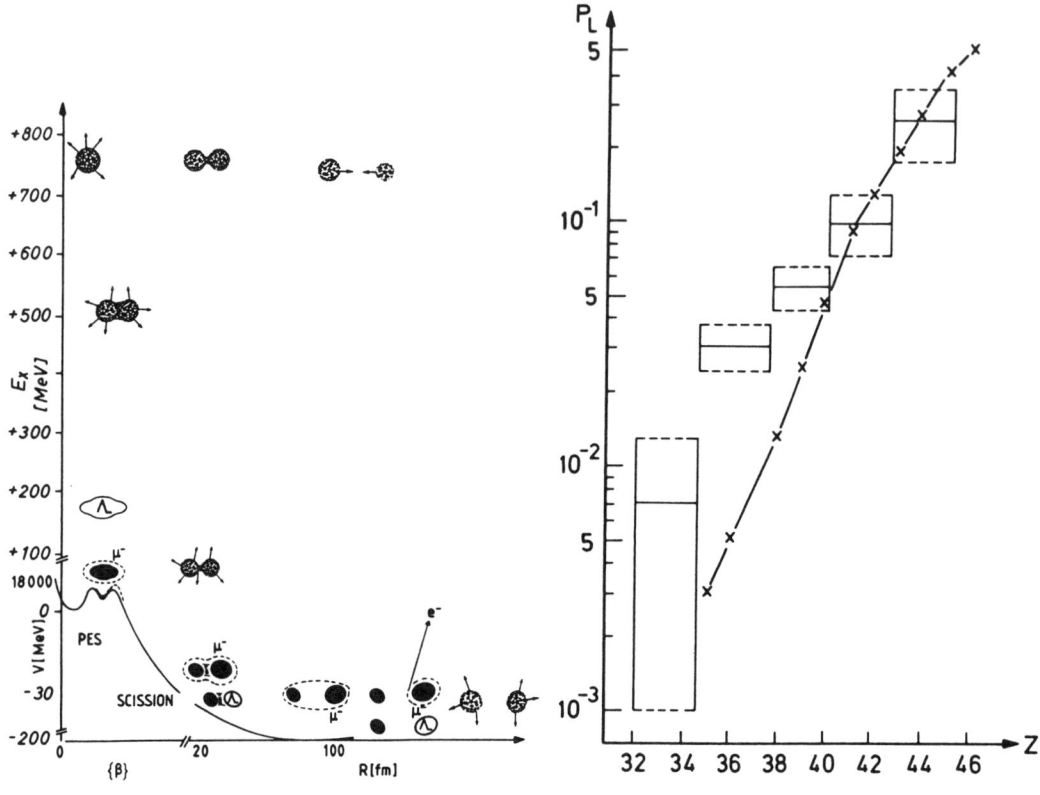

Figure 1:

Figure 2:

Table 1: Attachment probabilities P_L for some bins (Δm_L, and ΔTKE) of the light mass m_L and total kinetic energy (TKE) of fission fragments. The probabilities are corrected for conversion only.

Δm_L	ΔTKE	P_L(M,TKE) $[10^{-2}]$
104.5-111.5	100-170	9.5±3.7
	170-250	6.9±2.5
97.5-104.5	100-170	6.2±1.9
	170-250	4.9±1.3
90.5-97.5	100-170	5.2±1.3
	170-250	3.1±1.3

Muonic Atoms Spectroscopy

Laser Spectroscopy of Muonic Hydrogen

D. CHATELLARD[1], F. CIOCCI[2], G. DATTOLI[2], P. DE CECCO[3], F. DELLA VALLE[4], M. DENORÉAZ[1], A. DORIA[2], J.-P. EGGER[1], G.P. GALLERANO[2], L. GIANNESSI[2], G. GIUBILEO[2], P. HAUSER[3], E. JEANNET[1], F. KOTTMANN[3], G. MESSINA[2], E. MILOTTI[4], C. PETITJEAN[3], L. PICARDI[2], A. RENIERI[2], C. RIZZO[4], C. RONSIVALLE[2], L.M. SIMONS[3], D. TAQQU[3], A. VACCHI[4], A. VIGNATI[2], E. ZAVATTINI[4]

presented by E. Zavattini

[1]Institut de Physique de l'Université, Breguet 1, CH-2000, Neuchâtel, Switzerland
[2]ENEA - Area INN, Dipartimento Sviluppo Tecnologie di Punta, Centro Ricerche Energia, I-00044 Frascati, Italy
[3]Paul Scherrer Institut, CH-5232 Villigen PSI, Switzerland
[4]Dipartimento di Fisica dell'Università di Trieste and Istituto Nazionale di Fisica Nucleare, Sezione di Trieste, Via Valerio 2, I-34127 Trieste, Italy

Abstract.

We present here a discussion about the physics motivations and the experimental feasibility of our recent experimental proposal "Spectroscopy of Muonic Hydrogen" presented to the Dec. 1991 meeting of the PSI Benutzervesammlung (proposal R-9206-1).

1. Introduction.

In this paper we describe an experimental plan to measure, with high accuracy, the energy differences between levels with principal quantum number $n \leq 4$ of muonic hydrogen (μp). The experimental method is to irradiate the excited muonic atom, during its formation, with an intense, tuneable laser radiation, and detect at the same time the K_α and K_β X-rays emitted by the excited $(\mu p)^*$ system as it cascades electromagnetically toward the ground level. When the frequency of the laser radiation has the correct value to excite a transition between two of the levels involved in the cascade the intensities of some of the K lines will change. By detecting these intensity changes, while tuning the laser, we will be able to measure the energy difference between the two levels with great precision [1].

To perform such an experiment we need

1. A proper low-energy μ^--beam. Each incoming muon, which may either decay in flight or stop in the target, must be tagged by a proper scintillator counter approximately 1÷2 µsec before it stops in the hydrogen gas target to trigger in advance the necessarily intense electromagnetic radiation source.

2. A special cavity-target, containing in the stopping region a proper infrared cavity, where the hydrogen gas is at a rather low pressure. To reduce the interaction rate of collision of the $(\mu p)^*$ atom with the neighbouring hydrogen molecules H_2, the pressure must be kept around or below 30 mbar [2]. Moreover the spatial distribution of the stopped muons in the target must have sufficiently small dimensions so that we can achieve a high electromagnetic energy density in the stopping region.

3. A pulsed tuneable far-infrared source of high and stable intensity with a relative linewidth $\leq 0.5\%$.

4. High-resolution X-ray detectors: these detectors must be able to distinguish the K_α, K_β and K_γ lines of the (μp) system.

In what follows we wish to discuss first the physics motivations for such a program; then we will discuss its experimental feasibility.

2. Theoretical overview.

The energy levels of the muonic hydrogen atom can be computed within the framework of the Dirac theory: the hyperfine corrections, recoil corrections, etc., can be introduced as usual. To agree with the experimental results, one has to include also the Quantum Electrodynamics radiative corrections.

Thus the list of physics motivations in the field of muonic hydrogen spectroscopy includes the following items:

i) Test of the QED radiative corrections in a lepton-hadron bound system [1,3].

The QED radiative corrections, to the lowest order, can be classified in two broad categories: the vertex corrections and the photon propagator corrections. The first are associated to the quantum fluctuations of the electromagnetic field, whereas the second, called QED vacuum polarization corrections, are associated with the quantum fluctuations of the electron-positron field.

In general for a muonic atom the QED vertex corrections are extremely small compared to the QED vacuum polarization corrections [3].

The muonic atoms are the ideal systems to test experimentally these last corrections.

Laser Spectroscopy of Muonic Hydrogen

ii) Study of the corrections due to the proton finite size. We wish to reach sufficient precision in the measurements to be able to infer from the data significant informations about the proton form factor as seen by the muon.

iii) Test of μ-e universality, or more generally search for anomalous μ-P interactions.

iv) Experimental study of (μp)* atom formation and the ensuing cascade at different values of the hydrogen gas target pressure (3).

v) Extension of the above studies to the (μD) system.

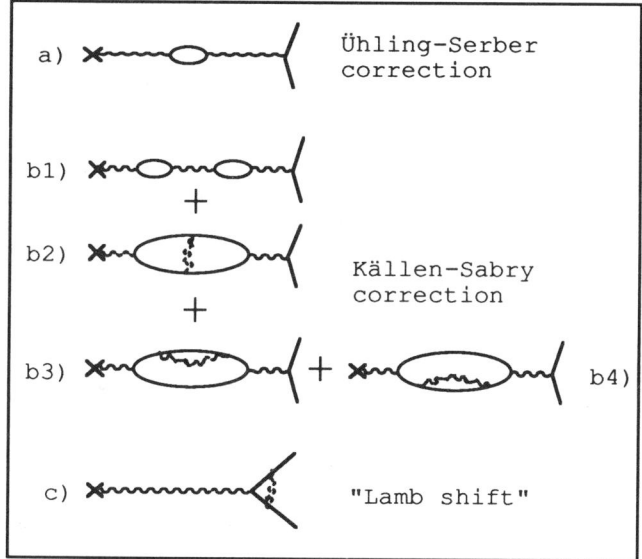

Figure 1: The lowest QED radiative corrections.
a) and b) vacuum polarization contributions; c) vertex correction.

To clarify point "i" it is especially important to review briefly the state of the art.
We recall that the values of the fine-structure constant α determined from condensed-matter physics (and independently of any QED corrections) are

from quantum Hall effect [4]:

$$\alpha_{QHE}^{-1} = 137.0359979 \ (32) \tag{1}$$

from AC Josephson effect [5]:

$$\alpha_{acJ}^{-1} = 137.0359770 \ (77) \tag{2}$$

The difference between these two values is approximately 3 standard deviations, a fairly large discrepancy if one recalls that both values should be free of "theoretical error". It has been argued [6] that the theoretical bases of the two determinations are not too clear and the theoretical consistency of the two methods has never been proved.

Obviously this discrepancy is crucial in all those cases in which the quantity measured depends very strongly on the value of α.

Let us now look at the results of the "g-2" experiments [7] as a source of the QED tests.

The electron anomaly $a_e = \frac{g-2}{2}$ can be computed theoretically as an expansion in terms of α:

$$a_e^{th} = c_1(\alpha/m) + c_2(\alpha/m)^2 + c_3(\alpha/m)^3 + c_4(\alpha/m)^4 + \ldots \delta_{ae} \qquad (3)$$

where $c_1, c_2, c_3, c_4, \delta_{ae}$ are quantities calculated using QED: using for α the value (1) one gets [6]

$$a_e^{th} \text{ (QHE)} = 1.1596521400 \, (53) \, (41) \, (271) \, 10^{-12} \qquad (4)$$

Using (2) would lead instead to a value a_e^{th} (acJ) $\approx 1.159651910 \, 10^{-12}$. This difference shows that the experimental values [7]

$$a_{e^+}^{exp} = 1.596521884 \, (43) \, 10^{-12}$$

$$a_{e^-}^{exp} = 1.596521879 \, (43) \, 10^{-12}$$
(5)

favour the value from the Quantum Hall effect rather than the one from AC Josephson effect.

Since the relative vacuum polarization correction to a_e^{th} is about 100 p.p.m. and the α theoretical determination discrepancy is 0.2 p.p.m. one sees that these measurements provide a test of this QED contribution at a level $2.0 \, 10^{-3}$.

Since a_e is so critically dependent on α equation (5) should rather be used to define α in the framework of QED; in fact assuming QED it is possible to derive the most precise value of α to date:

$$\alpha_{g-2}^{-1} = 137.03599222 \, (94) \qquad (6)$$

α_{g-2}^{-1} is an order of magnitude more accurate than α_{QHE}^{-1} and α_{acJ}^{-1}.

In conclusion to test QED vacuum polarization contributions an experiment that is less dependent on the value of α is absolutely needed.

We turn now to the ($2S_{1/2}$-$2P_{1/2}$) Lamb shift results for ordinary hydrogen as a QED test :

The latest experimental results are [8]

$$\Delta E_{L.S.}^{exp} = 1057.845 \ (9) \ \text{MHz}$$

$$\Delta E_{L.S.}^{exp} = 1057.851 \ (2) \ \text{MHz}$$

(7)

and these can be compared with the theoretical expectations:

$$\Delta E_{L.S.}^{th} = 1057.853 \ (13) \ \text{MHz for a proton rms radius} = 0.805 \ (11) \ \text{fm} \quad (8)$$

$$\Delta E_{L.S.}^{th} = 1057.871 \ (13) \ \text{MHz for a proton rms radius} = 0.862 \ (11) \ \text{fm} \quad (9)$$

Comparing (7), (8) and (9) and remembering that the QED contribution to (8) or (9) is ≈ 2.7 10^{-2} [6] one sees that the test of this term, due to the uncertainty of the proton rms radius, is at a level 1 10^{-3}.

Of course assuming QED to be valid values (8) and (9) can be used to deduce the proton rms radius as seen by the electron. Thus one obtains [9]

$$(\text{proton rms radius})^e = 0.785 \pm 0.007 \ \text{fm}$$

The agreement with (8) is barely acceptable, and there is a contradiction with (9).
This experiment is indeed less dependent on the α value as far as the QED vacuum polarization contributions test is concerned, however unfortunately it depends on the important empirical parameter that is the proton rms radius.

In table 1 we summarize the present experimental status of the QED vacuum polarization tests.

Now to support the main motivation of proposal R9206-1 presented at the P.S.I [1], a few comments are necessary:

a) It is interesting to see that today, the QED vacuum polarization corrections are checked to a level of about 10^{-3}. The situation is almost the same as we had at the Zurich conference in 1978 [10] where a similar discussion was held. Since then the experimental accuracies in the Lamb shift and g-2 experiments have improved by more than an order of magnitude.

b) Given the reasons of the limitation in the QED vacuum polarization tests it is clear that an accurate measurement on the elementary muonic system (μp)* can lead to a significant improvement and push the measurement accuracy below the 10^{-3} level.

In fact,
- the dependence on α of the energy level differences is sufficiently weak that one is allowed to do QED vacuum polarization contribution tests beyond 10^{-3};
- choosing transitions that do not involve S levels one can avoid corrections due to the proton rms radius;
- at low pressure in the H_2 gas target there are certainly no electron screening corrections for the neutral $(\mu p)^*$ system

Table 1

exp.	momentum transfer (keV/c)	limited by	QED v.p. test	quantity derived assuming QED	ref.
a_e	500	value of α	$2 \; 10^{-3}$	α_{g-2}	[6]
$\Delta E_{L.S.}$	2	proton form factor	$1 \; 10^{-3}$	proton rms radius = 0.785 ± 0.007 fm	[9]
(2s - 2p) in μ^-He	370	nuclear form factor	$1.7 \; 10^{-3}$	He rms radius = 1.673 ± 0.003 fm	[11]
$(3d_{5/2} - 2p_{3/2})$ in $^{28}Si\mu^-$ and $^{24}Mg\mu^-$	1000	electr. screening	$0.95 \; 10^{-3}$	-----------	[12]

In the final part of this paper we will show how the planned experiment might lead to an improved test of the QED vacuum polarization terms at a level of $(1 \div 2) \; 10^{-4}$. Such an improvement is possible and it can be done only at the PSI where muon beams that fulfil some of the necessary conditions for such an experiment already exist.

We will finish this section by looking more in detail at the energy levels for $n \leq 4$ for the (μp) system and at the lifetimes of the different levels.

The width Γ of the resonance will depend on the lifetimes of the levels involved: since we wish to reach a precision of $(2 \div 1) \; 10^{-4}$ in locating the resonances we think that the most appropriate transitions to look at are:

	line	$\Gamma/\Delta E$
1.	4S-4P	$4.2 \; 10^{-4}$
2.	4P-4D	$4.3 \; 10^{-3}$
3.	3P-3D	$5.3 \; 10^{-3}$

Transitions (2) and (3) are particularly suitable for a QED vacuum polarization test since no proton rms radius corrections are needed, whereas transition (1) is suitable to extract information about the (proton rms radius)$^\mu$.

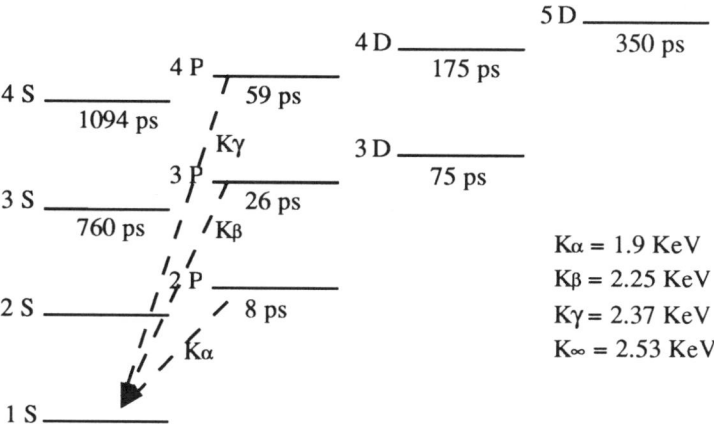

Figure 2: K lines and lifetimes of different levels for the (μp) system

If we turn again to figure 2 we may also note that each of the transitions schematically indicated actually represents many lines due to hyperfine sublevels.

Table 2 lists various energy level differences calculated from first order perturbation theory at the 1-2% level. Table 3 shows how the total energy is shared among the various contributions.

As it will be clarified later we plan to start the experimental program [1] measuring with accuracy the lines:

$$\Delta_3 = 3D_{5/2}{}^7 - 3P_{3/2}{}^5 , \qquad (10)$$

$$\Delta_4 = 4D_{5/2}{}^7 - 4P_{3/2}{}^5 .$$

As far as the measurements of the 4S - 4P line and the study of (μD) atom are concerned these should be a further step in the future program.

Table 2: The 4P-4D, 3P-3D and 4S-4P calculated energy differences for the various sublevels, arranged in order of increasing frequency (no vertex or finite size correction).

transition	hν	frequency	wavelength
$4P^5_{3/2} - 4D^3_{3/2}$	2.284 meV	552.2 GHz	543.275 μm
$4P^5_{3/2} - 4D^5_{3/2}$	2.566 meV	620.5 GHz	483.506 μm
$4P^5_{3/2} - 4D^5_{5/2}$	2.705 meV	654.1 GHz	458.676 μm
$4P^3_{3/2} - 4D^3_{3/2}$	2.754 meV	666.0 GHz	450.467 μm
$4P^5_{3/2} - 4D^7_{5/2}$	2.886 meV	697.9 GHz	429.837 μm
$4P^3_{3/2} - 4D^5_{3/2}$	3.037 meV	734.2 GHz	408.587 μm
$4P^3_{3/2} - 4D^5_{5/2}$	3.175 meV	767.8 GHz	390.714 μm
$4P^3_{1/2} - 4D^3_{3/2}$	3.218 meV	778.1 GHz	385.546 μm
$4P^3_{1/2} - 4D^5_{3/2}$	3.500 meV	846.4 GHz	354.451 μm
$4P^1_{1/2} - 4D^3_{3/2}$	4.394 meV	1063. GHz	282.342 μm
$3P^5_{3/2} - 3D^3_{3/2}$	5.307 meV	1283. GHz	233.803 μm
$3P^5_{3/2} - 3D^5_{3/2}$	5.976 meV	1445. GHz	207.621 μm
$3P^5_{3/2} - 3D^5_{5/2}$	6.305 meV	1525. GHz	196.779 μm
$3P^3_{3/2} - 3D^3_{3/2}$	6.422 meV	1553. GHz	193.198 μm
$3P^5_{3/2} - 3D^7_{5/2}$	6.735 meV	1629. GHz	184.210 μm
$3P^3_{3/2} - 3D^5_{3/2}$	7.091 meV	1715. GHz	174.966 μm
$3P^3_{3/2} - 3D^5_{5/2}$	7.42 meV	1794. GHz	167.203 μm
$3P^3_{1/2} - 3D^3_{3/2}$	7.521 meV	1819. GHz	164.960 μm
$3P^3_{1/2} - 3D^5_{3/2}$	8.190 meV	1980. GHz	151.482 μm
$3P^1_{1/2} - 3D^3_{3/2}$	10.31 meV	2493. GHz	120.346 μm
$4S^3_{1/2} - 4P^1_{1/2}$	25.89 meV	6261. GHz	47.9146 μm
$4S^3_{1/2} - 4P^3_{1/2}$	27.07 meV	6546. GHz	45.8326 μm
$4S^3_{1/2} - 4P^3_{3/2}$	27.53 meV	6658. GHz	45.0606 μm
$4S^3_{1/2} - 4P^5_{3/2}$	28.00 meV	6771. GHz	44.3035 μm
$4S^1_{1/2} - 4P^1_{1/2}$	29.42 meV	7114. GHz	42.1679 μm
$4S^1_{1/2} - 4P^3_{1/2}$	30.60 meV	7399. GHz	40.5469 μm
$4S^1_{1/2} - 4P^3_{3/2}$	31.06 meV	7511. GHz	39.9416 μm

Laser Spectroscopy of Muonic Hydrogen

Table 3: the main contributions to the 3P-3D energy differences

Transition	FS	HFS	Uhling	Källen	Total
$3P^5_{3/2} - 3D^3_{3/2}$	0. meV	0.8365 meV	-6.103 meV	-0.03975 meV	-5.307 meV
$3P^5_{3/2} - 3D^5_{3/2}$	0. meV	0.1673 meV	-6.103 meV	-0.03975 meV	-5.976 meV
$3P^5_{3/2} - 3D^5_{5/2}$	-0.8311 meV	0.6692 meV	-6.103 meV	-0.03975 meV	-6.305 meV
$3P^3_{3/2} - 3D^3_{3/2}$	0. meV	-0.2788 meV	-6.103 meV	-0.03975 meV	-6.422 meV
$3P^5_{3/2} - 3D^7_{5/2}$	-0.8311 meV	0.239 meV	-6.103 meV	-0.03975 meV	-6.735 meV
$3P^3_{3/2} - 3D^5_{3/2}$	0. meV	-0.948 meV	-6.103 meV	-0.03975 meV	-7.091 meV
$3P^3_{3/2} - 3D^5_{5/2}$	-0.8311 meV	-0.4461 meV	-6.103 meV	-0.03975 meV	-7.42 meV
$3P^3_{1/2} - 3D^3_{3/2}$	-2.493 meV	1.115 meV	-6.103 meV	-0.03975 meV	-7.521 meV
$3P^3_{1/2} - 3D^5_{3/2}$	-2.493 meV	0.4461 meV	-6.103 meV	-0.03975 meV	-8.19 meV
$3P^1_{1/2} - 3D^3_{3/2}$	-2.493 meV	-1.673 meV	-6.103 meV	-0.03975 meV	-10.31 meV

3. Requirements for the experimental apparatus.

When a μ^- is stopped in the H_2 gas target cavity it is captured by a proton and forms an atom with high n (n≈14), which subsequently cascades electromagnetically to the ground level. The natural transition probabilities combined with the statistical nature of the cascade process, yield well-defined intensities for the K_α, K_β, etc. lines at each gas pressure.

The cavity is also filled with FIR radiation from the tuneable source: when the frequency of this radiation has the right value to excite a transition between levels (like 3D -> 3P), the cascade process is changed and the relative intensities of the K lines also change.

We have already listed the most important parts of the apparatus that are necessary to implement this experimental program, and now we discuss in more detail points (2) and (3).

3.1 The beam, the target and the detector. As far as the low energy μ^- beam is concerned, the PSI accelerator complex has beams with momentum p less or equal to 30 MeV/c (πE1, πE5) and these beams are adequate when equipped with a high voltage separator to eliminate the electron contamination.

The x-ray detectors will be provided both by the Neuchatel group (commercial CCD's that have already been successfully tested in other measurements [13]) and by the Trieste group (silicon drift chambers, which are still at the research stage and have improved performances [14]).

As target chamber we plan to use the "Cyclotron Trap" developed by L. Simons and collaborators [15]. In this target the incoming muon is tagged by a thin scintillator, and then it spirals to the centre of the trap as it loses energy in the rarefied H_2 gas that fills it. Eventually the muons stop in the centre of the trap: it has been shown that the muon stop distribution is a cigar-shaped gaussian distribution with a radius of 1 cm and with a length of 4 cm. This allows one to have a high energy density electromagnetic field overlapping the muon stop distribution and contained inside a Fabry-Perot resonant cavity.

Moreover a sufficiently low H_2 pressure can be used to stop the negative muons in the target so that we can minimize (especially in the case of the Δ_3 line) the statistical population of the 3P level. This is important since the deexcitation of this level is a main source of constant background. Fig. 3 shows the expected populations of the 3D and 3P levels versus pressure [16]: clearly the low pressure is quite convenient since the 3D level is highly populated while the 3P is not. We plan to work in the range 10-30 mbar. Similar considerations apply to the n = 4 populations.

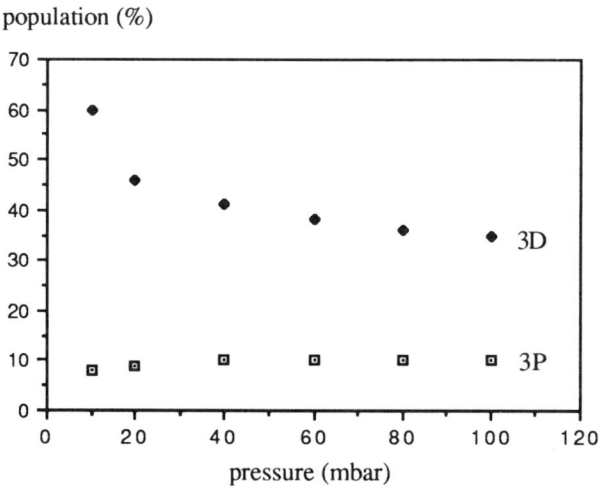

Figure 3: Expected populations of the 3D and 3P levels vs. pressure [16]

Using the tagging scintillator as a start signal it has been found experimentally that the stopping time distribution has a FWHM of about 300 nsec. We can also tune the pressure and the magnetic field in the trap so that the muons stop in the centre of the trap approximately 1.5 μsec after the start signal.

Thus the muons stop in a temporal window that is 300 nsec wide and appears 1.5 μsec after the scintillator signal: this allows a synchronization with the tuneable laser source (which has a discharge time approximately 1.5 μsec long).

Figure 4 shows how these parts might be arranged.

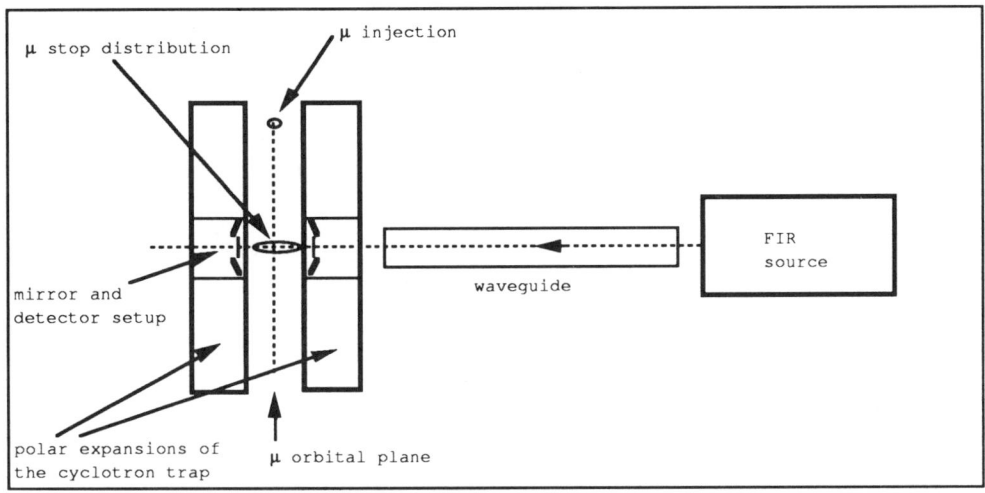

Figure 4: schematic experimental layout

3.2 The FIR source. After having studied various options for the intense and tuneable electromagnetic source, we have decided that the only possibility is to use a FEL (Free Electron Laser) [17].

This source has the following characteristics:

1. it is continuously tuneable in a rather wide range of frequency simply by varying the magnetic field or the undulator gap (major changes are also allowed by changing the undulator parameters or the energy of the LINAC that feeds the FEL)

2. it is triggerable within a time of less of 1.5 µs and gives a train of short pulses of radiation (for a total macropulse length greater than 0.5 µs), and its repetition rate can reach 100 to 200 Hz.

3. The short pulses in a macropulse are however sufficiently long ($\geq$ 100 ps) to give a sufficiently small linewidth. The lines (Δ_3 and Δ_4) that we wish to study have a width of about $5 \cdot 10^{-3}$, and the FEL has approximately the same linewidth.

4. The peak energy in each pulse is sufficiently high to induce an acceptable number of transitions above the background due to spontaneous deexcitation.

The FEL is presently under study at the ENEA laboratories (Frascati); table 4 gives an approximate set of FEL parameters.

Eventually we expect to have a $\Delta\nu/\nu \approx 0.4\ \%$ and a 60 kW average macropulse power.

Table 4: FEL parameters (3P - 3D line operation)

Electron energy	9 MeV
Radio Frequency	3 GHz
Macropulse duration	≥ 0.5 µs
Micropulse duration (after stretcher)	≥ 100 ps
Undulator period	5 cm
Number of periods	30
Output wavelength	160 µm - 320 µm
Radiation macropulse power	20 kW
Beam waist	≤ 2 mm
Repetition rate	100 - 200 Hz

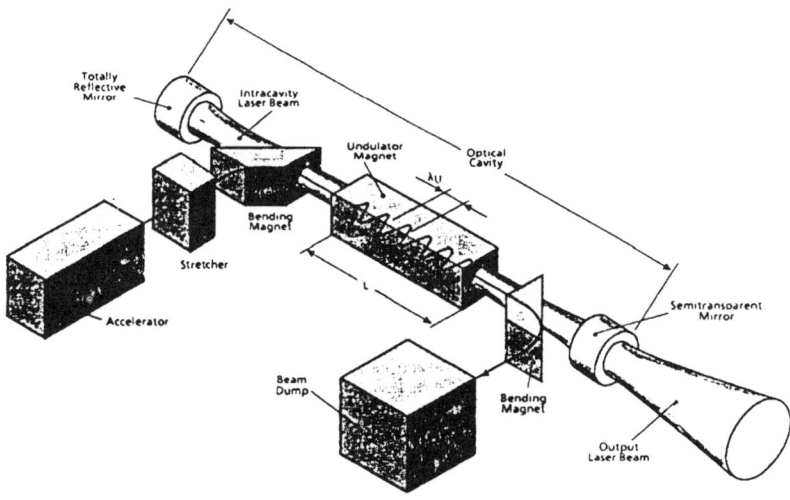

Figure 5: FEL layout

4. Rates

An accurate evaluation of the transition probability requires the exact quantum mechanical treatment of the two levels involved (this is given in P.Hauser's contribution to this conference). However (as shown by P.Hauser) a straightforward application of the Fermi golden rule is sufficient for a first evaluation of the transition probabilities, which are then given by the expression:

$$T_{a->b}(\omega = \omega_{res}) = \frac{16\pi^2\alpha I_0}{h} f \frac{1}{[\Gamma_a(\Gamma_a + \Gamma_b)]} R \qquad (11)$$

where $f \approx 0.7$ is a factor that takes into account the finite laser pulse length of about 100 ps, $\Gamma_a = 1/\tau_{3D}$, $\Gamma_b = 1/\tau_{3P}$ are the natural level linewidths and I_0 is the average intensity of radiation in the cavity that overlaps the muon stop distribution (≈ 6.7 kW/cm^2). R is the average number of reflections of the laser beam in the cavity (≈ 100).

Therefore if the 3D initial state population is $\approx 50\%$ of the stopped muons and the 3P population is $\approx 8\%$ of the stopped muons, and noting that the statistical population of the $3D_{5/2}{}^7$ level is 35% of the total 3D population, one obtains for the ratio $Q_{\beta,\alpha}$

$$Q_{\beta,\alpha} = \frac{\text{stimulated K}\beta \text{ intensity}}{\text{spontaneous K}\beta \text{ intensity}}$$

a value of $\approx 2\%$.

So tuning the laser around the resonance value of the Δ_3 transition one should see a peak 2% higher than the spontaneous emission background. Calling

$$SN = \frac{\text{number of stimulated K}\beta}{\sqrt{\text{number of spontaneous K}\beta}}$$

for a given total time T of counting per point we find that we need 10 points (around the peak) in which for each point one has $SN \geq 5$ to achieve the precision we request.

This means that for a useful solid angle fraction ≈ 0.15 a total time $T \approx 2 \cdot 10^6$ sec (i.e. about 5 weeks) is required.

5. References

[1] P.S.I. Proposal R-9206-1, "Spectroscopy of muonic hydrogen" (Dec. 1991)
[2] H.Anderhub et al., Phys.Lett. **143B** (1984) 65
[3] A. Di Giacomo, Nucl.Phys. **B11** (1969) 411. See also E.Borie and G.A.Rinker, Rev.Mod.Phys. **54** (1982) 67
[4] M.E.Cage et al., IEEE Trans.Instrum.Meas. **38** (1989) 233
[5] E.R.Williams, IEEE Trans.Instrum.Meas. **38** (1989) 284
[6] T.Kinoshita and D.R.Yennie: "High Precision Tests of Quantum Electrodynamics - An Overview", in T.Kinoshita (ed.) "Quantum Electrodynamics",World Scientific (Singapore, 1990)
[7] R.S.Van Dyck et al., Phys.Rev.Lett. **59** (1987) 26
[8] E.Zavattini: "Transitions in muonic helium", in G.F.Bassani,M.Inguscio and T.Hänsch (eds.) "The Hydrogen Atom", Springer-Verlag (New York, 1989)

[9] Yu.Sokolov: in G.F.Bassani, M.Inguscio and T.Hänsch (eds.) "The Hydrogen Atom", Springer-Verlag (New York, 1989)

[10] E.Zavattini "Quantum Electrodynamics Tests in Muonic Systems", in the Proceedings of the 7th International Conference on High-Energy Physics, M.Locher editor, Birkhäuser-Verlag (Basel, 1978)

[11] G.Carboni et al., Nucl.Phys. **A278** (1977) 381

[12] I.Beltrami et al., Nucl.Phys. **A451** (1986) 679

[13] See, e.g. J-P.Egger et al.: "Progress in X-ray detection", to be published in Physics World

[14] See, e.g. E.Gatti and P.Rehak, Nucl.Instr.and Meth. **225** (1984) 608, and P.Rehak et al., Nucl.Instr. and Meth. **A248** (1986) 367

[15] L.M.Simons, Phys.Scrip. **T22** (1988) 90

[16] Courtesy of L.M.Simons, from a cascade code developed by E.Borie and M.Leon, Phys.Rev. **A21** (1980) 1460

[17] See e.g. P.Sprangle and T.Coffey, Phys.Today March 1984, p.44 for a general overview of FEL's and other high-power sources. V.L.Granatstein et al., IEEE Trans.on Microwave Th.and Techn., **MTT-22** (1974) 1000 is a "historical" source on radiation from relativistic electron beams. A "sample" of the activity of the ENEA group is given e.g. in Nucl.Instr.and Meth. **A272** (1988) 132.

Laser Induced 3D→3P and 4S→4P Transition in Muonic Hydrogen

P. HAUSER
PSI-West, CH-5232 Villigen

Transition probabilities between atomic states in the external electromagnetic field of a FEL (free electron laser) are discussed, and requirements on the laser are evaluated. Special attention has to be given to the FEL substructure. If a sharp resonance line is needed, the micropulse length of the substructure has to exceed the natural lifetime of at least one of the involved states. Considering the 3D→3P and 4S→4P transitions in muonic hydrogen a minimal FEL-micropulse length of 100 ps is required.

1. Introduction

At the users meeting in January 1992 at PSI a spectroscopy program for light muonic atoms was proposed [1]. In a first step we considered 3D→3P transitions in muonic hydrogen to determine the vacuum polarization to an accuracy of $\sim$ 100 ppm, and 4S→4P transitions to evaluate the charge radius of the proton. To induce these atomic transitions a free electron laser (FEL) was presented. Meanwhile more detailed calculations on the requirements for the laser source were performed. In this paper we discuss the 3D→3P transition in detail and in a second part we apply these calculalations to the 4S→4P transition.

2. 3D-3P fine and hyperfine transition intensities

In the absence of a magnetic field the 3D and 3P states in muonic hydrogen are split each in 4 fine and hyperfine levels as shown in Fig. 1. The possible transitions between the 3D and 3P-sublevels are given by the selection rules of dipol transitions ($\Delta F, \Delta J = \pm 1$ or 0) leading to 10 possible radiative 3D-3P transitions. From these selection rules follows that the $3D_{5/2}^7$ state has only one radiative transition channel to the $3P_{3/2}^5$ state whereas the other 3D states have to distribute the transition intensity to two or more transition channels. The relative intensities can be calculated using the Clebsch-Gordan coefficients. As optimum transition we chose $3D_{5/2}^7 \rightarrow 3P_{3/2}^5$: the $3D_{5/2}^7$-level is the most populated 3D-state with a statistical population of 35%, and the mentioned transition is the most intense. In Fig. 2 the intensities of all possible 3D-3P transitions including also the statistical population of the initial 3D state are plotted against the frequency [2]. The $3D_{5/2}^7 \rightarrow 3P_{3/2}^5$ transition is well separated from other transitions. The wavelength of the mentioned transition lies in the fare infrared region ($\sim 200~\mu m$). In this wavelength region the FEL was found to be the most adapted light source which is tunable (to find the resonance), which could satisfy the requirements of high power (~ 1 MW peak power, as shown afterwards), and which offers the possibility of external triggering (to have a coincidence of the laser pulse with the muon stop).

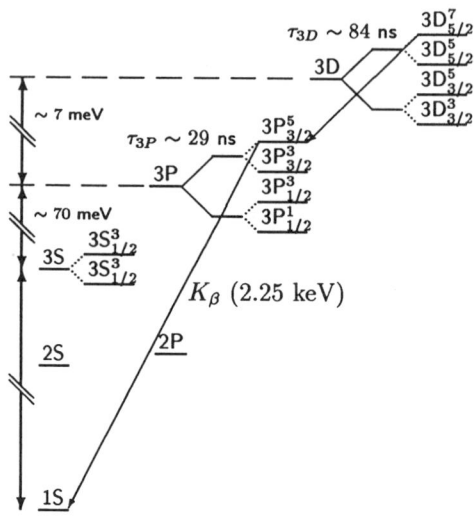

Fig. 1: Schematic level diagram for $\mu^- p$. The fine and hyperfine levels are denoted in the spectroscopic notation nX_J^{2F+1}, where F = total momentum and J = angular and spin momenta.

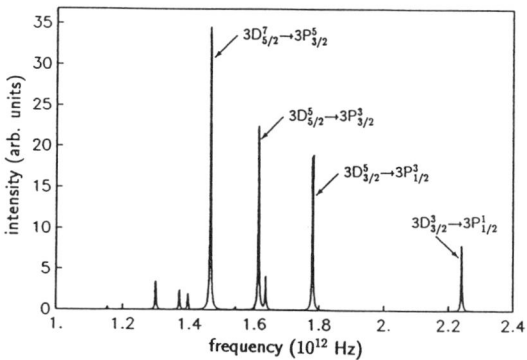

FIG. 2: Radiative 3D-3P-transition intensities including the statistical population of the initial 3D-state. The widths of the shown resonances correspond to the natural line widths.

2.1 FEL and induced 3D-3P transition probabilities

Before calculating the FEL-induced transition probabilities we have to mention some properties of this laser device. The main components of the FEL are an electron accelerator and a magnetic "wiggler" field inside an optical cavity. The laser light is produced by electrons oscillating in the magnetic field. The output of the laser contains macropulses of $\sim 1 \mu s$ duration separated by ~ 10 ms dead time corresponding to a repetition rate of $\sim 100~s^{-1}$(Fig. 3). These macropulses have a substructure, i.e. micropulses of length T separated by the inverse of the accelerator radio frequency ν_{RF}, with a duty cycle $T \cdot \nu_{RF} \sim 1/20$. The micropulse width T lies typically in the pico second region, i.e. in the same order of τ_{3D}, τ_{3P}. In this case, where for both involved atomic states the ratio T/τ_i is less than or about unity, the Golden Rule to calculate the transition probability fails. Therefore we have to investigate the time-dependent Schroedinger equation. Considering now the transition between two atomic states $|a> \to |b>$ with natural decay rates γ_a, γ_b in a monochromatic electromagnetic field of frequency ω and intensity I, the Schroedinger equation leads to two differential equations for the amplitudes a and b [3]:

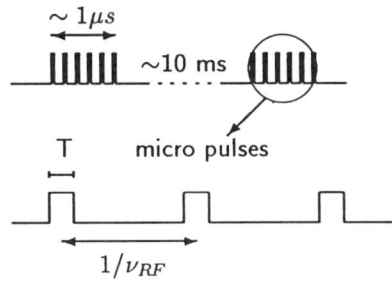

FIG. 3: Output of the FEL device

$$\dot{a} = -\frac{1}{2}\gamma_a a + \frac{V^*}{\hbar i}e^{i\Omega t}b \qquad (1)$$
$$\dot{b} = -\frac{1}{2}\gamma_b b + \frac{V}{\hbar i}e^{-i\Omega t}a$$

with

$$\omega_o = \omega_b - \omega_a = (E_b - E_a)/\hbar$$
$$\Omega = \omega - \omega_o$$
$$V = <b|H_{int}(0)|a>$$
$$= -i\sqrt{2\pi\hbar\alpha I_o}<b|\vec{\epsilon}\vec{x}|a>$$

The probability to find the atom after t in the state $|a>$ or $|b>$ is then $|a(t)|^2$ or, $|b(t)|^2$ respectively. In this paper we will discuss the solutions of eq. (1) for the case of a constant intensity I during T without giving the explicit mathematical expressions for the amplitudes a(t), b(t). For a fully occupied 3D state at t=0 (corresponding to initial conditions a(0)=1, b(0)=0) and the laser frequency on resonance ($\omega = \omega_o$) the 3D and 3P populations as a function of time are shown in Fig. 4 .The 3D-population drops nearly exponentially towards 0 whereas the 3P population reaches a maximum

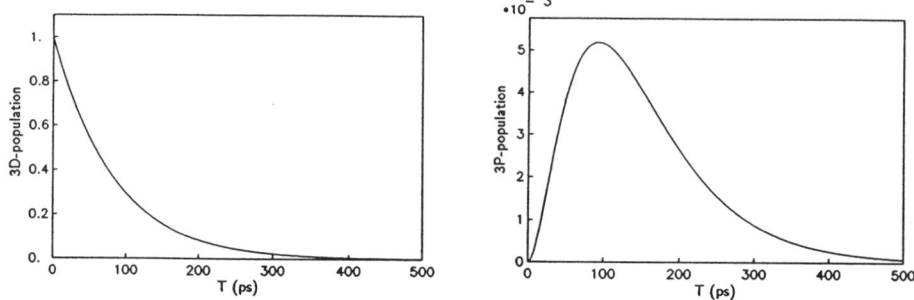

FIG. 4: $3D_{5/2}^7$ and $3P_{3/2}^5$ populations against the laser pulse duration starting at t=0 with a fully populated 3D-state. The laser is supposed to be on resonance with an intensity of I=1 MWcm^{-2}.

after $\approx$ 100 ps and decreases for longer t towards 0 due to the natural decays of both states. The transition probability per unit of time follows from the equation

$$dW_{|a\rangle \to |b\rangle}(t) = -\frac{d|a(t)|^2}{dt} - \gamma_a |a(t)|^2 \quad (2)$$

By integrating over the micropulse length T we get the total transition probability:

$$W_{|a\rangle \to |b\rangle}(T) = \int_0^T dW_{|a\rangle \to |b\rangle}(t) dt = 1 - |a(T)|^2 - \gamma_a \int_0^T |a(t)|^2 dt \quad (3)$$

In Fig. 5 the resulting transition probability (on resonance) against the micropulse length T is compared to the value obtained by using the Golden Rule which predicts a constant transition rate $\lambda_{GR} = 4|V|^2/(\hbar^2(\gamma_{3P} + \gamma_{3D}))$. For a short micropulse (T< 100 ps) the transition probability is considerably smaller than the Golden Rule value whereas for long T both results are nearly in agreement. In order to understand the suppressed transition probability for short la-

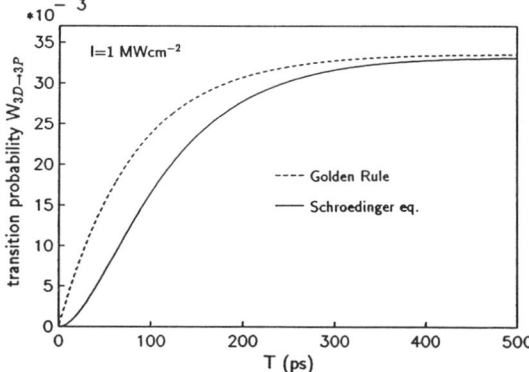

FIG. 5: Comparison between two different calculations of the $3D_{5/2}^7 \to 3P_{3/2}^5$ transition probability as a function of the micropulse length T. At t=0 the muon is supposed to be in the $3D_{5/2}^7$-state.

ser pulses we have to consider the region of laser frequencies outside of the resonance ω_o. Investigating the resonance curves for micropulses with different length T, constant intensity I during T and a monochromatic laser frequency ω we find a broader resonance curve with decreasing pulse duration T (Fig. 6). The line width of the 3D-3P-resonance corresponds for $T > 500$ ps to the natural linewidth $\Gamma = \gamma_{3D} + \gamma_{3P}$ (FWHM), for T=100 ps it is increased by a factor 1.5 whereas for T=10 ps the increase is a factor 10.

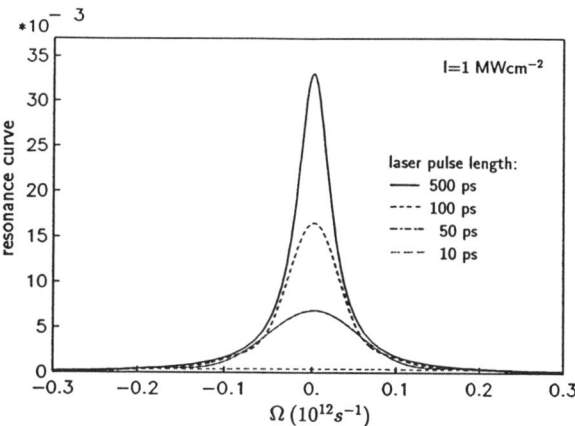

FIG. 6: Resonance curves for different laser pulse lengths T and a constant intensity I=1 MWcm^{-2} during the pulse. The laser is supposed to be monochromatic, *i.e.* with a vanishing intrinsic line width. The broadening of the resonance curve with decreasing T is due to the limited time for the development of the quantum mechanical state (uncertainty principle).

2.2 $3D_{5/2}^7 \to 3P_{3/2}^5$ - transition

The dipol matrix element of the $3D_{5/2}^7 \to 3P_{3/2}^5$ - transition in muonic hydrogen becomes

$$|<3P_{3/2}^5 | \hat{\epsilon}\vec{x} | 3D_{5/2}^7>|^2 = 13.5 a^2 \qquad (4)$$

where $a = 2.85 \cdot 10^{-13}$ m is the muonic Bohr radius.

To estimate the total increase of K_β x-rays due to the external electromagnetic perturbation we have to consider

- the probabilities f_{3D} and f_{3P}, that the muon reaches the 3D- or the 3P-state during the cascade
- the laser induced 3D→3P transition if the muon is in the 3D-state, and the 3P→3D-back-transition if it is in the 3P-state.
- the statistical population of the $3D_{5/2}^7$ (35%) and, for the back-transition, the statistical population of the $3P_{3/2}^5$ (41.7%).

From measurements and cascade calculations we get at low pressure (~ 0.1 atm) $f_{3D} \approx 0.5$ and $f_{3P} \approx 0.08$ [4]. Considering now a micropulse with duration T and assuming a flat muon stop time distribution ρ_t we get due to the 3D→3P-transition an induced K_β-x-ray yield during the micropulse of

$$W_{K_\beta}^+(T) = \rho_t \cdot 0.35 \cdot f_{3D}[\frac{1}{\gamma_{3D}} \cdot W_{3D \to 3P}(T) + \int_0^T W_{3D \to 3P}(t)dt] \qquad (5)$$

and a reduction of the expected K_β-yield due to the back-transition of

$$W_{K_\beta}^-(T) = \rho_t \cdot 0.42 \cdot f_{3D}[\frac{1}{\gamma_{3P}} \cdot W_{3P \to 3D}(T) + \int_0^T W_{3P \to 3D}(t)dt] \qquad (6)$$

with the natural decay rates $\gamma_{3D} = 1.19 \cdot 10^{10} s^{-1}$ and $\gamma_{3P} = 3.45 \cdot 10^{10} s^{-1}$. The first term in eqs. (5) and (6) considers muons reaching the 3D (3P) state before being crossed by the micropulse and the second term during the cross time of the micropulse. The back transition probability $W_{3P \to 3D}(T)$ is smaller than the forward transition probability $W_{3D \to 3P}(T)$ ($\gamma_{3P} > \gamma_{3D}$); thus the reduction of the K_β-yield due to the back-transition is less than 5%. The relative enhancement of the K_β-yield during the micropulse is

$$Q_{K_\beta}(T) = (W_{K_\beta}^+(T) - W_{K_\beta}^-(T))/\rho_t T f_{3P} \qquad (7)$$

For T=100 ps and I=1 MW/cm² we get $Q_{K_\beta} = 4.5\%$.

Until now we evaluated the K_β-enhancement during a micropulse without taking into account the background from cascade-K_β in between the FEL-micropulses. To calculate the K_β-enhancement at more realistic circumstances we have to consider the "optical" cavity in the target allowing an intense enlightening of the muon stop volume. We thought of two possible cavity arrangements: a small cavity where during the macropulse the stop volume is illuminated "homogeneously" in time by micropulses reflecting between the mirrors, or a larger cavity where the mirrors are placed in a distance such that each entering micropulse will overlap inside the cavity with the preceding micropulses (Fig. 7) and thus leading to one intense micropulse reflecting between the mirrors. In this paper I will briefly discuss the latter case without going into technical details. Assuming a micropulse of length T=100 ps and a duty cycle $T \cdot \nu_{RF} = 1/20$, i.e. micropulses separated by 20T=2 ns, we can achieve the mentioned arrangement by placing the cavity mirrors at a spacing of $\frac{c}{2\nu_{RF}} = 10cT$=30 cm.

Laser Induced 3D→3P and 4S→4P Transition in Muonic Hydrogen

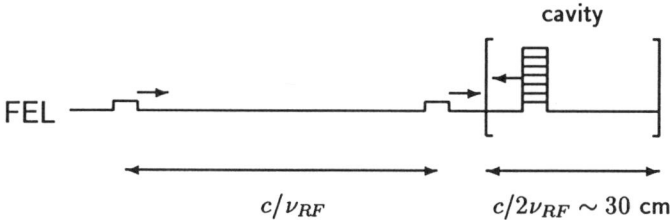

FIG. 7: Cavity arrangement with overlapping micropulses

Supposing now $N = 30$ reflections of each micropulse in the cavity, we have during the macro pulse $N/2 = 15$ overlapping micropulses inside the cavity. With a cross section $A = 3 cm^2$ for the muon stop volume and a laser peak power P=1 MW we achieve in the cavity an intensity $I = NP/2A \approx 5$ MWcm^{-2}. While this superposed micropulse is crossing the muon stop volume we have a K_β-enhancement. Using a fast X-ray detector with a time resolution ~100 ps in the 2 keV energy region [1] we may be able to distinguish K_β during the laser irradiation of the muon stop volume from the K_β where the micropulse is outside of the muon stop volume. For the 3 cm long stop volume the micropulse needs 100 ps to cross it. Hence we expect FEL-induced K_β during $\approx$ 200 ps each time when the micropulse is crossing the stop volume. This leads to a ratio "induced K_β/K_β from cascade" of

$$Q_{K_\beta} = (W^+_{K_\beta} - W^-_{K_\beta})/2\rho_t T f_{3P} \approx 1/10 \quad (8)$$

A ratio "effect/background" ~1/10 is fully sufficient to perform the experiment. Before making a rough rate estimate I have to mention some principles of the experiment. Every FEL macro pulse will be triggered by a muon entering the cyclotron trap. Because of technical reasons the FEL macro pulse will be generated after $\sim 1.5 \mu s$ relative to this trigger. Muon momentum and target pressure have therefore to be chosen such that the muon stop falls within the interval of the macro pulse, i.e. within the interval $[1.5 \mu s, 2.5 \mu s]$ after the trigger. With the mentioned cavity arrangement the probability for a FEL induced K_β per muon stop is then

$$P_{K_\beta} = 0.35 \cdot f_{3D} \cdot 0.95 [\frac{1}{\gamma_{3D}T} \cdot W_{3D \to 3P}(T) + \frac{1}{T} \int_0^T W_{3D \to 3P}(t)dt] \cdot \frac{cT}{L_{cav}} \quad (9)$$
$$= 1.7 \cdot 10^{-3} \quad \text{(for T=100 ps)}$$

where the factor 0.95 corresponds to the decreasing K_β-intensity due to the 3P-3D

---
[1] with this requirement on time resolution only a bad energy resolution is possible. Therefore suitable x-ray absorbers have to be used to identify the K_β x-rays.

back transition and the ratio $\frac{cT}{L_{cav}}$ is about 1/10, due to the cavity duty cycle. As the muon stop rate in the cyclotron trap is by an order of magnitude larger than the FEL macro pulse repetition rate $f_{laser} \sim 100 s^{-1}$, the induced K_β rate is limited by the FEL, leading to the following rate estimate:

$$Y_{K_\beta} = f_{laser} \times 1/3 \times (\Omega \cdot \varepsilon)_{K_\beta} \times P_{K_\beta} \approx 8 \cdot 10^{-3} s^{-1} \approx 30 h^{-1} \qquad (10)$$

where the factor 1/3 is the trigger quality which is mainly limited by the muon decay while slowing down. The product of "solid angle × efficiency" $(\Omega \cdot \varepsilon)_{K_\beta}$ is supposed to be $\sim 15\%$. Based on this estimated rate and an "effect/background" ratio of 1/10 we performed a computer simulation of the experiment (Fig. 8). With 400 h measuring time and 23 measuring points arround the resonance a relative accuracy of 240 ppm resulted. The arrangement with a smaller cavity without overlaping micropulses gives similar results (higher K_β rate but smaller ratio "effect/ background"). For the 4S-4P transition however the cavity with overlapping micropulses is of special interest and was therefore also mentioned as a possible arrangement for the 3D-3P transition.

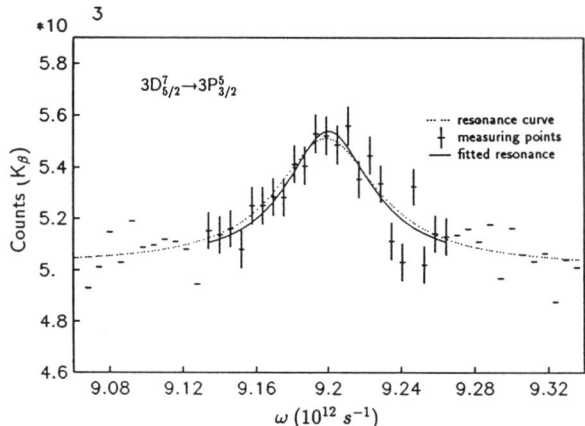

FIG. 8: Simulation of the 3D-3P resonance measurement. The dotted line represents the expected resonance curve for a micropulse length T=100 ps corresponding to a relative resonance line width of $7.5 \cdot 10^{-3}$(FWHM) and the solid line to the fit of the simulated measuring points.

3. $4S^3_{1/2} \rightarrow 4P^5_{3/2}$ - transition

The $4S^3_{1/2} \rightarrow 4P^5_{3/2}$ - transition will be discussed as for the $3D^7_{5/2} \rightarrow 3P^5_{3/2}$ - transition, but the involved parameters are considerably different. The dipol matrix element is larger by a factor ~ 7.5 ($|<4P^5_{3/2} | \vec{\epsilon} \cdot \vec{\tilde{x}} | 4S^3_{1/2}>|^2 = 100a^2$). The statistical $4S^3_{1/2}$ population is 75% and the probabilty for the muon to reach the 4S (4P) state is $f_{4S} \sim 0.5\%$ ($f_{4P} \sim 3\%$)[5]. The 4S-state is metastable ($\tau_{4S} = 1.24$ ns) whereas the 4P state decays rapidly ($\tau_{4P} = 67$ ps). The $4S^3_{1/2} \rightarrow 4P^5_{3/2}$ energy difference is 4 times larger than the $3P^5_{3/2} \rightarrow 3D^7_{5/2}$ energy difference, the relative natural resonance line width is about a factor 10 smaller ($4.2 \cdot 10^{-4}$). The forward and backward transition probabilities for a muon which is at t=0 in the 4S (4P) state are shown in Fig. 9 for a FEL pulse with constant intensity during the micropulse length T. Considering the values for f_{4S} and f_{4P} one may think that the 4P→4S back transition exceeds the forward transition, giving a decrease of the K_γ yield. However the 4S state is metastable and thus the muons which reach the 4S state in between the laser pulses give a large contribution to the forward transition. For the mentioned cavity where one pulse (15 superposed micropulses) of length T and intensity I=5 MWcm^{-2} is reflecting between the cavity mirrors spaced at a distance 10cT (see Fig. 7) the induced K_γ yield per muon stop is illustrated in Fig. 10 against the pulse length T. For T=100 ps the induced yield per muon stop is $1.64 \cdot 10^{-3}$. This value has to be compared to the expected K_γ-background from the cascade which is $f_{4P} \approx 3 \cdot 10^{-2}$ giving a ratio "effect/background" of 5%. Using a rate estimate similar to eq. (10) we get an expected in-

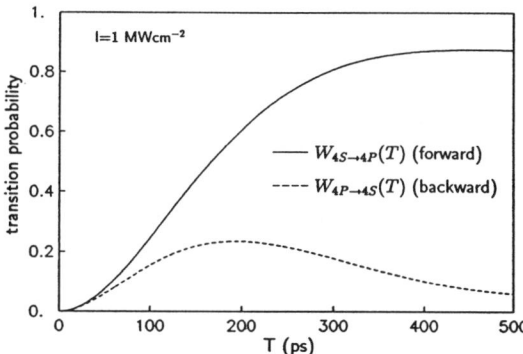

FIG. 9: 4S→4P transition (forward) and 4P→4S back-transition for a constant intensity I=1 MWcm^{-2} during the pulse length T and the laser frequency on resonance.

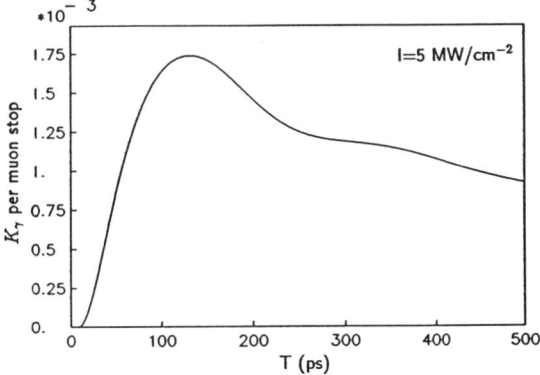

FIG. 10: Laser induced K_γ enhancement per muon stop due to $4S^3_{1/2} \to 4P^5_{3/2}$ transitions in a cavity with overlapping micropulses and a FEL duty cycle $T\nu_{RF} = 1/20$ (cavity mirrors spaced at a distance 10cT).

duced K_γ rate of $\sim 30 h^{-1}$. The broadening of the resonance curve due to the micropulse of length T=100 ps is of a factor 2.5. Fig. 11 shows a computer simulation for the 4S-4P experiment. A measuring time of 400 h gives an accuracy of the 4S-4P energy difference of 50 ppm (23 measuring points).

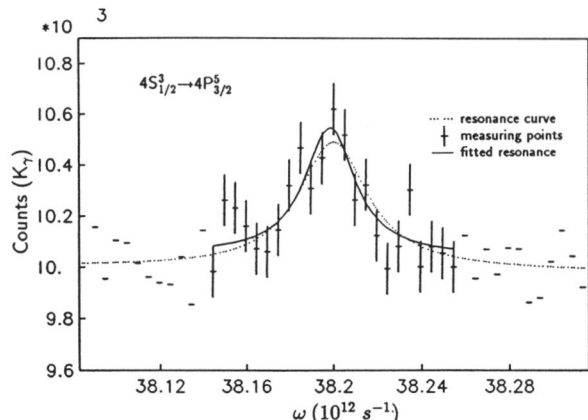

FIG. 11: Simulation of the 4S-4P resonance measurement. The dotted line represents the expected resonance curve for a micro pulse length T=100 ps corresponding to a relative resonance line width of $1.0 \cdot 10^{-3}$ (FWHM) and the solid line to the fit of the 23 simulated measuring points.

4. Conclusions and requirements for the FEL

With our present calculations on the proposed measurement of the 3D-3P and 4S-4P resonance in muonic hydrogen we now know the requirements for the free electron laser which are - concerning the peak power and substructure for both transitions- the same. To keep the broadening of the resonance within acceptable limits we need a FEL with a minimal micropulse length of 100 ps. To achieve a feasible transition probability, *i.e.* $\sim 30\hbar^{-1}$ induced K x-rays, a peak power of 1-2 MW together with a macropulse repetition rate of ~ 100 is required. Such a device would allow us a QED-test of the vacuum polarization with an accuracy of $\sim 10^{-4}$ in a first step via the 3D-3P resonance. In addition, if the 4S-4P resonance can be included, the proton charge radius could be determined to an accuracy of $< 0.5\%$.

References

[1] PSI proposal R-92-06
[2] E. Milotti, internal report
[3] R. C. Retherford and W. E. Lamb, Phys. Rev **79**, 549 (1950)
[4] H. Anderhub et al., Phys. Lett. **143B**, 65 (1984); E. C. Aschenauer and L. Simons, private communication.

Muonium Spectroscopy

Klaus P. JUNGMANN
Physikalisches Institut der Universität Heidelberg
Philosophenweg 12, D-6900 Heidelberg, Germany

Abstract

Highly precise spectroscopic studies of the purely leptonic muonium atom are of interest for tests of the standard theory of the electroweak interaction and for the accurate determination of fundamental constants. The ground state hyperfine splitting, the n=2 Lamb shift and the 1S-2S interval are particularly interesting for both theory and experiment.

1. Introduction

The hydrogen-like muonium atom $(\mu^+ e^-)$ is the bound state of a positive muon (μ^+) and an electron (e^-) [1,2]. Muonium completes the chain of natural hydrogen isotopes hydrogen (pe^-), deuterium (de^-) and tritium (te^-) together with other exotic atoms like positronium $(e^+ e^-)$, pionium (πe^-), and also muonic helium $(\alpha \mu^- e^-)$. The term energies of the isotopes with hadronic nuclei (p, d, t, π^+, α) are influenced by the not arbitrarily well known charge radii, polarizability, and inner structure of these particles. In atomic hydrogen, for example, the accuracy of Lamb shift experiments has reached the same order of magnitude (10^{-5}) as the uncertainty of theoretical calculations due to the influence of the proton's internal structure [3]. The hyperfine splitting of the hydrogen ground state can be measured 10^6 times more accurate than theoretical calculations, because they are limited by the knowledge of the proton's polarizability.

Positronium and muonium consist each of two leptons only for which no finite size has been found, yet. The theoretical description reaches very high precision by using exclusively quantum electrodynamics(QED). Contributions due to the strong interaction arising from vacuum polarization loops with hadrons, pions for example, are of the order of magnitude of 10^{-11} eV [4]. The weak interaction contributes to the order of $6 \cdot 10^{-13}/n^3$, where n is the principal quantum number of the state under consideration [5]. Muonium and positronium are ideal systems for experimental investigations of QED and fundamental principles in modern physics. In particular, muonium spectroscopy tests the nature of the muon as a heavy, point-like leptonic particle.

Detailed theoretical work has been performed on the QED description of muonium [6]. The ground state hyperfine splitting and the Lamb shift of the first excited state were the only quantities accessible for precision experiments in the past [7,8,9,10], since muonium atoms at thermal energies could be produced in gases only [1] and muonium in vacuum was available from a beam foil technique [11] at keV energies only. Most of the theoretical efforts went into calculations of these quantities. Intense sources of thermal muonium in vacuum from SiO_2 powder [12,13] and SiO_2 aerogel targets [14] made laser experiments and measurements of the ground state hyperfine splitting in vacuum possible.

2. Hyperfine structure splitting $\Delta\nu_{HFS}$ in the ground state

The ground state hyperfine splitting allows the most sensitive tests of QED for the muon-electron interaction. An unambiguous and very precise atomic physics value for the fine structure constant α can be derived from $\Delta\nu_{HFS}$. The experimental precision for $\Delta\nu_{HFS}$ has reached the level of 0.036 ppm or 160 Hz which is just a factor of two above the estimated contribution of 70 Hz (16ppb) from the weak interaction arising from an axial vector – axial vector coupling via Z-Boson exchange. With increased experimental accuracy muonium can be the first atom where a shift in atomic energy levels due to the weak interaction will be observed. The theoretical calculations for $\Delta\nu_{HFS}$ can be improved to the necessary precision. Right now they are limited by the accuracy of the measured value of magnetic moment μ_μ of the positive muon.

Table 1: Results extracted from the muonium ground state hyperfine structure splitting in comparison with theory and relevant quantities from independent experiments.

$\Delta\nu_{HFS}$(theory)	4 463 303.11(1.33)(1.0) kHz	(0.4 ppm)	[6]
$\Delta\nu_{HFS}$(expt.)	4 463 302.88(0.16) kHz	(0.036 ppm)	[7]
$\Delta\nu_{HFS}(theory) - \Delta\nu_{HFS}(expt.)$	(-0.23 ± 1.4) kHz		
$\alpha^{-1}(\Delta\nu_{HFS})$	137.035 988(20)	(0.15 ppm)	[15]
α^{-1}(electron g-2)	137.035 992 22(94)	(0.007 ppm)	[16]
α^{-1}(condensed matter)	137.035 9979(32)	(0.024 ppm)	[17]
$\mu_\mu/\mu_p(\Delta\nu_{HFS})$	3.183 346 1(11)	(0.36 ppm)	[7]
$\mu_\mu/\mu_p(\mu$SR in lq.bromine)	3.183 344 1(17)	(0.54 ppm)	[18]

All precision experiments to date have been carried out with muonium formed by charge exchange after stopping μ^+ in a suitable gas mixture [1]. Currently such an experiment is under way at LAMPF [19]. In a homogeneous magnetic field of about 1.8 Tesla one can expect significant improvements, in particular, for the value of the magnetic moment μ_μ which can be obtained from the Zeeman splitting. The experiment will employ the technique of "old muonium". The linewidth of the signals can be reduced below the "natural" linewidth $\delta\nu_{nat} = (\pi \cdot \tau_\mu)^{-1}=144$ kHz by detecting atoms only that have been interacting with the microwave fields for periods longer than several muon lifetimes. These measurements, however, suffer from the interaction of the muonium atoms with the foreign gas atoms.

With the discovery of polarized thermal muonium emerging from SiO_2 powder targets into vacuum [13] measurements of the $\Delta\nu_{HFS}$ are now feasible in vacuum in the absence of a perturbing foreign gas. Corrections for density effects are obsolete. In a preliminary experiment at PSI transitions between the $1^2S_{1/2}, F = 1$ and the $1^2S_{1/2}, F = 0$ hyperfine levels could be induced in zero magnetic field (see Fig. 1). A rectangular rf resonator operated in TEM_{301} mode was placed directly above the muonium production target. The atoms entered the cavity with thermal velocities through a wall opening. The e^+ from the parity violating muon decay $\mu^+ \to e^+ + \nu_e + \bar{\nu}_\mu$ were registered in two scintillator telescopes which were mounted close to the side walls of the resonator. The telescope axes have been oriented parallel respectively antiparallel to the spin of the incoming muons which had been rotated transverse to the muon propagation direction by an **ExB** separator in the muon beam line. At the resonance frequency of 4.46329(3) GHz a reduction of the muon polarization of 16(2)% has been observed as a signal from the hyperfine transitions [20].

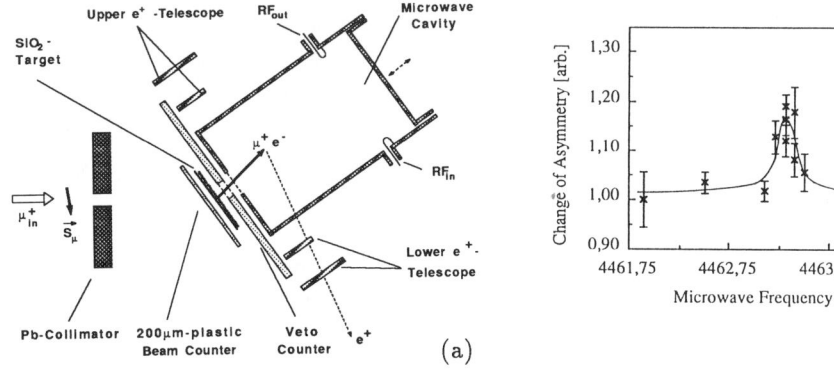

Figure 1: (a) Principle of the setup for a measurement of $\Delta\nu_{HFS}$ in vacuum, (b) first observed muonium $\Delta\nu_{HFS}$ signal in vacuum.

Certainly the method needs a lot of development work in order to improve the signal strength. However, ultimately one expects higher precision from experiments on muonium in vacuum than from measurements in gases.

3. Lamb shift in the first excited state

The classical Lamb shift in the atomic hydrogen atom between the metastable $2^2S_{1/2}$ and the $2^2P_{1/2}$ states is totally QED in nature and seems to be ideal for testing radiative QED corrections. An important contribution arises from the internal proton structure. Its uncertainty limits any theoretical calculation. Today, experiment and theory agree on the 10 ppm level. The purely leptonic muonium atom is free from nuclear structure problems. In addition, the relativistic reduced mass and recoil contributions are about one order of magnitude larger compared to hydrogen. Lamb shift measurements at TRIUMF [8] and LAMPF [9,10] have reached the 10^{-2} level of precision:

Table 2: n=2 Lamb shift in muonium. Comparison between experiment and theory.

Experiment	$\Delta\nu_{2^2S_{1/2}-2^2P_{1/2}}$[MHz]	experimental method	Ref.
TRIUMPF	1070(+12)(-15)	direct $2^2S_{1/2} - 2^2P_{1/2}$ trans.	[8]
LAMPF	1042(+21)(-23)	direct $2^2S_{1/2} - 2^2P_{1/2}$ trans.	[9]
LAMPF	1027(+30)(-35)	extracted from $2^2S_{1/2} - 2^2P_{3/2}$ trans.	[10]
Theory	1047.49(1)(9)		[6]

All these measurements were carried out using fast muonium produced by a beam foil method [11] which is the only method known to date that produces muonium in the metastable 2S state. Significant improvements are expected from the utilization of muon beams with an increased phase space density for low energy μ^+, for example a PSC beam at PSI, and from laser excited metastable muonium [21].

4. Optical excitation of the 1S-2S twophoton transition

A precision measurement of the 1S-2S energy splitting ($\Delta\nu_{1S-2S}$) can be interpreted in different ways. Firstly, it can be regarded as test for QED calculations. Secondly, one may want to interpret the result as a measurement of the Rydberg constant in a purely leptonic system. Thirdly, one can extract a precise value for the muon mass.

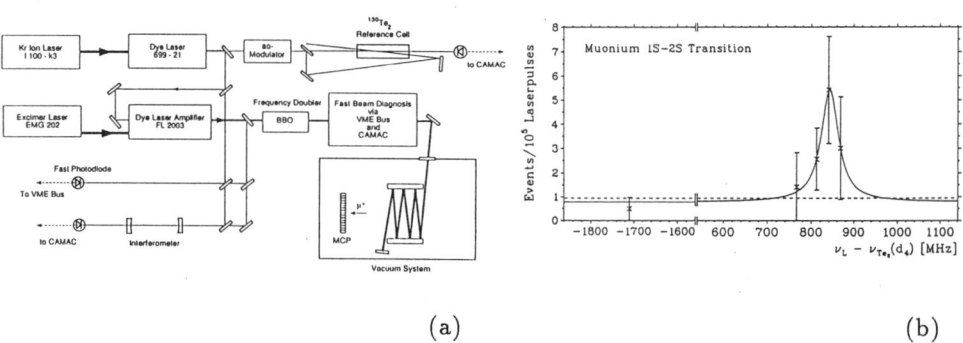

(a)   (b)

Figure 2: (a) Principle of the setup of the lasers and the optical system in the RAL muonium 1S-2S experiment, (b) Muonium $1^2S_{1/2}, F = 1 - 2^2S_{1/2}, F = 1$ signal.

At the Rutherford Appleton Laboratory (RAL) an experiment is under way which aims for a precise determination of $\Delta\nu_{1S-2S}$ using Doppler-free twophoton spectroscopy. The muonium atoms are created in a SiO_2 powder target and interact close to the target with two counterpropagating laser beams at λ=244 nm wavelength. The laser system (see Fig. 1) consists of a Krypton ion laser pumped cw single mode ring dye laser at

488 nm wavelength and a pulsed excimer laser pumped dye laser amplifier. The pulsed blue light is frequency doubled in a BBO crystal (see Fig. 2). A fraction of the cw laser light is frequency downshifted by approximately 840 MHz by double passing through a pair of acoustooptic modulators. For the frequency calibration of the system the cw laser is locked to the d_4 line in the spectrum of $^{130}Te_2$ using saturation spectroscopy with the frequency downshifted beam. The d_4 secondary frequency standard is accurate to 0.8 ppb [22]. The 1S-2S transition is detected by the photoionization of the 2S state with a third photon from the same laser field. The slow muon released in this process is electrostatically accelerated to 2 keV and guided through an energy and momentum selective path to a microchannel plate (MCP) particle detector. The detection of the Michel positron from the muon decaying on the MCP in a double layer of plastic scintillator around the MCP is required as part of the signature.

Attention must be paid to systematic effects. The largest contributions are related to the intense laser fields. Rapid changes of the refractive index in the dye cells of the pulsed laser amplifier cause time dependent chirping of the laser frequency. The effect amounts to several 10 MHz. The dynamic Stark shift of the muonium energy levels is of the same order of magnitude and depends strongly on the temporal and spatial properties of the laser pulse. The residual linear Doppler effect due to a finite crossing angle between the counterpropagating laser beams and the nonisotropic velocity distribution contributes less than about 10 MHz and the relativistic quadratic Doppler shift accounts for less than 1 MHz. All these effects can be well controlled by a careful laser beam diagnosis running in parallel with the data taking for the 1S-2S resonance.

Table 3: Muonium 1S-2S transitions. Comparison between experiment and theory.

	$\Delta\nu_{1^2S_{1/2}-2^2S_{1/2}}$ [MHz]	Reference
KEK-Experiment	2 455 527 936(120)(140)	[24]
RAL-Experiment	2 455 528 016(58)(43)	[23]
Theory	2 455 527 958.5(4.3)	[6]

The results of the RAL experiment are in good agreement with earlier measurements at KEK and with theory (see Table 2). The accuracy of the theoretical value is limited by the precision of the muon mass of 0.36 ppm. The Lamb shift contribution to $\Delta\nu_{1S-2S}$ of 7 056.2(1.2) MHz is tested in the experiment at the 1% level. Even in its first stage the RAL experiment happens to be the best Lamb shift measurement in muonium so far.

A measurement of the muonium-hydrogen isotope shift using the same setup for both atomic species is particularly interesting. Firstly, most of the systematic effects cancel and, secondly, for the isotope shift measurement the accuracy of the Rydberg constant does not limit the precision with which one can determine the QED corrections. From an experiment using available laser technology at todays most intense pulsed muon source at RAL one can expect an accuracy of 1 MHz for $\Delta\nu_{1S-2S}$ in the near future. An improved value for the μ^+ mass can be extracted.

Acknowledgements

We wish to thank J.R.M. Barr, P.E.G. Baird, A.I. Ferguson, V.W. Hughes, F. Maas, B. Matthias, G. zu Putlitz, P.G.H. Sandars, W.T. Toner, W. Schwarz, K.A. Woodle and D. Yennie for stimulating discussions and the German BMFT for financial support.

References

[1] V.W. Hughes, in *Muon Physics* p. 78ff, ed. V.W. Hughes and C.S. Wu (Academic Press, 1977), and V.W. Hughes and G zu Putlitz, in *Quantum Electrodynamics* p. 822ff, ed. T. Kinoshita (World Scientific, 1990).
[2] V.W. Hughes and G. zu Putlitz, Com.Nucl.Part.Phys.**12**,259(1984).
[3] V.W. Hughes in *Atomic Physics 10* p. 1ff,ed. H. Narumi and I. Shimamura (World Scientific, 1986), and references therein.
[4] S.S. Brodsky and S.D. Drell, Ann.Rev.Nucl.Sci.**20**,147(1970).
[5] M.A.B. Bég and G. Feinberg, Phys.Rev.Lett.**33**,606(1974) and Phys.Rev.Lett. **35**,130(1975).
[6] J.R. Sapirstein and D.R. Yennie, in *Quantum Electrodynamics* p. 560ff, ed. T. Kinoshita (World Scientific, 1990), and references therein.
[7] F.G. Mariam et al., Phys.Rev.Lett.**49**,993(1982).
[8] C.J. Oram et al., Phys.Rev.Lett.**52**,910(1984).
[9] A. Badertscher et al., Phys.Rev.Lett.**52**,914(1984), *see also* K.A. Woodle et al., Phys.Rev.**41**,94(1990).
[10] S.H. Kettell et al.,Bull.Am.Soc.Phys.**36**,1258(1991).
[11] D.A. Bolton et al.,Phys.Rev.Lett.**47**,1441(1981).
[12] G.A. Beer et al.,Phys.Rev.Lett.**57**,671(1986).
[13] K.A. Woodle et al., Z.Phys.D**9**,59(1989)
[14] W. Schwarz et al., J.Non-Cryst.Solids, in print (1992).
[15] J.R.Sapirstein, Phys.Rev.A**51**,985(1983).
[16] T. Kinoshita and D.R. Yennie, in *Quantum Electrodynamics* p. 1ff, ed. T. Kinoshita (World Scientific, 1990), and references therein.
[17] M.E. Cage et al., IEEE Trans.Instr.Meas.**38**,284(1989), *see also* E.R. Williams et al., IEEE Trans.Instr.Meas.**38**,233(1989).
[18] E. Klempt et al.,Phys.Rev.D**25**,652(1982).
[19] LAMPF research proposal 1054,*Ultrahigh Precision Measurements on Muonium Ground State: Hyperfine Structure and Magnetic Moment*, V.W. Hughes, G. zu Putlitz, P.A. Souder et.al. (1986).
[20] V. Ebert et al.,to be published, and PSI Newsl.**20**,910(1990).
[21] K. Jungmann, in *Proceedings of the Workshop on The Future of Muon Physics, Heidelberg, 1991* ed. K. Jungmann,G. zu Putlitz and V.W. Hughes, Z.Phys.C in print (1992).
[22] G.P. Barwood et al., Phys.Rev.A**43**,4783(1991).
[23] K. Jungmann et al., Z.Phys.D**21**,241(1991).
[24] S. Chu et al., Phys.Rev.Lett.**60**,101(1988).

Relativistic Corrections to Particle Overlap in Atoms and Muon Capture in Hydrogen

D. BAKALOV[1]

Istituto Nazionale di Fisica Nucleare
Via Valerio 2, I-34017, Trieste, ITALY

Abstract. It is demonstrated that δ-functionlike singularities of the relativistic interaction Hamiltonian of two spin particles with electromagnetic interactions have an effect on the muon-proton overlap in the ground state of $p\mu$-atoms of the order of 1–2% that may be important for the interpretation of high-accuracy measurements of the rate of muon capture by protons in hydrogen.

1. Introduction

Relativistic Schrödinger equations in coordinate space have proved to be very efficient in high precision two-body bound state calculations. Although the Hamiltonians in the various approaches look different they all produce energy spectra that coincide up to order $O(\alpha^6)$ or even higher [1]; the corresponding 3-dimensional relativistic wave functions may significantly differ, however. I shall focus my attention here on the short-distance singularities of the relativistic Hamiltonian of two spin particles with electromagnetic interactions and the corresponding corrections to the wave function near the origin. In the particular case of muonic hydrogen the effect of the Fermi spin-spin interaction on particle overlap is shown to be observable in precise muon capture rate measurements.

---

[1] on leave from the Institute for Nuclear Research and Nuclear Energy, boul. Trakia 72, Sofia 1784, BULGARIA

2. The relativistic Schrödinger equation

Two particular relativistic equations for the two-body Coulomb problem will be used in the calculations to follow: the Breit equation (BE) [2] and Todorov's quasipotential equation (QPE) [3]; this choice was influenced by the previous experience of the author. The Hamiltonian H is the sum of a "nonrelativistic" part H^{nr} that contains the kinetic energy and the Coulomb potential, and a relativistic part H^{rel} that includes all the remaining terms: $H = H^{nr} + H^{rel}$. The explicit form of H^{rel} depends on the approach adopted even in the lowest nonvanishing order $O(\alpha^2)$ (see [4,5]) and includes terms that grow up at least as fast as $1/r$ when $r \to 0$. Such terms will be called "singular"; when added to the Coulomb potential, they modify the shape of the wave function at short distances, while only slightly perturbing it for larger values of r. The change of the wave function near the origin is of particular interest in S-states for which $\psi(0) \neq 0$. For sake of simplicity only $^1S_0$-states (without any admixture of D-wave) will be considered in what follows; the Schrödinger equation then reduces to a single uncoupled equation for the radial function $u(r)$:

$$u'' + \frac{2}{r}u' + 2\left(\frac{1}{r} - \langle l=0, S=0 | V^{rel}(r) | l=0, S=0 \rangle + \lambda\right)u = 0 \qquad (1)$$

where $V^{rel}(r) = V^{(0)}(r) + V^{(F)}(r)$ is the sum of the surviving relativistic terms. Denote by μ_i the magnetic dipole moment of particle i. In BE and QPE $V^{(0)}(r)$ has the form:

$$V^{(0)}_{BR} = \frac{\alpha^2}{2m_1 m_2}\left(\frac{1}{r}\Delta + \frac{1}{r^3}(\mathbf{r}.(\mathbf{r}.\nabla)\nabla)\right) + \alpha^2 \sum_{i=1,2} \frac{2\mu_i - 1}{2m_i^2}\pi\delta^{(3)}(\mathbf{r})$$
$$-\frac{\alpha^2}{8m^3}\left(1 - 3\frac{m}{(m_1+m_2)}\right)\Delta^2 + U^{(VP)} + o(\alpha^2), \qquad (2)$$

where $e = \hbar = m = 1$, $m^{-1} = m_1^{-1} + m_2^{-1}$ and $U^{(VP)}$ is the Ühling potential [6],

$$V^{(0)}_{QP} = -\frac{\alpha^2}{2r^2} + \alpha^2\left(\frac{1}{mM} + \frac{2\mu_1 - 1}{2m_1^2} + \frac{2\mu_2 - 1}{2m_2^2}\right)\pi\delta^{(3)}(\mathbf{r}) + U^{(VP)} + o(\alpha^2) \qquad (3)$$

For both BE and QPE the matrix element of the spin-spin interaction $V^{(F)}(r)$ is:

$$\langle l=0, S=0 | V^{(F)} | l=0, S=0 \rangle = -\frac{\alpha^2 \mu_1 \mu_2}{m_1 m_2} 4\pi\delta(\mathbf{r}) \qquad (4)$$

The $O(\alpha^2)$-relativistic Hamiltonian involves the following singularities:

1) $V_1(r) = U^{(VP)}(r)$, dominated by $const \cdot (\ln r)^2/r$ at small r [6,7];
2) $V_2(r) = \alpha^2/2r^2$ – "squared Coulomb" – the first term of Eq.3;
3) $V_3(r) = \left(\frac{1}{r}\Delta + \frac{1}{r^3}(\mathbf{r}.(\mathbf{r}.\nabla)\nabla)\right)$ in the BE;
4) $V_4(r) = \pi\delta(\mathbf{r})$ in $V^{(0)}$ and $V^{(F)}$ of both BE and QPE.

It is quite natural to also include in this list the "finite size" term $V^{(FSZ)}(r)$ [5], i.e. the difference between the Coulomb potential of particles with finite charge distribution (that remains bound at short distances), and the "bare" Coulomb:

5) $V_5(r) = V^{(FSZ)}(r) \sim 1/r$ for $r \to 0$.

The effect of the singular interaction operators $V_i, i = 1, \ldots, 5$ on the two-body wave functions was studied numerically. The $O(\alpha^2)$-relativistic terms in Eq.1 are "small" enough and have approximately additive effects on the wave function $u(r)$; that is why the singular potentials were considered separately, i.e.

$$u'' + \frac{2}{r}u' + 2\left(\frac{1}{r} + \lambda - V(r)\right)u = 0 \qquad (5)$$

was solved by identifying consecutively the potential $V(r)$ with any of the items $V_i(r)$ listed above:

$$V(r) = \beta V_i(r), i = 1, \ldots, 4 \text{ or } V_5(r). \qquad (6)$$

always keeping the "variable intensity" β small enough to avoid any restructuring of the energy spectrum in the spirit of [8]. Some of the singular terms were regularized: e.g. the Dirac δ-function in V_4 was replaced by the δ-like function $\hat{f}_\Lambda(r)$ (Λ being a cutoff parameter) of the particular type encountered when considering the electromagnetic interactions of particles with finite size charge distribution of radius $\langle r_{ch} \rangle$ (in this case the cutoff being simply $\Lambda = \sqrt{12/\langle r_{ch}^2 \rangle}$) [5]:

$$\hat{f}_\Lambda(r) = \frac{\Lambda^3}{2}e^{-\Lambda r}, \quad 4\pi\delta(\mathbf{r}) = \lim_{\Lambda \to \infty} \hat{f}_\Lambda(r) \qquad (7)$$

Similarly, the problems with the term V_3 (the coefficient in front of the d^2/dr^2-operator in Eq.5 changes sign at short distances) were eliminated by using the regularized expressions of V_3, derived in [5] for nonpointlike particles. The main purpose of the numerical calculations was to study the "response" of $u(r)$ to the various singular perturbations $V(r) = \beta V_i(r)$ in Eq.5. This "response" is best described by the function:

$$R(r) = -\lim_{\beta \to 0} \frac{\delta\left(u(r)/u^{(0)}(r)\right)}{\delta \lambda} \qquad (8)$$

where $\delta u(r)$, $\delta\lambda$ are the corrections to the wave function and to the energy level and $u^{(0)}$ is the nonperturbed $^1S_0$-wave function obtained for $\beta = 0$. For "small" perturbations $R(r)$ does not depend on β and only shows "how much singular" the corresponding potential $V(r)$ is. Table shows the values of $R(r)$ for the $^1S_0$-state wave function. The "response" $R(r)$ of the $^3S_1$-state wave function also depends on the strength of the spin-orbit and tensor coupling; for strengths of order of magnitude $O(\alpha^2)$, however, the ortho-values of $R(r)$ only slightly differ from the para-values of Table .

Table 1: Values of the "response" function $R(r)$ (defined in Eq. 8) for the singular potentials V_i listed above, calculated with cutoff parameter $\Lambda = 1000$ for a set of interparticle distances r (in dimensionless units)

r	V_1	V_2	V_3	V_4	V_5
0.00001	5.3	13.2	18.3	253.2	381.9
0.0001	5.3	10.9	18.2	252.8	370.8
0.001	5.3	8.6	17.4	227.1	277.8
0.01	5.1	6.3	12.6	51.1	55.9
0.1	4.2	3.9	5.6	3.7	8.5

As expected, the most pronounced effects on particle overlap $u(r)$ have the particle-finite-size correction to the Coulomb potential and the δ-like contact terms in the Fermi spin-spin and relativistic spin-independent interactions. The following empiric formulae describe their dependence on the cutoff: $R_4 \sim 0.2516 \cdot \Lambda + 1.6$, $R_5 \sim 0.372 \cdot \Lambda + 10.0$. Of particular interest is the dependence on the cutoff Λ of the expectation value of operators that describe short-range interactions of radius R_0. Consider the integral:

$$I(R_0) = \int_0^\infty u^2(r) e^{-r/R_0} r^2 dr \qquad (9)$$

For Λ such that $\Lambda R_0 \ll 1$ this integral is proportional to Λ, while for $\Lambda R_0 \gg 1$ it does not depend on Λ. Here now is the rule for choosing the value of Λ in numerical calculations: if the particles have finite charge distribution, set $\Lambda = \sqrt{12/\langle r_{ch}^2 \rangle}$; if the particles are pointlike, set instead $\Lambda \gg 1/R_0$, where R_0 is of the order of magnitude of the range of the particle interaction.

All the numerical results of Table involve only dimensionless quantities and can be directly applied to any hydrogenlike system if rescaled in an appropriate way.

3. Muon capture in hydrogen

Muonic hydrogen atoms provide a good opportunity for the experimental study of the relativistic corrections to the wave functions at short distances: the muon-proton overlap here is relatively large, and experimental data on the rate of muon capture allow (in principle) for measuring $\psi(0)$ in both the singlet and triplet ground states.

If the dipole approximation is used for the proton weak formfactors [9], the probability for the muon to be captured by a proton in a $p\mu$-atom is proportional to the integral $I(R_0)$

of Eq. 9 (the range of the weak forces, R_0, being the inverse of the cutoff parameter Λ_d of the dipole approximation). The numerical calculations show that the relativistic effects augment the capture probability for the singlet state, but reduce it for the triplet state. Table contains the values of these corrections for a set of typical values of the cutoff Λ_d.

Table 2: Relativistic corrections, in %, to the integrals $I(R_0)$ of Eq. 9, for a set of values of R_0, corresponding to cutoffs Λ_d

R_0, fm	Λ, (MeV/c)	BE		QPE	
		$^1S_0$	$^3S_1$	$^1S_0$	$^3S_1$
0.569	346	0.40	-0.43	0.20	-0.63
0.285	693	0.40	-0.94	0.12	-1.20
0.212	929	0.39	-1.17	0.08	-1.46
0.190	1039	0.39	-1.26	0.06	-1.56
0.142	1386	0.38	-1.46	0.02	-1.80

When determining the constants $\Lambda_{\uparrow\downarrow}$ and $\Lambda_{\uparrow\uparrow}$ of the reactions $\mu_\uparrow^- + p_\downarrow \to n + \nu_\mu$ and $\mu_\uparrow^- + p_\uparrow \to n + \nu_\mu$ from the experimental values Λ_s and Λ_t of the rate of muon capture in the singlet and triplet ground states of $p\mu$, measured in gaseous hydrogen, the commonly used relations $\Lambda_s = \Lambda_{\uparrow\downarrow} \cdot D_{1S}$, $\Lambda_t = \Lambda_{\uparrow\uparrow} \cdot D_{1S}$ (D_{1S} being the muon density at the proton in the ground state of $p\mu$) should be replaced by

$$\Lambda_s = \Lambda_{\uparrow\downarrow} \cdot D_{1^1S_0}, \quad \Lambda_t = \Lambda_{\uparrow\uparrow} \cdot D_{1^3S_1} \qquad (10)$$

$D_{1^1S_0}$ and $D_{1^3S_1}$ differing from D_{1S} as shown on Table .

In liquid hydrogen, the observable rate of muon capture Λ_c depends also on the rates of formation of $pp\mu$-molecules $\lambda_{pp\mu}$, of ortho-para transition λ_{op}, of muon decay λ_0 and of muon capture in the ortho- and para-states of the $pp\mu$-molecules Λ_{om} and Λ_{pm} [10]. Similarly to Eq. 10, the expression of [11] $\Lambda_{om} = \sum_{n=1}^{5} \rho_n 2\gamma_o (Q_s^{(n)}\Lambda_s + Q_t^{(n)}\Lambda_t)$ (where ρ_n is the initial population of the n-th hfs level of the ortho-state of $pp\mu$, γ_o is the ratio of the muon density at the proton in the $pp\mu$ to that in $p\mu$, and $Q_{t,s}^n$ is the probability for the "left" proton and muon spins in the n-th state of $pp\mu$ to be parallel or antiparallel [12]) should now be replaced by

$$\Lambda_{om} = \sum_{n=1}^{5} \rho_n 2\gamma_o^{(n)} (Q_s^{(n)}\Lambda_s + Q_t^{(n)}\Lambda_t) \qquad (11)$$

because in any of the 5 hfs states the muon-proton overlap will undergo a specific correction from the singular interaction terms in the 3-body relativistic Hamiltonian [13]. The deviation of $\gamma_o^{(n)}$ from γ_o is expected to be of the order of 1–2%. Comparison of the theoretical predictions for the rate of muon capture in liquid hydrogen with the experimental data of [14] should necessarily be based on the results of the new more accurate calculations of the γ-factor and the rate of ortho-para transitions in $pp\mu$ that are now in progress.

Acknowledgements

The author is grateful to Profs. E.Zavattini and G.Barbiellini from I.N.F.N. – Trieste, for the hospitality and financial support, to Prof. L.Ponomarev - for initiating the work on the problem and help in all further stages, and to the Direction of SISSA – for the wonderful working conditions.

References

[1] G.T.Bodwin, D.R.Yennie, M.A.Gregorio. Rev.Mod.Phys. **57** (1985) p.723.

[2] H.Bethe, E.Salpeter. Quantum Mechanics of One- and Two- Electron Atoms. Springer, Heidelberg, 1957.

[3] V.Rizov, I.Todorov. Sov.J.Part.Nucl.,**6**,3 (1976) p.269.

[4] D.Bakalov. Phys.Lett.**93B**,3 (1980)p.265.

[5] D.Bakalov. Yad.Fiz.**48** (1988)p.335.

[6] A.E.Uehling. Phys.Rev.**48** (1935)p.55.

[7] L.W.Fullerton, G.A.Rinker. Phys.Rev. **A13** (1976) p.1283.

[8] A.O.Barut, J.Kraus. Jour.Math.Phys. **17** (1976) p.506.

[9] L.B.Okun'. Leptons and Quarks. Amsterdam, North-Holland, 1982.

[10] D.Bakalov et al. Comm.JINR, R4-82-633, Dubna, 1982.

[11] D.Bakalov, M.Faifman, L.Ponomarev, S.Vinitsky. Nucl.Phys.**A384** (1982)p.302.

[12] D.Bakalov, S.Vinitsky, V.Melezhik. Sov.Phys.JETP,**52**(4), (1980)p.820.

[13] D.Bakalov, V.Melezhik. Phys.Lett.**B161** (1985)p.5.

[14] A.Bertin et al. Nuovo Cim.**86A**,2 (1985)p.123.

Muonic Atoms: Charge Radii and Nuclear Polarization

Lukas A. SCHALLER
Institut de Physique de l'Université, Pérolles
CH-1700 Fribourg, Switzerland

Abstract: Muonic atom spectroscopy is reviewed in the nuclear range from Z,N = 6 up to Z = 64 and N = 86. Special emphasis is given to the systematics of isotope and isotone shifts. The obtainable accuracy in the transition energies is of the order of 10 ppm, which translates to about 0.1 am for the nuclear charge moments. However, this presupposes a precise knowledge of the nuclear polarization corrections. Since these corrections are important, but badly known corrections to the low-lying muonic s- and p-states in heavy nuclei, they should be determined in "complete" experiments. Results for muonic Zr and muonic Pb show that the nuclear polarization 2p (and 3p) fine structure splittings are in disagreement with theory.

1. Introduction; Theory

Muonic atoms are "exotic" atoms in the sense that an electron in the atomic cloud is replaced by a negatively charged muon. Since the muonic Bohr radius is roughly 200 times smaller than the electronic Bohr radius, the muon-nucleus system for muonic orbits with principal quantum numbers n between about 4 and 10 can be treated as a hydrogen-like system. Also, with the muon being a charged second generation lepton, its interaction with the nucleus is purely electromagnetic, in complete analogy to the electron. Contrary to the electron however, there is a large overlap of the muonic with the nuclear wave function in low-lying orbits (n < 5) of medium-heavy and heavy muonic atoms. Hence, muonic atoms provide a very sensitive and accurate electromagnetic probe of the nuclear charge distribution. Muonic atom results on the nuclear charge moments are complimentary to elastic electron scattering measurements and to optical laser spectroscopy. Using the latter technique, also unstable isotopes can be examined. The great advantage of muonic atoms however as compared to optical measurements of the nuclear charge distribution is the high absolute accuracy which can be obtained. Muonic atom data for only a few isotopes may therefore provide an absolute calibration for laser-spectroscopic data on long chains of isotopes.

The present work presents in its first part an overview of precision measurements on nuclear charge moments using the muonic-atom method for nuclei as light as carbon and as heavy as Nd or Sm. If the obtained precision of about 10 ppm in the transition energies could be fully translated to the nuclear charge extension, an accuracy of about 0.1 am for the latter quantity would be attainable. However, in order to find out what precision is realistically possible, we have to have a closer look regarding all corrections to our muon-nucleus system. Firstly, in all but the lightest nuclei, the finite nuclear charge extension in low-lying muonic orbits cannot be treated as a perturbation. Hence, in the Dirac equation, the Coulomb potential between two pointlike charges has to be replaced by a potential with adjustable parameters, which describe the size of the nuclear charge extension, and the wave equation has to be numerically solved. Such a procedure is only weakly dependent on the form of the chosen nuclear charge distribution (normally a two-parameter Fermi distribution), due to a double integration process involved. The final result is thus not appreciably changed, if the nuclear charge density is varied. This is particularly true if generalized nuclear charge moments are employed, that is the socalled Barrett moments [1]

$$\langle r^k e^{-\alpha r} \rangle = \int_0^\infty \rho(r)\, r^k\, e^{-\alpha r}\, 4\pi r^2\, dr$$

Each muonic atom transition sensitive to the nuclear charge distribution determines a different and model-independent generalized moment. Instead of the Barrett moment, the radius of a homogeneously charged sphere yielding the same generalized moment as the real charge distribution is often introduced and defined as the equivalent radius $R_{k\alpha}$. For more details, see e.g. ref.[2].

Besides the potential for the finite nuclear charge, the first order vacuum polarization (VP) potential is also included before numerically solving the Dirac equation. The corresponding energy shift is of the order of the fine structure constant α smaller than the binding energy. All other corrections are added as perturbations to the Dirac hamiltonian. The least-known of these corrections is the nuclear polarization (NP) correction. The NP shift is particularly large in the 1s levels of heavy muonic atoms. In μ^- - ^{208}Pb e.g., it amounts to about 4 keV or 600 ppm of the 1s binding energy. The nuclear polarization is the analogon of the dispersion correction in electron scattering. By "muon double scattering", nuclear states are virtually excited up to energies close to the muon mass, in any case quite up into the giant resonance regions. This leads to real energy shifts in the muonic energy levels. Due to incomplete knowledge of the nuclear excitation spectrum, the NP shifts can only be calculated approximately, i.e. to an absolute error of typically 20 to 30% [2,3]. In relative determinations of nuclear charge radii., the NP error is reduced to about 10 %. It is clear that such an uncertainty limits the ultimate accuracy with which nuclear charge moments can be determined. QED corrections on the other hand are now reliably known to the order of 0.1%, at least concerning the largest one of these corrections, i.e. the first order VP [4,5]. They do therefore not appreciably increase the statistical error of 0.1 am reached in charge radii measurements using the muonic atom method. But due to the NP errors, the radii uncertainties are of the order of 0.5 to 1 am, depending on whether we are interested in relative or absolute charge radii. The absolute precisions however are generally still superior to the accuracies obtained in elastic electron scattering, and considerably higher than in laser spectroscopic work. Note also that no isotone shifts can be determined when using optical transitions.

In the following, we shall first present results on nuclear charge radii differences (chapter 2). Then, we discuss what has been done up til now regarding an experimental determination of nuclear polarization shifts, and what problems still remain (chapter 3).

2. Isotope and isotone shifts

During the past two decades, a lot of nuclear charge radii have been measured using the muonic atom method. Since isotope and isotone shifts are particularly sensitive to nuclear structure effects, we shall concentrate on such results, i.e. on charge radii differences when adding pairs of neutrons or protons. Fig.1 shows isotope shifts for $6 \leq N \leq 86$ and isotone shifts for $6 \leq Z \leq 62$. The results, taken from our collaboration between Fribourg, Mainz and Los Alamos [6-13], are given in terms of equivalent radii differences $\Delta R_{k\alpha}$ [in am].

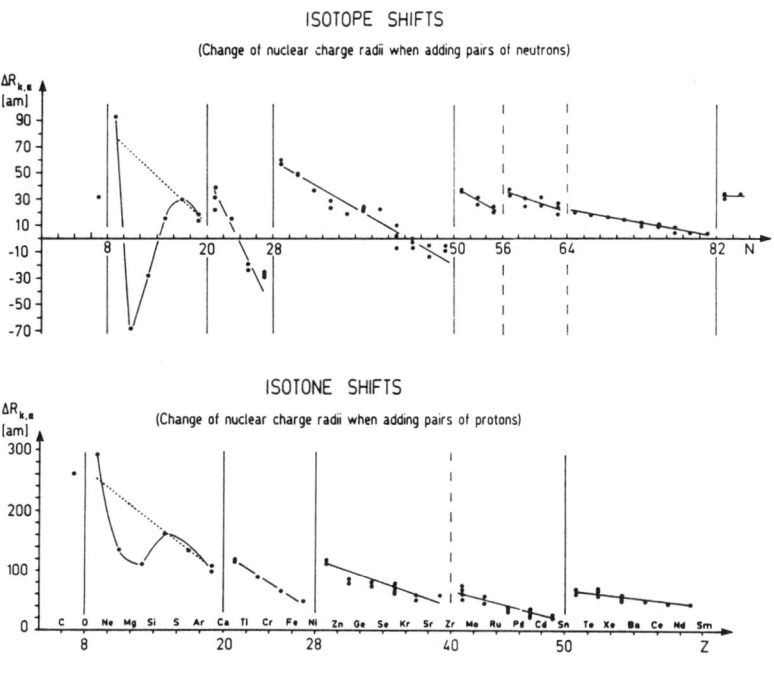

Figure 1: Isotope and isotone shifts measured with muonic atoms

For nuclei with N,Z > 20, the trends are similar and can be described by the following characteristica:
(1) After the closing of a major shell, there is a sudden increase in the radii differences, reflecting the shell structure of the nuclei.
(2) In-between these magic numbers, the addition of a pair of neutrons or a pair of protons results in an almost linear decrease of the radii differences.
(3) The change of nuclear charge radii when adding a pair of neutrons (protons) is only slightly dependent on the proton (neutron) configuration.

Precision measurements may also reveal subshell effects, as has e.g. been demonstrated by C.Piller et al.[12]. Subshell effects, together with a rapidly changing deformation, are also at the origin of the "anomalous" behaviour of the light s-d-shell nuclei ($8 \leq N,Z \leq 20$), where the 2nd characteristicum is completely absent, and where certain nuclear charge radii may even diminish when a pair of neutrons is added (see figure 1 and refs.[13, 14]).

3. Nuclear Polarization

Nuclear polarization corrections in the innermost muonic orbits limit the accuracy with which nuclear charge radii can be determined (see also chapter 1). It is therefore important to check the theoretical estimations experimentally, whenever this is possible. In order to arrive at this aim, all muonic atom transitions sensitive to the nuclear charge extension have to be measured with the highest possible precision. This has been done for muonic ^{208}Pb, ^{140}Ce and ^{90}Zr [14 - 17]. Fig.2 shows the energy spectrum of μ^{-}-^{140}Ce in the region of the weakly populated 2s - 2p transitions, with (ACS on) and without (ACS off) Compton suppression.

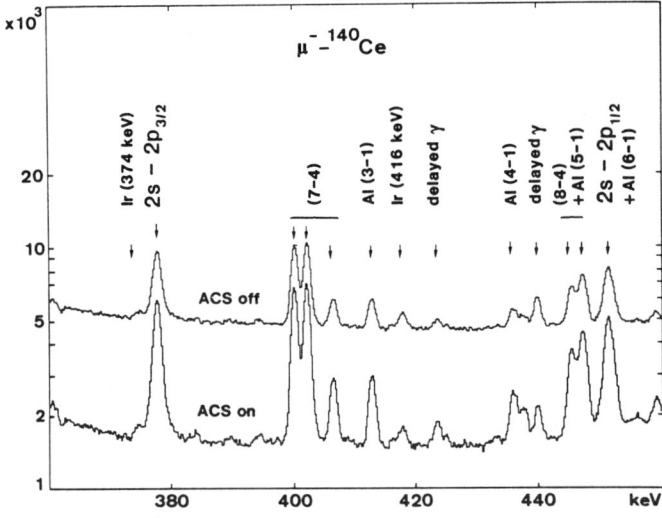

Figure 2: Energy spectrum of muonic ^{140}Ce between 360 and 460 keV

When comparing experiment with theory, correlations between different states should be made. The theoretically well predictable ratio of the NP correction in the 2s state to the 1s state [3] e.g. is in accordance with experiment [15,16]. However, there is a serious discrepancy when looking at the NP splittings in the p levels. Since the $2p_{1/2}$ and the $2p_{3/2}$ states lead to practically the same nuclear charge moment, while their sensitivity to the nuclear polarization is quite different, a plot of the 2pNP fine structure splitting $\Delta(NP2p) =$ = $NP2p_{1/2} - NP2p_{3/2}$ versus the NP correction in the 1s level should represent a particularly sensitive test. All current NP theories yield for $\Delta(NPnp)$ a positive value, since, simply stated, the $2p_{1/2}$ state overlaps more strongly with the nuclear charge than the $2p_{3/2}$ state. The experiment however leads both in μ^- - ^{90}Zr and in μ^- - ^{208}Pb to negative values, if the NP1s correction is taken to be about 1 keV in Zr and 4 keV in Pb. In order to obtain agreement with theory, NP1s values about three times larger than predicted would be needed (see figure 3). Such high values can practically be excluded if the muonic atom data are fitted with a nuclear charge density taken from elastic electron scattering data [17].

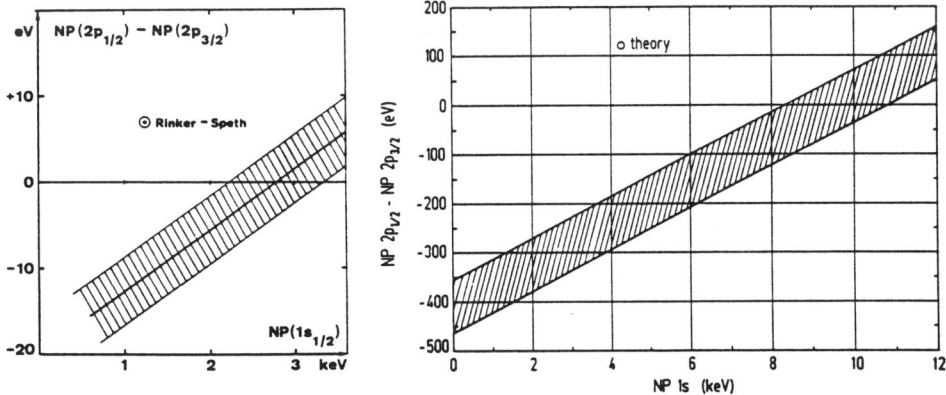

Figure 3: The difference of the nuclear polarization (NP) values in the $2p_{1/2}$ state minus the $2p_{3/2}$ state versus the NP 1s value in μ^- - ^{90}Zr (left) and μ^- - ^{208}Pb (right).

More evidence should be obtained when the μ^- - ^{140}Ce data are analysed. In addition, one could have a closer look at the 2p fine structure splittings in our measurements of isotope and isotone shifts. In these cases however, only one parameter of the nuclear charge density, i.e. the half-density radius c, can then be varied, the theoretical NP1s result has to be taken at face value, and no consistency checks are possible. It is not surprising, that such an analysis, where performed, does not allow to draw definitive conclusions.

It is not clear what effects could cause the discrepancies in the nuclear polarization corrections. Uncertainties in the NP values of higher-lying states, or in higher-order vacuum polarization corrections, would influence the μ^- - ^{208}Pb data considerably more than the μ^- - ^{90}Zr data. The effect however is in both cases the same. Also, recent calculations about transverse polarization effects do not seem to resolve the binding-energy anomaly in μ^- - ^{208}Pb [18]. The next talk by Dr. Rosenfelder [19] may give us some hints, how the nuclear polarization problem could be solved or at least "retackled" from a theoretical point of view.

To conclude, the author would like to thank all members of the muonic atom collaboration, especially in Fribourg and Mainz, without their work this presentation would not have been possible. He also acknowledges the financial support of the Swiss National Foundation.

References

[1] R.C.Barrett, Phys.Lett. B**33**, 388 (1970)
[2] E.F.Borie and G.A.Rinker, Revs.Mod.Phys. **54**, 67 (1982)
[3] G.A.Rinker and J.Speth, Nucl.Phys. A**306**, 397 (1978)
[4] T.Dubler, K.Kaeser, B.Robert-Tissot, L.A.Schaller, L.Schellenberg, and H.Schneuwly, Nucl.Phys. A**294**, 397 (1978)
[5] B.Aas et al., Nucl.Phys. A**375**, 405 (1982)
[6] L.A.Schaller, T.Dubler, K.Kaeser, G.A.Rinker, B.Robert-Tissot, L.Schellenberg, and H.Schneuwly, Nucl.Phys. A**300**, 225 (1978)
[7] H.J.Emrich et al., in *Proceedings of the 4th Int. Conf. on Nuclei far from Stability*, Helsingor 1981, ed. by L.O.Skolen, p.33
[8] H.D.Wohlfahrt, E.B.Shera, M.V.Hoehn, Y.Yamazaki, and R.M.Steffen, Phys.Rev. C**23**, 533 (1981)
[9] L.A.Schaller, L.Schellenberg, T.Q.Phan, G.Piller, A.Rüetschi, and H.Schneuwly, Nucl.Phys.A**379**, 523 (1982)
[10] L.A.Schaller et al., Phys.Rev. C**31**, 1007 (1985)
[11] E.B.Shera, M.V.Hoehn, G.Fricke, and G.Mallot, Phys.Rev. C**39**, 195 (1989)
[12] C.Piller, C.Gugler, R.Jacot-Guillarmod, L.A.Schaller, L.Schellenberg, H.Schneuwly, G.Fricke, T.Hennemann, and J.Herberz, Phys.Rev. C**42**, 182 (1990)
[13] G.Fricke, J.Herberz, T.Hennemann, G.Mallot, L.A.Schaller, L.Schellenberg, C.Piller, and R.Jacot-Guillarmod, Phys.Rev. C**45**, 80 (1992)
[14] L.A.Schaller, in *Proceedings of the Int.Conf on the Future of Muon Physics*, Heidelberg 1991, ed. by K.P.Jungmann, to be published in Z.Phys C
[15] T.Q.Phan, P.Bergem, A.Rüetschi, L.A.Schaller, and L.Schellenberg, Phys.Rev. C**32**, 609 (1985)
[16] P.Bergem, G.Piller, A.Rüetschi, L.A.Schaller, L.Schellenberg, and H.Schneuwly, Phys.Rev. C**37**, 2821 (1988)
[17] L.A.Schaller, in *Proc. of the Int. Symp. on Collective Phenomena in Nuclear and Subnuclear Long Range Interactions in Nuclei, Bad Honnef, 1987*, ed. by P.David (World Scientific Publ. Co., Singapore 1988, p.145)
[18] Y.Tanaka and Y.Horikawa, Phys.Lett. B**281**, 191 (1992)
[19] R.Rosenfelder, Nucl.Phys. A**393**, 301 (1983), and this conference

Nuclear Polarization in Muonic Atoms

R. ROSENFELDER
Paul Scherrer Institute
CH-5232 Villigen PSI, Switzerland

Abstract. A survey is given on the theoretical evaluation of the energy shift in muonic atoms due to virtual excitations of the nucleus. In light muonic atoms the shift can be expressed by the forward Compton amplitude and it is therefore related to measured inelastic cross sections. This is not the case for heavy nuclei where detailed models of transition densities and currents are required. Comparison is made with a recent analysis of electron scattering data from ^{12}C which includes dispersion corrections. Finally the persistent anomaly in the nuclear polarization shifts of p–levels in heavy atoms is discussed. It is shown that virtual Coulomb excitations favour larger shifts for the $p_{1/2}$-level than for the $p_{3/2}$-level, i.e. most probably cannot account for this anomaly.

1. Introduction

The spectra of muonic atoms and the elastic cross sections measured in electron scattering from nuclei are our main sources of information about the nuclear charge distributions. With the increasing accuracy of the experiments it is necessary to take into account all sorts of corrections to the simple picture of Coulomb interaction between lepton and static nuclear charge distribution. One of these – and the least well understood – is the contribution arising from virtual excitations of the target : in muonic atoms these corrections are called *nuclear polarization* (NP) shifts while in electron scattering they traditionally go under the name of *dispersion corrections*

(DC). From the vast literature on NP shifts some calculations which are relevant for the following are listed in refs. [1 - 8]. For ordinary (electronic) atoms see e.g. ref. [9].

Before going into more detail it is worthwhile to consider the classical analogue of this process : replacing

$$\text{nucleus} \longrightarrow \text{earth}$$

$$\text{muon} \longrightarrow \text{moon}$$

one has to deal with the century-old (and difficult) problem of *tides* [10]. Besides the central potential for the moon there is a residual interaction which deforms the surface of the earth. This slows down the rotation of the earth until the rotational period will be synchronized with the orbital period. In addition, the moon retreats from the earth (i.e. it's energy increases) as rotational energy from the earth is transfered to the moon's orbital motion.

When a muon is captured into a high orbit around a spinless nucleus similar things happen : the Coulomb interaction between the muon and the Z protons in a nucleus is

$$V = -e^2 \sum_{i=1}^{Z} \frac{1}{|\mathbf{r} - \mathbf{r}_i|} = V_0 + \Delta V . \tag{1}$$

Here

$$V_0(r) = <0|V|0> = -e^2 \int d^3 r' \, \frac{\rho(r')}{|\mathbf{r} - \mathbf{r}'|} \tag{2}$$

is the Coulomb potential generated by the static charge distribution $\rho(r)$ in the ground state $|0>$ of the nucleus. As in the earth-moon system, the residual interaction ΔV will induce tidal forces in the target and deform it which in turn acts back on the muon. In quantum mechanics a deformation is described by an admixture of excited states to the ground state of the nucleus. For light nuclei the muon will be far away from the nucleus and these excitations will be dominated by the longest-ranged multipole, i.e. the dipole. However, two important differences to the classical analogue exist: first, the muon energy will be lowered and second, only *virtual* excitations [1] are considered in the following : the nucleus remains in its ground-state. In addition, for large Z the muon can penetrate the nucleus which obviously is not possible in the classical analogue ...

---

[1] Real excitations correspond to what is called dynamic nuclear polarization [11,12] occuring when muonic transition energies are nearly degenerate with nuclear excitation energies.

The usual procedure is that one takes into account the static Coulomb interaction $V_0(r)$ to all orders by solving the Dirac equation for the muon moving in that field. This yields a spectrum of energy levels ϵ_i. The residual interaction is then treated in second-order perturbation theory which gives a shift for the muon energy

$$\Delta \epsilon_i = \sum_{(n,N) \neq (i,0)} \frac{|<i,0|\Delta V|n,N>|^2}{\epsilon_i + E_0 - \epsilon_n - E_N}. \tag{3}$$

Here capital letters are used for nuclear and small letters for muonic states; E_N are the nuclear excitation energies. It is evident that for the ground state of the muon ($i = 0$) the nuclear polarization shift is always negative, i.e. the muon binding is increased:

$$\Delta \epsilon_{1s} < 0. \tag{4}$$

Eq. (3) illustrates the basic problem of all NP calculations: a knowledge of the complete excitation spectrum of the nucleus is required. However, eq. (3) is not the complete answer: in principle one also has to take into account the transverse electromagnetic interaction between muon and nucleus. This will be derived in the next section. Section 3 deals with light muonic atoms where some simplifications are possible. High Z atoms (section 4) are theoretically more difficult and the calculated NP shifts are more uncertain because the sum over excited states has to be performed in some specific nuclear model. In the section 5 I will discuss the anomaly in the p-level NP shifts of heavy nuclei where empirical fits demand more binding for the $2p_{3/2}$ level than for the $2p_{1/2}$ level whereas all theoretical calculations predict the opposite behaviour. Finally the conclusions are summarized in section 6.

2. Energy shift

NP shifts including the *full* electromagnetic interaction between muon and nucleus can be derived from the Feynman rules for scattering of the muon in the external field V_0 of the nucleus [2]. In second order one has the box graph, the crossed graph and the "seagull" graph depicted in Fig. 1. The seagull graph is essential to maintain gauge invariance [13] in a nonrelativistic system like the nucleus [3]. Note that it also contributes to the two-photon recoil correction [14].

---

[2] For simplicity, I assume in the following that the nuclear mass ($\simeq AM$) is much larger than the muon mass (m) so that recoil effects can be neglected.

[3] It arises from the minimal substitution $\sum_i \mathbf{p}_i^2/2M \to \sum_i (\mathbf{p}_i - e\mathbf{A}(\mathbf{x}_i))^2/2M$ in the free Hamiltonian.

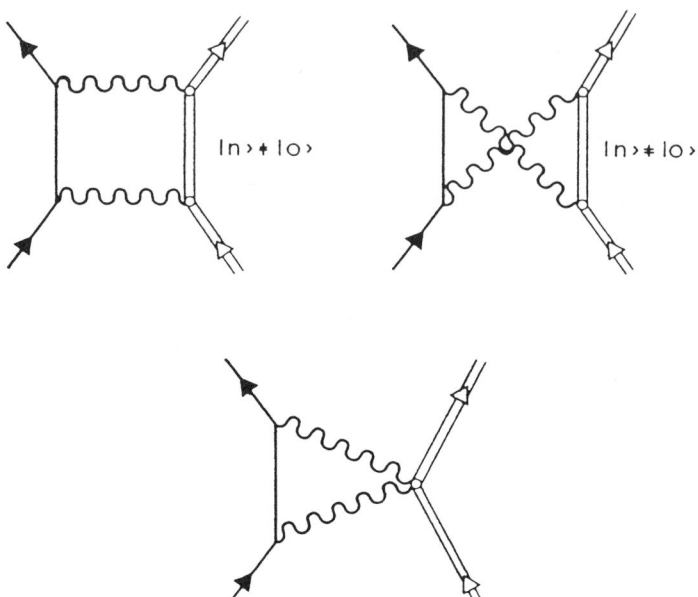

Figure 1: Second-order nonstatic contributions to lepton–nucleus scattering.

Evaluating the second-order S-matrix from the Feynman diagrams of Fig. 1

$$S^{(2)}_{(i,I)\to(f,F)} = -2\pi i\, \delta(\epsilon_i + E_I - \epsilon_f - E_F)\, T^{(2)}_{i\to f} \quad (5)$$

one can define an effective potential ΔV which produces $T^{(2)}$

$$T^{(2)}_{i\to f} = <\phi_f|\Delta V|\phi_i> \quad (6)$$

where $|\phi_i>, |\phi_f>$ are *scattering* states. The second-order shift due to virtual excitations is then obtained by calculating the expectation value of ΔV between *bound* state muon wavefunctions

$$\Delta\epsilon_i = <\phi_f|\Delta V|\phi_i> \quad (7)$$

where now $E_I = E_F = E_0$. This so-called scattering approximation explicitly gives

$$\Delta\epsilon_i = -(4\pi\alpha)^2\, \mathrm{Im}\int \frac{d^4q_1\, d^4q_2}{(2\pi)^7}\, \delta(q_1^0 - q_2^0)\, t^{(i)}_{\mu\nu}(q_1, q_2)\, D^{\mu\rho}(-q_2)\, D^{\nu\tau}(q_1)\, T_{\rho\tau}(q_1, q_2) \quad (8)$$

where $\alpha \simeq 1/137$ is the fine-structure constant and

$$t^{(i)}_{\mu\nu}(q_1, q_2) = <\phi_i|\, \gamma_0\gamma_\mu\, e^{i\mathbf{q}_2\cdot\mathbf{x}}\, \frac{1}{\epsilon_i - q_1^0 - h}\, e^{-i\mathbf{q}_1\cdot\mathbf{x}}\, \gamma_0\gamma_\nu\, |\phi_i> \quad (9)$$

denotes the virtual Compton amplitude for the muon ($h|\phi_i> = \epsilon_i|\phi_i>$). Although this looks complicated the real complexity of the NP problem resides in the virtual Compton amplitude for the nucleus [13]

$$T_{\rho\tau} = \text{seagull} + \sum_{N\neq 0}\left[\frac{<0,\mathbf{q}_1-\mathbf{q}_2|J_\rho(0)|N,\mathbf{q}_1><N,\mathbf{q}_1|J_\tau(0)|0,\mathbf{0}>}{E_0 - E_N + q_1^0 + i\eta}\right. \quad (10)$$

$$+ \left.\frac{<0,\mathbf{q}_1-\mathbf{q}_2|J_\tau(0)|N,-\mathbf{q}_2><N,-\mathbf{q}_2|J_\rho(0)|0,\mathbf{0}>}{E_0 - E_N - q_2^0 + i\eta}\right].$$

Here $J_\tau = (\rho, \mathbf{J})$ is the nuclear current operator. It can be shown [8] that the second-order shift $\Delta\epsilon_i$ is invariant under a general gauge transformation of the photon propagators

$$D^{\nu\tau}(q_1) \longrightarrow D^{\nu\tau}(q_1) + q_1^\nu \chi^\tau(q_1) + \chi^\nu(q_1) q_1^\tau \quad (11)$$

where χ^ν are arbitrary functions of q_1 provided the Compton amplitudes fulfill the condition

$$q_1^\mu t_{\mu\nu}^{(i)}(q_1, q_2) = q_1^\mu T_{\mu\nu}(q_1, q_2) = 0 \quad (12)$$

As mentioned before the seagull term in the nuclear tensor is indispensible for the current conservation expressed in eq. (12).

3. Light muonic atoms : ^{12}C

Considerable simplifications arise for light muonic atoms where the muon orbits far outside the nuclear radius R

$$a_B = \frac{1}{mZ\alpha} \gg R. \quad (13)$$

The muon wavefunction is then approximately constant over the nuclear volume

$$\phi(\mathbf{x}) \simeq \phi_{\text{av}} \quad (14)$$

and the muon acts as a static source transfering no momentum to the target

$$t_{\mu\nu} \simeq -\delta^{(3)}(\mathbf{q}_1 - \mathbf{q}_2)\,\bar\phi_{\text{av}}\,\gamma_\mu \frac{1}{q_1^0 + h_0(\mathbf{q}_1)}\gamma_0\gamma_\nu\,\phi_{\text{av}}. \quad (15)$$

In the spirit of this "wave-function approximation" [5] the binding energy of the muon has been neglected and the muon Hamiltonian replaced by the free one. It is then seen that only the *forward* nuclear Compton amplitude enters in the energy shift of eq. (8). This quantity is determined by structure functions which can be measured in inelastic electron scattering from nuclei. Retaining only the Coulomb interaction one easily obtains

$$\Delta\epsilon_{1s} = -8\alpha^2 \bar{R}\,\phi_{1s}^2(0)\int_0^\infty d\omega \int_0^\infty dq \frac{1}{q^2}\frac{S_L(\omega,q)}{\omega + q^2/2m} \quad (16)$$

where $\bar{R} < 1$ is a factor correcting for the change of the muon wavefunction over the nuclear radius and $S_L(\omega, q)$ is the longitudinal structure function. The transverse electromagnetic interaction and the seagull term can be included in a similar way [8] but they tend to cancel each other and their net effect (at least in a nucleus like ^{12}C) is small. The main nuclear information is contained in S_L for which an analytical model was used [15] which was fitted to experimental γ-absorption and quasielastic electron scattering data.

The final result (including transverse and seagull contributions) for the NP shift in ^{12}C is

$$\Delta \epsilon_{1s} = (-2.5 \pm 0.3) \text{ eV} \tag{17}$$

where the error has been estimated from the neglected Coulomb corrections in the muon propagator. This has to be compared with the 5 ppm accuracy of the measured $2p_{3/2} - 1s_{1/2}$ transition energy [17]

$$\epsilon_{2p_{3/2}} - \epsilon_{1s_{1/2}} = (75261.4 \pm 0.4) \text{ eV}. \tag{18}$$

Although small on a absolute scale the uncertainty in the NP shift is the dominant theoretical error as the detailed listing of QED and other corrections demonstrates [16]. It also turns out [8] that the main contribution to the double integral (16) comes from the low-energy – low-momentum transfer region which is dominated by the giant dipole resonance. The contribution from the quasielastic peak region is relatively small. The rms radius of ^{12}C extracted after applying QED and NP corrections is [17]

$$<r^2>_\mu^{1/2} = (2.483 \pm 0.002) \text{ fm}. \tag{19}$$

Previously there was a 2.4 standard deviation from the value derived from elastic electron scattering. A new analysis of the data from NBS, Stanford and Mainz together with new scattering data from NIKHEF recently gave [18]

$$<r^2>_e^{1/2} = (2.478 \pm 0.009) \text{ fm} \tag{20}$$

in excellent agreement with the value (19). This was achieved by fitting the experimental cross sections with *free* normalizations using a static charge distribution and applying the dispersion corrections (DC) calculated by Friar and Rosen [13] outside the diffraction minima. Inclusion of DC raised the rms radius by 0.007 fm to the value given in eq. (20)[4]. However, the case for an unambigous proof of dispersion

---

[4]In contrast the NP correction (17) lowers the muonic rms radius by the tiny amount of 0.0003 fm as can be estimated from the perturbative finite-size shift $\Delta \epsilon \simeq 2 <r^2>/(3a_B^3)$.

corrections is rather weak since in the minima the deviations from the static best fit is 10 times larger than theoretically predicted. This is also reflected in the total χ^2 of the fit

$$\chi^2/\text{dof} = \begin{cases} 280.2/252 & \text{without DC} \\ 279.0/252 & \text{with DC} \end{cases} \qquad (21)$$

which is not improved significantly by the inclusion of the theoretical dispersion corrections.

4. Heavy muonic atoms : ^{208}Pb

Heavy nuclei pose a greater challenge for a reliable calculation of NP shifts. This is not because of the muonic Compton amplitude $t^{(i)}_{\mu\nu}$ which can also be calculated exactly in the static field V_0 of a heavy nucleus – it is because of the full Compton amplitude $T_{\rho\tau}(\omega, \mathbf{q}_1, \mathbf{q}_2)$ of the nucleus given in eq. (10) which is needed now. To phrase it differently : the muonic part is a *one*-body problem with a well-known interaction whereas the nuclear part is linked to all the complications and uncertainties of a *many*-body problem with not very well understood interactions.

The standard approach (see e.g. ref. [7]) for which general computer codes are available [19] consists of the following steps :

- Keep only the Coulomb interaction.

- Evaluate

$$T_{00}(\omega, \mathbf{q}_1, \mathbf{q}_2) = \sum_{N \neq 0} \frac{<0|\rho^\dagger(\mathbf{q}_2)|N><N|\rho(\mathbf{q}_1)|0>}{E_0 - E_N + \omega + i\eta} \qquad (22)$$

  by including *low-lying* levels (using empirical transition densities and B(EL)-values) and some *high-lying* levels. For the higher levels the sum-rule strength is concentrated into a single giant resonance and a model for the transition strength is assumed.

- Perform the sum over muonic excitations exactly using the Dalgarno-Lewis [20] technique of solving inhomogenous equations of motion.

An example for the outcome of such a calculation in ^{208}Pb is shown in Fig. 2 where binding energies are listed, i.e. one defines

$$\text{NP(nl)} \equiv -\Delta\epsilon_{\text{nl}}. \qquad (23)$$

TABLE II. Theoretical nuclear polarization corrections in ^{208}Pb.

Energy (MeV)	I^π	$B(E\lambda)\uparrow$ ($e^2b^{2\lambda}$)	$1s_{1/2}$ (eV)	$2s_{1/2}$ (eV)	$2p_{1/2}$ (eV)	$2p_{3/2}$ (eV)	$3p_{1/2}$ (eV)	$3p_{3/2}$ (eV)	$3d_{3/2}$ (eV)	$3d_{5/2}$ (eV)
2.615	3^-	0.612	135	12	90	84	26	26	111	-63
4.085	2^+	0.318	198	20	182	180	76	84	6	4
4.324	4^+	0.155	14	1	8	7	2	2	1	1
4.842	1^-	0.00156	7	1	-9	-8	0	0	1	1
5.240	3^-	0.130	27	2	16	15	5	5	2	2
5.293	1^-	0.00204	9	2	-27	-19	0	-1	1	1
5.512	1^-	0.00380	16	3	-90	-53	-1	-1	1	1
5.946	1^-	0.00007	0	0	3	-30	0	0	0	0
6.193	2^+	0.0505	29	3	22	21	7	7	0	0
6.262	1^-	0.00024	1	0	3	5	0	0	0	0
6.312	1^-	0.00022	1	0	3	4	0	0	0	0
6.363	1^-	0.00014	1	0	2	2	0	0	0	0
6.721	1^-	0.00075	3	1	6	7	0	-1	0	0
7.064	1^-	0.00156	6	1	9	11	-1	-1	0	0
7.083	1^-	0.00075	3	1	4	5	-1	-1	0	0
7.332	1^-	0.00204	8	1	10	11	-2	-2	0	0
Total low-lying states			458	48	233	242	111	117	123	-53
13.5	0^+	0.047872	906	315	64	38	24	15	1	0
22.8	0^+	0.043658	546	147	43	26	15	10	0	0
13.7	1^-	0.537672	1454	221	786	738	255	258	66	54
10.6	2^+	0.761038	375	37	237	222	67	68	33	30
21.9	2^+	0.566709	207	21	108	99	29	29	8	7
18.6	3^-	0.497596	77	7	40	36	11	11	3	2
33.1	3^-	0.429112	53	5	25	23	7	7	2	1
	$>3^a$		176	15	80	71	21	21	4	4
Total high-lying states			3794	768	1383	1253	429	419	117	98
Total			4252	816	1616	1495	540	536	240	45

^{a}Values from Ref. 7. Positive NP values mean that the respective binding energies are increased.

Figure 2: Theoretical nuclear polarization shifts in muonic lead (from ref. [21]).

It is seen that the giant dipole resonance at 13.7 MeV still gives a large contribution to the total NP shift but that monopole (for the s-levels) and quadrupole excitations are now also important. The total theoretical NP shifts

$$\text{NP}(1s_{1/2}) = 4.25 \text{ keV} \tag{24}$$

$$\text{NP}(2p_{1/2}) = 1.62 \text{ keV} \tag{25}$$

$$\text{NP}(2p_{3/2}) = 1.50 \text{ keV} \tag{26}$$

should be contrasted with the 15 ppm accuracy of the measured $2p_{3/2}$ - $1s_{1/2}$ transition energy [21]

$$\epsilon_{2p_{3/2}} - \epsilon_{1s_{1/2}} = (5962.854 \pm 0.090) \text{ keV} . \tag{27}$$

Compared with the results given in eqs. (17) and (18) one sees that NP shifts are much more important in heavy nuclei. Unfortunately it is practically impossible to

5. p-level anomalies

Yamazaki et al. [22] have introduced a method for experimentally determining the NP shifts by a χ^2-fit to the measured transition energies. Since the NP shifts of higher levels turn out to be strongly correlated with the 1s-shift they plotted them as a function of NP(1s) and found good agreement in ^{208}Pb with the theoretical prediction for the case of NP(2s) vs. NP(1s). However, it turned out that

$$\Delta \text{NP}(2p) \equiv \text{NP}(2p_{1/2}) - \text{NP}(2p_{3/2}) < 0 \tag{28}$$

for "reasonable" values of the NP(1s) shift whereas the theoretical prediction from eqs. (25) and (26) is positive ! This has been confirmed by an analysis of more precise lead data in ref. [21] where the same type of anomaly has been found also for the fine-structure partners of the 3p-level (see Fig. 3). The same type of p-level NP anomaly was later found in ^{90}Zr [23] and there are indications for a similar anomaly also in Sn isotopes [24]. This suggests that *the anomaly is independent of the nuclear structure.*

To understand the sign of the difference in the NP shifts of fine-structure partners it is helpful to treat the spin-orbit coupling of the muon as a perturbation. As is well known from standard textbooks (see, e.g. ref. [25]) the nonrelativistic reduction of the muon Hamiltonian gives

$$h = \frac{\mathbf{p}^2}{2m} + V_0(r) + V_{\text{LS}}(r)\, \boldsymbol{\sigma} \cdot \mathbf{l} + ... \tag{29}$$

with

$$V_{\text{LS}}(r) = \frac{1}{4m^2} \frac{1}{r} \frac{\partial V_0(r)}{\partial r}. \tag{30}$$

In perturbation theory this gives rise to the fine structure splitting

$$\epsilon_{2p_{1/2}} - \epsilon_{2p_{3/2}} = \left(\lambda_{3/2} - \lambda_{1/2}\right) < \phi_{2p}|V_{\text{LS}}|\phi_{2p} > \tag{31}$$

where

$$\lambda_j = <\boldsymbol{\sigma} \cdot \mathbf{l}> = j(j+1) - l(l+1) - \frac{3}{4} = \begin{cases} -2 & \text{for } p_{1/2} \\ +1 & \text{for } p_{3/2} \end{cases}. \tag{32}$$

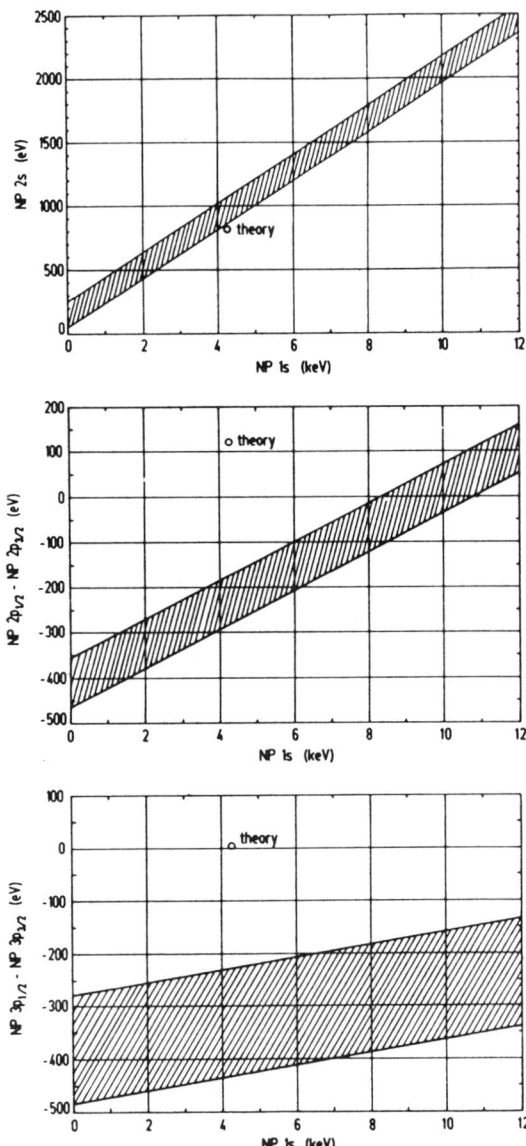

Figure 3: Empirically determined nuclear polarization shifts in ^{208}Pb as a function of the NP(1s) shift together with the theoretical predictions (from ref. [21]).

Evaluating the matrix element $< \phi_{2p}|V_{LS}|\phi_{2p} >$ with point Coulomb wave functions but with the finite size static potential $V_0(r)$ one obtains

$$\epsilon_{2p_{1/2}} - \epsilon_{2p_{3/2}} = -3\frac{Z\alpha}{96m^2 a_B^3} 4\pi \int_0^\infty dr\, r^2 \rho(r) e^{-\frac{r}{a_B}}(1+\frac{r}{a_B}) \simeq -\frac{Z\alpha}{32m^2 a_B^3} e^{-x}(1+x) \quad (33)$$

with $x = <r^2>^{1/2}/a_B$. Numerically this estimate gives $\simeq -200$ keV for lead which is in good agreement with the -185 keV obtained by solving the Dirac equation exactly (see Fig. 2). This shows that a perturbative treatment of the spin-orbit coupling makes sense even in a heavy nucleus like ^{208}Pb.

Let us now apply this perturbative treatment to the *difference* of NP shifts in the 2p-level. For simplicity I only keep the Coulomb interaction and use Coulomb gauge. According to eq. (8) and keeping in mind the sign convention (23) one then has

$$\Delta\mathrm{NP}(2p) = (4\pi\alpha)^2 \operatorname{Im} \int_0^\infty d\omega \int \frac{d^3q_1\, d^3q_2}{(2\pi)^7} \Delta t_{00}^{(2p)}(\omega, \mathbf{q}_1, \mathbf{q}_2) \frac{1}{\mathbf{q}_1^2 \mathbf{q}_2^2} T_{00}(\omega, \mathbf{q}_1, \mathbf{q}_2). \quad (34)$$

The difference of muon Compton amplitudes in the 2p-level is now expanded to first order in V_{LS} *both* for the muon wavefunction and for the muon Green function (see eq. (9)). If, in addition, one employs the closure approximation

$$\frac{1}{\epsilon - h} \simeq \frac{1}{-\bar{\epsilon}} \quad (35)$$

one obtains a very simple and transparent result

$$\Delta t_{00}^{(2p)}(\omega, \mathbf{q}_1, \mathbf{q}_2) \simeq -3 \left[\frac{<V_{LS}>_q}{(\omega+\bar{\epsilon}_{2p})^2} + \frac{<V_{LS}>_q - <V_{LS}> f_{2p}(q)}{\bar{\epsilon}_{2p}(\omega+\bar{\epsilon}_{2p})} \right]. \quad (36)$$

Here $\mathbf{q} = \mathbf{q}_1 - \mathbf{q}_2$ and the following quantities have been defined

$$<V_{LS}>_q = <\phi_{2p}|V_{LS}\, e^{-i\mathbf{q}\cdot\mathbf{x}}|\phi_{2p}> \quad (37)$$
$$f_{2p}(q) = <\phi_{2p}|e^{-i\mathbf{q}\cdot\mathbf{x}}|\phi_{2p}>. \quad (38)$$

Using the same kind of approximation which led to the fine-structure splitting given in eq. (33) one can show [26] that all these quantities are positive for $q \lesssim (1/a_B)$. Therefore the sign of $\Delta t_{00}^{(2p)}$ is determined by the same factor $\lambda_{1/2} - \lambda_{3/2} = -3$ in front of eq. (36) which also determines the sign of fine-structure splitting. On the nuclear side

$$\operatorname{Im} T_{00}(\omega, \mathbf{q}_1, \mathbf{q}_2) = -\sum_{N \neq 0} \delta(\omega - (E_N - E_0)) <0|\rho^\dagger(\mathbf{q}_2)|N><N|\rho(\mathbf{q}_1)|0> \quad (39)$$

is mostly negative at least for momenta $\mathbf{q}_1 \simeq \mathbf{q}_2$. Due to the finite size of the nucleus these momenta are restricted to values less than $\mathcal{O}(1/R) \simeq \mathcal{O}(1/a_B)$ and therefore the dominant contribution to the integrand of eq. (34) will be positively weighted. In other words we expect

$$\Delta \text{NP}(2p) \gtrsim 0 \,. \tag{40}$$

This is intuitively clear from the fact that the spin-orbit interaction pulls the $2p_{1/2}$ orbit closer to the nucleus where it has a greater chance to polarize the nucleus. It is obvious that some drastic approximations have been made to arrive at eq. (40). However, since here only the *sign* of the NP difference in the 2p-levels matters it seems rather plausible that virtual Coulomb excitations of the nucleus do not provide a solution to the 2p NP-anomaly.

Recently Tanaka and Horikawa [27] have calculated the NP shifts in ^{208}Pb including also transverse electric excitations up to 13 MeV excitation energy. Compared to the results shown in Fig. 3 they obtain a slight shift of the theoretical point (NP(1s) = 4.10 keV and Δ NP(2p) = 0.10 keV) but the discrepancy to the empirically determined NP values remains.

6. Conclusions

I have given a rather selected survey on the problem of energy shifts in muonic atoms due to virtual polarization of the nucleus. These nuclear polarization shifts

- are only calculated in a gauge invariant way if transverse and sea gull terms are included (usually these terms are neglected);

- can be related to the forward nuclear Compton amplitude in *light* muonic atoms. This allows measured photo-absorption and inelastic electron scattering cross sections to be used as an input in the theoretical calculation. There seems to be good agreement now between the rms radii deduced from muonic atoms and electron scattering if in the latter case also dispersion corrections are included in the analysis;

- require detailed knowledge of the nuclear excitation spectrum in *heavy* muonic atoms which leads to a considerable model dependence of the theoretically calculated shifts;

- limit the accuracy with which nuclear radii can be extracted from the increasingly more precise experimental data;

- show consistent *anomalies* in the p-level splittings of heavy muonic atoms : contrary to all theoretical predictions empirical fits require stronger $p_{3/2}$ NP attraction than for the $p_{1/2}$ level. Simple arguments which treat the spin-orbit interaction of the muon as a perturbation suggest that this cannot be explained by the Coulomb interaction alone which polarizes the nucleus.

Accepting the empirical determination of nuclear polarization shifts there are several possibilities to explain theoretically this anomaly :

- Mixing with nearby 1^- states has been proposed for the case of lead [7, 11], but no appropriate candidates have been found. As the p-level anomaly shows up in other heavy nuclei as well such an accidental degeneracy is very unlikely.

- Nuclear polarization shifts for the 1s -level are much larger than predicted by the most reliable theoretical calculations up to now. This would allow a positive empirical Δ NP(2p) shift (see Fig. 3). In view of the agreement of theory and experiment for the NP(2s) vs. NP(1s) shifts this is not very plausible.

- Higher nuclear excitations (e.g into the continuum) may lead to an inversion of Δ NP(2p).

- Spin-dependent excitations could give different contributions to the fine structure partners. Possible candidates for all nuclei are quasifree excitations induced by the spin current of nucleons which make a sizable contribution in quasielastic electron scattering.

Further theoretical work and additional experimental evidence for the anomaly are needed to clarify these points.

Acknowledgements

I would like to thank the organizers for arranging this successful workshop and, in particular, Lukas Schaller for rekindling my interest in nuclear polarization shifts. I am grateful to Dina Alexandrou for a critical reading of the manuscript.

References

[1] R. K. Cole, Phys. Rev. **177**, 169 (1969).

[2] M. - Y. Chen, Phys. Rev. C **1**, 1167 (1970).

[3] H. F. Skardhamar, Nucl. Phys. A **151**, 151 (1970).

[4] G. A. Rinker, Phys. Rev. A **14**, 18 (1976).

[5] J. L. Friar, Phys. Rev. C **16**, 1540 (1977).

[6] G. A. Rinker and J. Speth, Nucl. Phys. A **306**, 360 (1978).

[7] G. A. Rinker and J. Speth, Nucl. Phys. A **306**, 397 (1978).

[8] R. Rosenfelder, Nucl. Phys. A **393**, 301 (1983).

[9] G. Plunien et al., Phys. Rev. A **43**, 5853 (1991).

[10] J. Bartels, in *Handb. d. Physik*, Bd. XLVIII (Geophysik II), p. 734.

[11] E. Borie and G. A. Rinker, Rev. Mod. Phys. **54**, 67 (1982).

[12] J. Hüfner, F. Scheck and C. S. Wu, in *Muon Physics*, vol. I, p. 201, edited by V. W. Hughes and C. S. Wu, Academic Press, New York (1977).

[13] J. L. Friar and M. Rosen, Ann. Phys. **87**, 289 (1974).

[14] H. Grotch and D. R. Yennie, Rev. Mod. Phys. **41**, 350 (1969).

[15] R. Rosenfelder, Nucl. Phys. A **377**, 518 (1982).

[16] W. Ruckstuhl et al., Phys. Rev. Lett. **49**, 859 (1982).

[17] W. Ruckstuhl et al., Nucl. Phys. A **430**, 685 (1984).

[18] E. A. J. M. Offerman et al., Phys. Rev. C **44**, 1096 (1991).

[19] G. A. Rinker, Comp. Phys. Commun. **16**, 221 (1979).

[20] A. Dalgarno and J. T. Lewis, Proc. R. Soc. London A **233**, 70 (1955).

[21] P. Bergem et al., Phys. Rev. C **37**, 2821 (1988).

[22] Y. Yamazaki et al., Phys. Rev. Lett. **42**, 1470 (1979).

[23] T. Q. Phan et al., Phys. Rev. C **32**, 609 (1985).

[24] C. Piller et al., Phys. Rev. C **42**, 182 (1990).

[25] J. D. Bjorken and S. D. Drell, *Relativistic Quantum Mechanics*, p. 51, McGraw-Hill (1964).

[26] R. Rosenfelder, to be published.

[27] Y. Tanaka and Y. Horikawa, Phys. Lett. **281**, 191 (1992).

Muon Catalyzed Fusion and Cascade

Muon Catalyzed Fusion and Basic Muon Reactions in Deuterium and Hydrogen

Peter KAMMEL
Institute for Medium Energy Physics
Austrian Academy of Sciences
Boltzmanngasse 3
A-1090 Vienna
Austria

Abstract. By now, the kinetics of muon catalyzed fusion in pure D_2 and in H–D mixtures is understood in terms of the basic underlying processes. It provides rich information about muon induced few–body reactions. The current status and future directions of the field are discussed.

1. Introduction

Muon catalyzed fusion (μCF) has experienced a dramatic revival [1] since a resonant mechanism was found in the formation rate of $dd\mu$ and $dt\mu$ molecules which greatly enhances the fusion yield per muon. Yields of more than 100 fusions/μ^- have been observed in the most efficient D–T mixtures, bringing μCF close to practical applications (e.g., as an intense source of 14 MeV neutrons for fusion material research [2]).

In the present survey, however, we will not discuss the spectacular field of D–T μCF, but concentrate on the rich physics of muon induced processes in D_2 and H–D mixtures. In particular, we will investigate the relation

$$\mu CF \text{ kinetics} \longleftrightarrow \text{basic processes}.$$

On the one hand, the diversity of physical processes involved in the μCF kinetics is one of the major intellectual attractions of this field. On the other hand, this close interplay between different forces can be – and has been in the past – an obstacle to single out a specific basic process. It is just this respect where major achievements in the understanding of the muon induced kinetics in D_2 and H–D have been reached due to the ongoing intensified μCF research. To a large extent

- a quantitative analysis in terms of basic processes has been achieved,
- new techniques (experimental and theoretical) have been developed,
- and an important prototype for the more complex D–T μCF has been established.

Of course, there is still important work waiting for completion, but chances are excellent that the necessary experiments can be performed in the next years, in particular considering the upcoming high–intensity operation of the upgraded PSI accelerator.

In the present work we will follow the traditional practice of discussing the different steps of the muon's life story in chronological order (fig.1). As usual all rates are normalized to the density $N = 4.25 \times 10^{22}$ atoms/cm^3 of liquid H_2 and densities Φ are given relative to this value. The presentation will be mainly by example (selected according to personal and general current interest) and does not aim at completeness. While discussing the phenomenology a rigorous separation of physical fields is not always possible since often several effects are needed for the explanation of experimental observables. However, in order to prove our point that such a distinction is possible, the results concerning the different fields are separately collected in the final summary section.

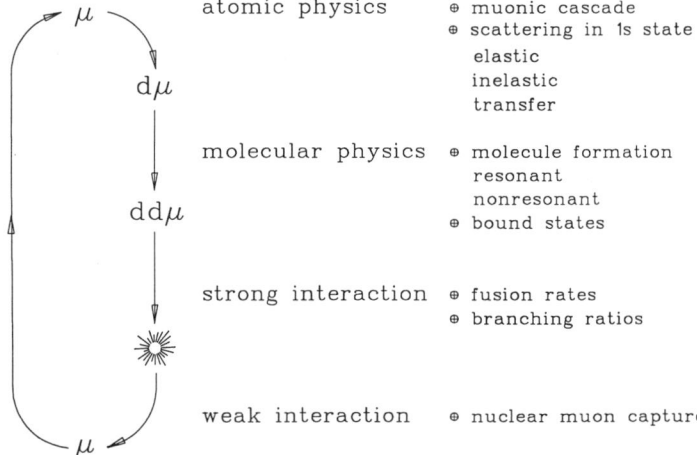

Figure 1: simplified phenomenology of muon induced reactions

2. Atomic Physics

2.1 Muonic Cascade. The characteristic times of initial Coulomb capture of the μ^- by the hydrogen nucleus x (x= p, d or t) and the subsequent deexcitation cascade are much shorter than the reaction times in the ground state of the muonic atom $x\mu$ [1]. Nevertheless, this early stage is important since it defines the initial conditions for the following slower reaction sequence. In particular, two related questions are of current interest:

(a) *What is the energy distribution f(E) of $x\mu'$s upon arrival in their ground state?* According to the standard theoretical picture $(x\mu)_{n\sim 14}$ atoms are initially formed with a kinetic energy of $\sim 1 eV$. It is expected that this energy is significantly modified by accelerating and decelerating processes during deexcitation to the ground state. Though there have been attempts [3] the complexity of the muonic cascade has prevented quantitative predictions for the distribution of the initial $(x\mu)_{1s}$ energies. In particular, theoretical results concerning Coulomb deexcitation rates vary considerably [4] and no complete calculation of elastic cross sections in excited states it yet available. Experimental information is based mainly on two recent experiments. In experiment [5] the diffusion of muonic hydrogen atoms is studied in the density range $\Phi \sim 10^{-3} - 10^{-4}$. As discussed by R.Siegel at this workshop, the observed distributions f(E) for $(d\mu)_{1s}$ atoms are well described by either rectangular or gaussian velocity distributions with mean energies of 1.3-1.5 eV. The energies of $(p\mu)_{1s}$ atoms appear to be about twice as high. These results are preliminary, partly because differential scattering cross sections required in the analysis were not yet available (see section 2.2). In the second type of experiment [6] the neutron energy from the reaction $\pi^- + p \rightarrow \pi^0 + n$ is measured. This determines the energy distribution of the excited $(p\pi)_{n=3,4}$ states where the reaction mainly occurs. In liquid hydrogen a distribution with a mean energy of ~ 16.2 eV was found with a flat tail extending up to 71.5 eV. An improved understanding of the cascade processes responsible for these surprisingly high energies is expected from investigating the dependence of these energy distributions on the hydrogen density [7].

(b) *How many μ's transfer during the muonic cascade from excited states according to*

$$(p\mu)_n + d \rightarrow (d\mu)_n + p + 135/n^2 eV \qquad (1)$$

The analogous transfer probability in D–T mixtures is a critical and controversial problem for understanding μCF there [1]. The probability for reaching $(p\mu)_{1s}$ after the initial formation of an excited $p\mu$ atom is usually expressed by a function $q_{1s}(\Phi, c_d)$. Theory [1] predicts a strong dependence of q_{1s} on both density and deuterium concentration resulting from the competition of Φ–independent radiative transition rates with the rates of Auger and Coulomb deexcitation ($\propto \Phi$) and the transfer rates according to Eq.1. ($\propto c_d\Phi$). Moreover, transfer rates are strongly energy dependent [8]. Thus, $x\mu$ energies during the cascade have to be known for quantitative predictions of q_{1s} (fig.2). Experimentally, muon transfer has been studied in the *pionic cascade* by detecting the two gamma's from π^0 decay as a clean signal that the stopped π^- was absorbed in the $\pi^- + p$ charge exchange reaction mentioned above [9]. The q_{1s} curve derived from experiments in the density range $\Phi = 0.04 - 0.12$ characterizes pion transfer in the upper part of the cascade because pion absorption dominates for n≤3. No strong Φ dependence is expected since an early experiment in liquid H–D gave similar results [10]. Only recently, a new technique was proposed to directly measure X–ray yields from the *muonic cascade* in H–D mixtures and, thus, to determine transfer also in the lowest atomic levels [11]. The main experimental challenge of separating the closely spaced muonic K_α peaks of the isotopic atoms was solved with the help of CCD detectors. First preliminary results [12] show a much stronger q_{1s} effect than observed with pionic atoms (fig.2). It is planned to increase the detection efficiency by an order of magnitude and to systematically measure q_{1s} for liquid and gaseous targets in 1993.

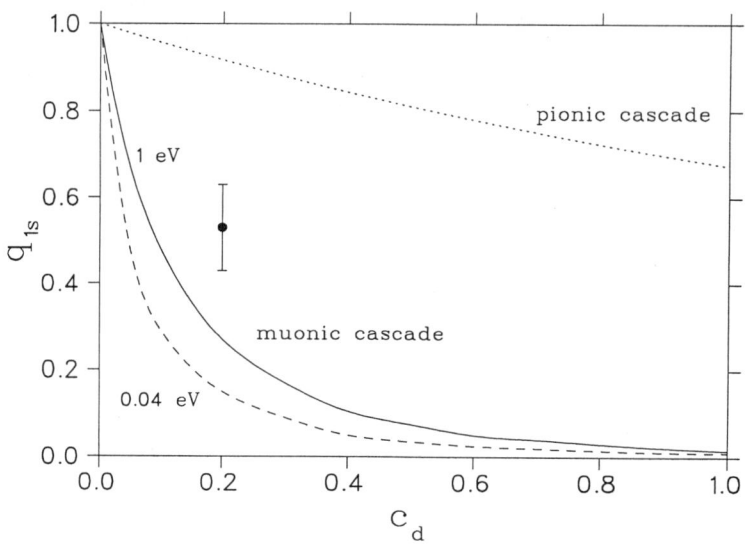

Figure 2: q_{1s} in H–D mixtures. Pionic cascade: experimental results [9] (dotted line). Muonic cascade in liquid H_2: theory [8] for different $p\mu$ energies (dashed and solid line) and new preliminary result from the X–ray experiment [12] (circle).

2.2 Scattering of Muonic Atoms in their Ground State.
Scattering determines the *energy, spacial and hyperfine* distribution of $(x\mu)_{1s}$ atoms as a function of time. Therefore scattering processes are of practical importance for a variety of experimental topics, several of which have been discussed at this workshop: diffusion of muonic hydrogen atoms [5], efficiency of solid H–D–T emission sources [13], thermalization during the μCF cycle [14], hyperfine effects in μCF, QED and nuclear muon capture, lifetimes of the $p\mu(F=1)$ and $d\mu(F=3/2)$ hyperfine states, etc.

Of course, scattering processes are of intrinsic theoretical interest as well. Only recently have the most advanced calculations based on an adiabatic expansion of the Coulomb 3-body scattering problem been completed which provide a systematic atlas of all cross sections for scattering between the 3 isotopic species of muonic hydrogen ([15,16,17]). It is impossible to present this wealth of information here, so we select two examples.

The elastic scattering process

$$p\mu(F=0) + H_2 \rightarrow p\mu(F=0) + H_2 \qquad (2)$$

has been notoriously difficult to calculate (cf. [18]). Fig. 3a shows that – by now – three recent calculations [20,15,21] of the pure "nuclear" scattering problem (i.e. $p\mu + p$ scattering) agree within 30%. These cross sections are larger by a factor of 2–3 than those extracted from most experiments [22,23]. Further experimental information will soon become available [5]. The discrepancy between experiment and theory appears to be dramatic if the effect of screening and molecular structure is included in the calculations [16]. However, for the detailed analysis of these diffusion experiments not only total cross sections but also *differential cross sections* are required (e.g., pure forward scattering would have negligible effect on the experimental observables). Indeed, first preliminary results [24] emphasize that screening and molecular structure effects strongly favor scattering in the forward direction (see fig.3b) (remember that the basic $p\mu + p$ cross sections are isotropic).

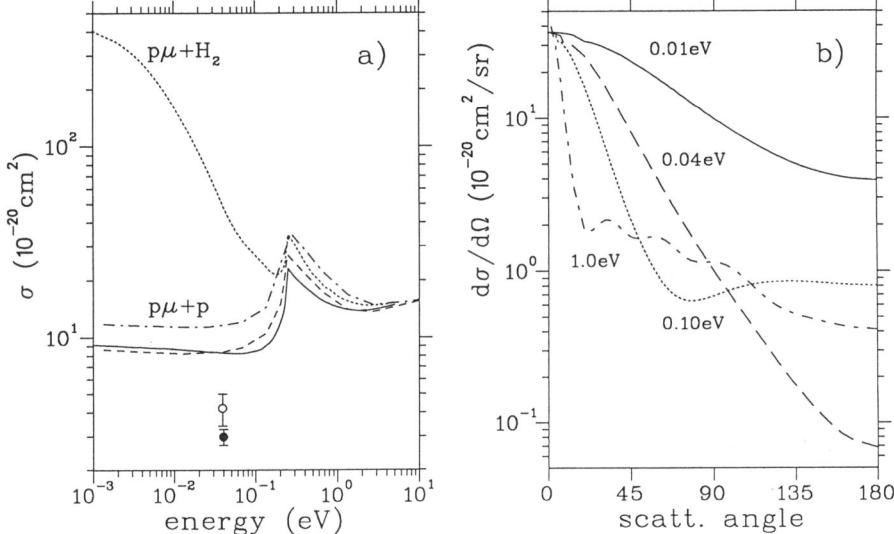

Figure 3: cross sections for $p\mu(F = 0) + H_2$ scattering as function of CMS energy. For comparison cross sections for $p\mu + p$ scattering are multiplied by two. Theory (solid [15], dashed [20], dashed–dotted [21] and dotted curve [16,24]), experiments since 1980 (open circle [23], filled circle [22]) b) differential cross sections for several CMS energies, kindly provided by A.Adamczak [24].

A case where accurate experimental results are available is the rate $\tilde{\lambda}_{3/2,1/2}$ for the hyperfine transition

$$d\mu(F = 3/2) + D_2 \to d\mu(F = 1/2) + D_2 \qquad (3)$$

because μCF provides a precise monitor of the $d\mu$ hyperfine populations (cf. section 3.2). Using this technique the temperature dependence of $\tilde{\lambda}_{3/2,1/2}$ has been studied by the Vienna–PSI group [25] and recently also by groups at LNPI [26] and JINR [27]. According to the present understanding of the $dd\mu$ cycle the total theoretical rate for process (3) consists of a scattering contribution [16,19] and a contribution from intermediate resonant $dd\mu$ formation [28,29] (cf. Eq.7). The total theoretical prediction, which is the sum of these two contributions, exceeds the experimental results by ~ 40% (see fig.4). This puzzle in the otherwise well understood $dd\mu$ μCF points to the fact that improvements in the quantitative theoretical description of scattering or resonant formation are still required.

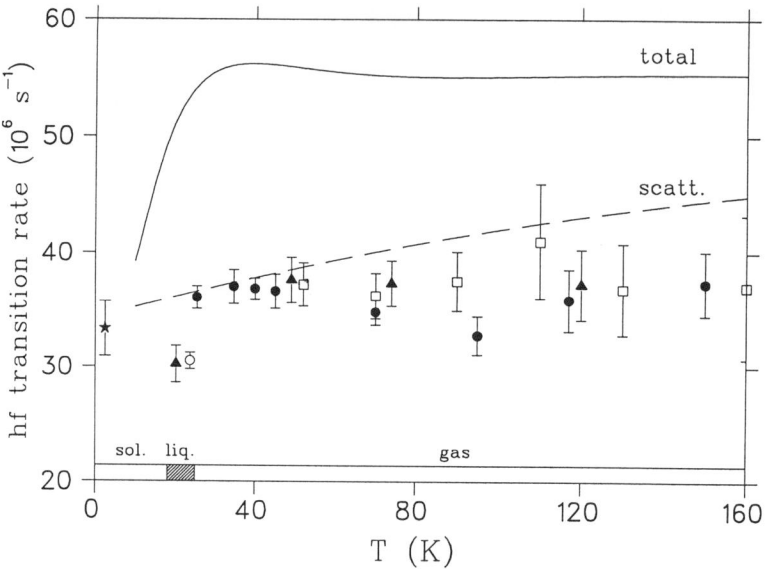

Figure 4: $d\mu(F = 3/2) + d$ hyperfine transition rates. Theory see text, experiments (circles [25], squares [26], triangles [27], star [13] –preliminary).

3. Molecular Physics

3.1. Muonic molecule formation.
Formation of muonic hydrogen molecules occurs in collisions according to two basic reaction types [1] (e.g., for D_2 molecules)

$$x\mu + D_2 \rightarrow [(xd\mu)^+ de] + e \qquad (4)$$
$$x\mu + D_2 \rightarrow [(xd\mu)^+ dee]^* \qquad (5)$$

The energy release in these reactions is dominated by the $xd\mu$ binding energies (of the order of 100 eV), which are large compared to the vibrational excitations of the electronic molecules (of the order of 100 meV). Therefore the usual formation mechanism is reaction (4), i.e. ejection of an Auger electron. Only if the $xd\mu$ molecule has an extremely weakly bound state is reaction (5) possible, where the released energy is absorbed in internal ro-vibrational excitations of the molecular complex. Such states exist only for $dd\mu$ and $dt\mu$ (with quantum numbers J=1, v=1). The simplified level scheme in fig.5 emphasizes the resonant character of this reaction, which is only possible if the resonance condition is exactly fulfilled. In addition to dramatically increasing the formation rates, the strong sensitivity of the resonant rate to small variations of ϵ_T provides a spectroscopic tool to reveal tiny energy splittings which otherwise would be completely negligible on the muonic scale.

$$d\mu + D_2 \rightarrow [(dd\mu)dee]$$

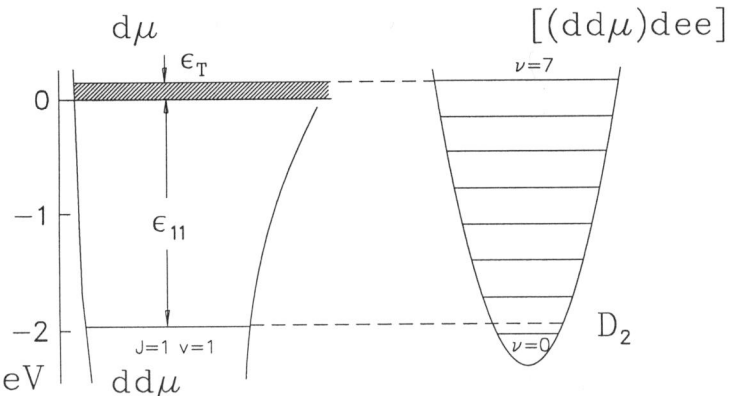

Figure 5: level scheme of resonant $dd\mu$ formation. ϵ_T denotes $d\mu$ kinetic energy.

3.2. Pure D_2. We refer to the extended literature [25,29] for a discussion of the μCF kinetics sketched in fig.6 and only want to draw your attention to two points.

(a) Due to the hyperfine splitting $\Delta E_{hfs} = 0.0485 eV$ the resonance behaviour for $dd\mu$ formation from the two $d\mu$ hyperfine states is very different. In particular at $T \leq 200K$ molecule formation from the $d\mu$ quartet state is strongly dominant, which leads to two distinct exponential components in the time spectra of $dd\mu$ fusion events (as first observed in experiment [30]). From these time spectra the temperature dependence of $d\mu$ hyperfine transition rates as well as $dd\mu$ molecular formation rates from both $d\mu$ hyperfine states have been determined, which allow for a comprehensive test of the theory of resonant formation.

(b) An *ab initio* theoretical description of all *electromagnetic* effects contributing to resonant molecule formation has been completed recently, including high precision calculations of the molecular binding energies [31] and calculations of all formation matrix elements $|V_{if}|$ [29]. In the comparison between theory and experiment one has to take into account that once the excited complex is formed, back decay into the entrance channel (rate $\Gamma_S \sim 1.5 \times 10^9 s^{-1}$) competes with fusion (rate λ_f, see table 2) and Auger deexcitation (rate $\lambda_e = 0.022 \times 10^9 s^{-1}$ [32]). Thus the experimentally observed formation rates are [29]

$$\tilde{\lambda}_F = \lambda^{nr} + \Sigma_S \lambda_{FS} \frac{\tilde{\lambda}_f}{\tilde{\lambda}_f + \Gamma_S} \tag{6}$$

In a similar way, the effective hyperfine transition rate contains an additional term due to intermediate molecule formation and subsequent back decay [28,29]

$$\tilde{\lambda}_{FF'} = \lambda_{FF'}^{scat} + \Sigma_S \lambda_{FS} \frac{\Gamma_{SF'}}{\tilde{\lambda}_f + \Gamma_S} \qquad (7)$$

Here S, F denote the spins of the $dd\mu$ and $d\mu$, respectively, λ^{nr} and λ_{FS} are the nonresonant [33] and resonant formation rate, respectively, $\tilde{\lambda}_f = \lambda_f + \lambda_e$, and $\Gamma_S = \sum_F \Gamma_{SF}$ [29]. Accordingly, also the effective fusion rate $\tilde{\lambda}_f$ critically enters the size of the observed rates.

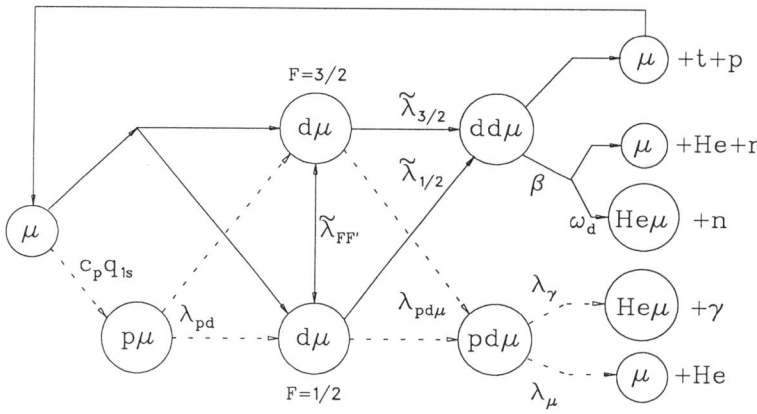

Figure 6: simplified μCF kinetics for H–D mixtures. The reaction path due to ^1H admixtures is indicated by dotted lines. Only reduced rates are shown, whereas observed rates depend on density, atomic and molecular concentrations.

In fig.7 the present state of the worldwide work on $dd\mu$ formation is summarized. The most striking feature is the strong variation of the observed rates with target temperature characteristic of the resonance mechanism. In the experiments of the Vienna-PSI group the molecule formation rates $\tilde{\lambda}_F$ from both $d\mu$ hyperfine states were systematically measured in gaseous and liquid targets [30,35,34,25] (fig.7a). The sensitivity to the resonance energies is most pronounced for $\tilde{\lambda}_{3/2}$ because for the quartet state the strongest resonances are located at low energies where the target Maxwell distribution is very narrow. The solid line shows that present *ab initio* theory [29], after minor adjustments of the input parameters (see section 5), agrees well with the data. In the majority of other experiments [36,26,27] the main observable was the steady state $dd\mu$ formation rate $\tilde{\lambda}_{dd\mu}$ corresponding to an average over $\tilde{\lambda}_F$ according to the kinetic equilibrium of hyperfine states in the steady state (Fig.7b). In case the kinetic conditions of the experiments are known, the corresponding rates $\tilde{\lambda}_{dd\mu}$ can also be calculated from the best theoretical fit to the basic rates $\tilde{\lambda}_F$ (fig.7a) and then extrapolated to higher temperatures. By and large the experimental results are in good agreement although some discrepancies at low temperature should still be investigated. Comparison with theory demonstrates that a quantitative theoretical description of all experimental $dd\mu$ formation rates measured in the range 20–600 K has been achieved.

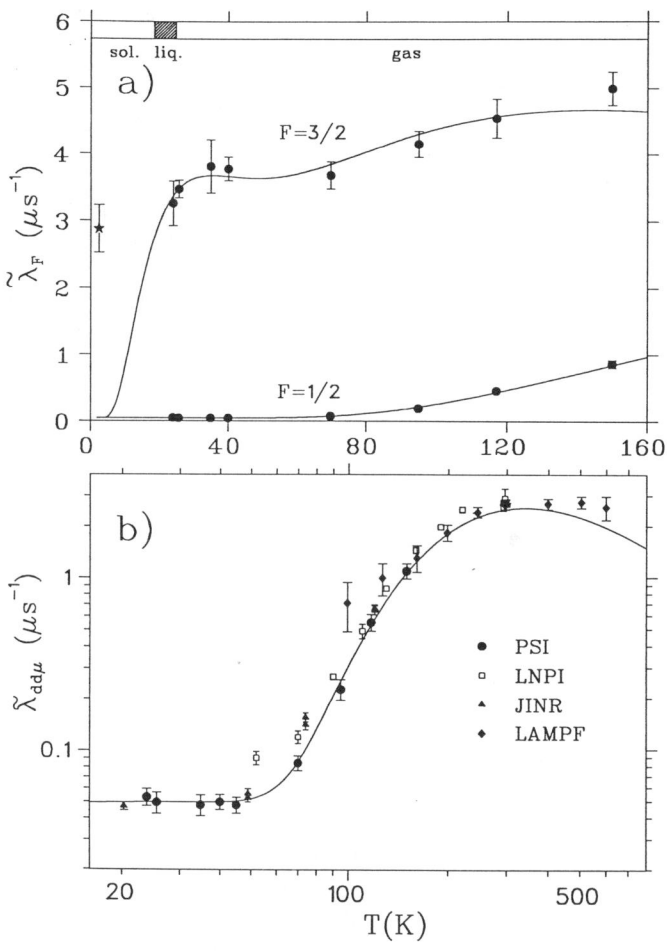

Figure 7: $dd\mu$ formation rates, solid curve represents best theoretical fit to Vienna-PSI data [25] a) $\tilde{\lambda}_F$: liquid and gaseous D_2 [30,35,34,25](circles), solid D_2 [13](star) – preliminary; b) steady state formation rates $\tilde{\lambda}_{dd\mu}$(from Ref. [25]): experiments [25,26,27,36]

Very recently $dd\mu$ formation has been investigated also in solid targets of T $\sim$ 2.5 K by observing 2.5 MeV neutrons and 3.0 MeV protons from dd fusion [13]. A time spectrum with an intense but quickly decaying initial spike was observed, which closely resembled the structure seen in liquid targets [34]. This observation suggests that $dd\mu$ formation from the quartet state is still resonant although the available thermal energies of $3/2kT \sim 0.3$ meV are far too low to reach the lowest resonance energy of ~ 4 meV as determined from measurements in gaseous and liquid targets [25]. Preliminary results for $\tilde{\lambda}_{3/2,1/2}$ and $\tilde{\lambda}_{3/2}$ in solid D_2 are included in fig.4 and 7, respectively. As an explanation for this very different resonance behaviour of $\tilde{\lambda}_{3/2}$ in the solid phase, e.g. the possibility of incomplete thermalization due to reduced elastic cross sections in $d\mu$ collisions with the solid lattice [37] as well as the modification of electronic energy levels due to the intermolecular interaction [38] should be considered.

3.2. H–D mixtures. Schematically, the μCF kinetics in H–D mixtures (fig.6) is very similar to D_2, only $p\mu \to d\mu$ transfer, $pd\mu$ formation and the subsequent pd fusion reactions are added. There are, however, several distinctive features not visible in this simplified diagram. (a) Resonant reactions are expected to depend on the participating different molecules H_2, HD and D_2. Thus, not only the atomic target concentrations but also the proportions of the molecular species have to be known, which can be varied between $H_2 + D_2$, statistical equilibrium $H_2 : HD : D_2 = c_p^2 : 2c_p c_d : c_d^2$ and – given sufficient effort – (nearly) pure HD. (b) Compared to the elastic $d\mu + d$ cross section, the $d\mu + p$ cross section has a deep minimum at collision energies around 2 eV due to the Ramsauer–Townsend effect [17]. Thus the thermalization times of $d\mu$'s formed with an energy of ~43 eV by muon transfer from hydrogen are comparatively long and lead to an enhanced population of epithermal $d\mu$'s in H–D. As discussed by Marshall [13] the anomalous transparency of solid H_2 layers can also be utilized to produce a beam of $d\mu$ atoms. (c) The physics of $pd\mu$ fusion is much richer than shown in fig.6 and will be discussed in the next section.

In spite of the fact that μCF has been discovered in H–D mixtures, the present understanding is less complete than that of pure D_2 (cf. section 5). This situation, however, can and should be improved in the next years. The following example shows that these studies can provide crucial supplementary information enabling us to fully utilize results obtained with pure D_2.

Precision measurements [11] of the time distribution of dd fusion events in H–D can determine the amount of back decay in resonant formation and hyperfine transition. According to the accurately verified picture of resonant formation only the process $d\mu(3/2) + D_2$ is resonant at $T \leq 50K$. Thus the effective quartet formation rate can be written as (cf. Eq.6)

$$\tilde{\Lambda}_{3/2} = c_d \lambda^{nr} + c_{D_2} \lambda_{3/2}^{res} \qquad (8)$$

while the doublet rate has only a non resonant contribution. The effective hyperfine transition rate is (cf. Eq.7)

$$\tilde{\Lambda}_{3/2,1/2} = c_d \lambda_{3/2,1/2}^{scat} + c_{D_2} \lambda_{3/2,1/2}^{res} \qquad (9)$$

neglecting the small influence of the molecular structure on the scattering rate ($\lambda_{3/2}^{res}$ and $\lambda_{3/2,1/2}^{res}$ denote the second term in Eq.6 and 7, respectively). Both rates $\tilde{\Lambda}_{3/2}$ and $\tilde{\Lambda}_{3/2,1/2}$ should be measured simultaneously in two mixtures of *constant atomic and different molecular concentrations* (see fig.8). Obviously the differences $\Delta\tilde{\Lambda}_{3/2}$ and $\Delta\tilde{\Lambda}_{3/2,1/2}$ of the rates at the two conditions is given by the resonant contribution. The back decay contribution can be determined directly from $\Delta\tilde{\Lambda}_{3/2,1/2}$ or, alternatively, the ratio of fusion to back decay can be fixed by

$$\frac{\Delta \tilde{\Lambda}_{3/2}}{\Delta \tilde{\Lambda}_{3/2,1/2}} \approx \frac{\tilde{\lambda}_f}{\Gamma_{1/2,1/2}} \qquad (10)$$

(This approximate expression follows from Eqs.6 and 7, if the summation is limited to the $dd\mu$ S=1/2 state, which dominates resonant formation at low T [25])

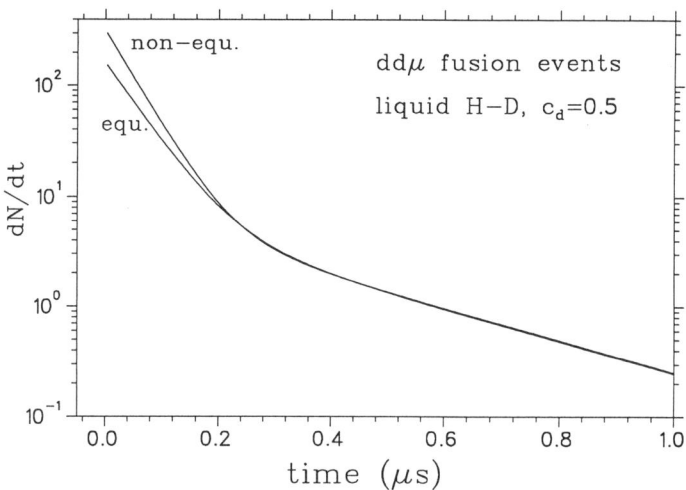

Figure 8: calculated time distribution of dd fusion events in equilibrated ($H_2 + HD + D_2$) and non equilibrated ($H_2 + D_2$) mixtures of $c_d = 0.5$. In order to demonstrate the expected effects a ratio of $\lambda^{scat}_{3/2,1/2} : \lambda^{res}_{3/2,1/2} = 2 : 1$ was assumed.

4. Nuclear Physics

We select two examples to show the fascinating possibilities for studying nuclear few–body reactions by μCF. For a more systematic summary we refer to section 5.3.

The relative orbital (L) and spin (S) quantum numbers of the two deuterons bound in a $dd\mu$ molecule depend on the molecule formation mechanism. At temperatures ~300 K molecule formation from both hyperfine states is completely dominated by resonant formation, which exclusively populates L=1 (S=1) d–d states. At these conditions muon catalyzed fusion occurs by pure p-wave interaction. By detecting all charged fusion products simultaneously in a high–pressure ionisation chamber at LNPI, the ratio between the ($^3He + n$) and the ($t + p$) fusion channels was accurately measured as R= 1.39 ± 0.04, indicating a surprisingly large difference between the two charge–conjugate reactions [39]. Recently, these measurements were extended to lower target temperatures [26] (Fig.9). For the experimental cuts used the observed events correspond predominantly to $dd\mu$ formation from the doublet $d\mu$ state, which increasingly populates L=0 states at lower temperatures.

Thus, with some assistance from theory [33], such measurements can provide valuable information about s–wave fusion too.

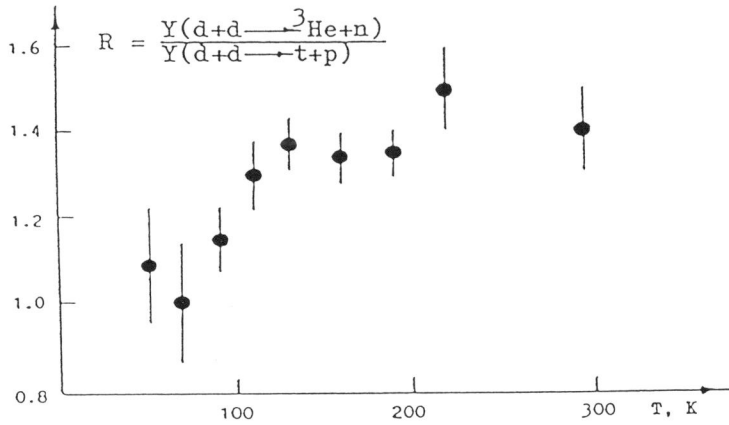

Figure 9: branching ratios for dd fusion observed in the time interval 0.5–5 μs after muon stop [26].

In the case of $pd\mu$, fusion takes place from four incoherent hyperfine states of the molecule. Every state corresponds to a definite population of the possible pd spin states with S=1/2 and 3/2. The rates for nuclear reactions depend on these nuclear spins according to table 2. The population of the molecular spin states (and thus the population of the nuclear spin states) is determined by $d\mu$ hyperfine transitions before $pd\mu$ formation. This basic idea, namely controlling the fusion rates in the molecule by atomic physics processes, is the essence of the *Wolfenstein–Gershtein effect* [40] and has lead to numerous experiments in the past. However, its detailed interpretation has undergone a major revision only recently. Up to 1988 complete dominance of pd doublet fusion was assumed. This hypothesis led to inconsistencies in the interpretation of the observed c_d dependence of the gamma yield [41]. By now, the importance of quartet pd fusion [42] is proven by an *ab inito* calculation [43] and by a new experiment [44] (where the fusion rates were directly determined from the observed time distribution, thus eliminating ambiguities in the interpretation of earlier experiments).

5. Summary and Outlook

In this summary we will try to compile the relevant information separately for each field of physics connected to muon catalyzed fusion in deuterium and hydrogen. It contains results as well as present and future research topics. We conclude from the detailed list given below that significant progress has been achieved over the last years. For many of the basic processes quantitative results exist. For others research programs have been formulated which, we hope, will solve the remaining problems within the next few years. Increasingly, the effort is shifting from establishing general features of the subject towards investigating especially attractive problems with high precision. This trend is strongly benefitting from the basic techniques developed in the course of systematic μCF research.

5.1 Atomic Physics. Muon induced processes in mixtures of hydrogen isotopes provide a rich field for studying the 3–body Coulomb interaction for a case where – due to the considerable size of the mass ratio $m_\mu/m_p \sim 0.1$ – nonadiabatic effects are essential. Current research efforts are concentrated on the following topics:

(a) elastic scattering and muon transfer in excited states of muonic hydrogen. Due to the complexity of the muonic cascade it is mandatory to use different techniques to provide sufficient experimental constraints for the development of an accurate theoretical description. Ongoing experimental work includes the study of $p\mu$ and $d\mu$ diffusion at low densities [5], the observation of the neutron kinetic energies from the process $\pi^-p \to n+\pi^0$ at $\Phi = 1$ [6] and Φ=0.02–0.05 [7] and the measurement of muonic X rays in H–D mixtures [12]. Important theoretical work consists of quasiclassical and quantum mechanical calculations of elastic and transfer cross sections in excited states, calculations of Auger and Coulomb de-excitation rates where results are still controversial and the development of a Monte Carlo program which microscopically follows the atom's energy evolution during the muonic cascade.

(b) elastic scattering in the muonic ground state. Two experiments are in the analysis stage (muonic atom diffusion [5] and $d\mu$ emission from solid hydrogen [13]) and additional information comes from epithermal μCF [14]. By now theoretical calculations use sophisticated methods to solve the three–body scattering problem [15,17] and include molecular and screening effects [16]. At the level of accuracy reached the double differential cross sections $\frac{\partial^2 \sigma}{\partial\Omega\partial E_f}$ (E_f is the final $x\mu$ energy) is required for the analysis of experiments [24].

(c) hyperfine transitions. Hyperfine transition rates have been calculated for all isotopic cases [16,19]. With the final analysis for $p\mu$ scattering still pending [5], the best experimental test case is $d\mu(3/2) + D_2$ hyperfine transitions. Surprisingly the theoretical expectation for the rate $\tilde\lambda_{3/2,1/2}$ significantly exceeds the experimental values (see fig.4). It is important to clarify the origin of this discrepancy by experiments with H–D mixtures (cf.section 3.2). Tighter limits on the asymmetric spin–flip reaction $d\mu(3/2) + p \to d\mu(1/2) + p$, which has been controversial, should be established by direct experiments (cf. Ref. [44]).

5.2 Molecular Physics

(a) binding energies of muonic molecules. In order to predict resonant formation rates, binding energies of muonic molecules have to be known with extremely high precision (i.e. with uncertainties of some 0.1 meV corresponding to $\sim 10^{-7}$ of the muonic Rydberg energy). By now variational calculations have achieved these stringent requirements for the nonrelativistic 3-body Coulomb problem [31] and corrections [45,46] (relativistic, finite

Table 1: binding energy ϵ_{11} (meV) of the $dd\mu$ excited state with J=v=1

theory		experiment [25]
non relativistic ϵ_{11} [31]	-1974.9	
relativistic shift [45]	1.44	
vacuum polarization [45]	8.66	
nuclear finite size [45]	-1.54	
deuteron polarizability [45]	-0.1	
$dd\mu$ finite size [1,46]	0.24	
total ϵ_{11}	-1966.2±0.5	-1966.1±0.3

size effects etc.) have been carefully evaluated. Experimentally, the beautiful test case of resonant $dd\mu$ formation has been systematically studied and a high-precision value for ϵ_{11} was derived from the data, nearly independent of the assumption used in the analysis [25]. The result is in excellent agreement with theory (see table 1). Results from the analysis of other recent experiments [26,27] are expected soon.

(b) muonic molecule formation.

$dd\mu$ formation. In pure D_2 theory and experiments covering a wide range of conditions (Fig.7) are in good agreement. However, the formation matrix elements $|V_{if}|$ [29], which are a challenging theoretical problem, are *not tested well by experiment*, since the effect of fusion and back decay given explicitly in Eq.6 cannot be unambiguously separated. As shown in Refs. [47,25] observed rates only define a range of pairs $(|V_{if}|, \tilde{\lambda}_f)$ compatible with experiment. An additional constraint is necessary, which could be provided by experiments in H-D mixtures described in section 3.2. The surprising effect of resonant $dd\mu$ formation in cold solid D_2 needs quantitative explanations [13].

Concerning $dd\mu$ formation on HD molecules data is scarce and partially inconsistent [48] with theory and further experimental effort is required. Moreover, the significant temperature dependence predicted for nonresonant formation [33] could be tested in H-D, since resonant formation turns on at higher temperatures in $dd\mu$ formation with HD molecules (compared to D_2 molecules).

$pd\mu$ formation. Further careful experiments are needed to clarify the remaining puzzles (cf. Ref. [48,44]). For the analysis of these experiments a Monte Carlo study of epithermal effects would be helpful which accounts for the exceptionally small $d\mu + p$ cross sections.

5.3 Nuclear physics. Fusion reactions in muonic molecules occur at conditions which usually cannot be studied in nuclear few-body physics: the collision energy is extremely low and the two nuclei are prepared with selected relative spin (S) and orbital (L) quantum numbers in the molecular environment. Moreover, exotic reaction channels (e.g. muon conversion) exist.

dd fusion. Resonant $dd\mu$ formation provides a unique possibility to study low-energy L=1 (S=1) dd fusion. Unfortunately, so far only a range of allowed fusion rates can be extracted from experiments [47,25] and additional effort is required (cf. section 3.2). An update of theoretical results is given in table 2. The second interesting observable is the branching

Table 2: fusion rates and branching ratio $R = (^3He+n)/(t+p)$ in experimentally relevant states of $dd\mu$ and $pd\mu$ molecules. S denotes the total nuclear spin.

state(J,v)	reaction	S	symbol	theory	experiment
$dd\mu(1,1)$	$^3He + n + \mu$	1	$\lambda_f\,(10^8 s^{-1})$	4.4^a, 3.8^b	$\sim$2.3–6.5^c
	$t + p + \mu$		R	1.43 [51]	1.39±0.04 [39]
$dd\mu(0,0)$		0,2	R	0.886 [51]	
$pd\mu(0,0)$	$^3He + \mu$	1/2	$\lambda_\mu^{1/2}\,(10^4 s^{-1})$	6.2±0.2 [43]	5.6±0.6^d
	$^3He\mu + \gamma$	3/2	$\lambda_\gamma^{3/2}\,(10^4 s^{-1})$	10.7±0.6 [43]	11.0±1.0 [44]
	$^3He\mu + \gamma$	1/2	$\lambda_\gamma^{1/2}\,(10^4 s^{-1})$	37±1 [43]	35±2 [44]

^{a} from [49] with molecular density at nuclear coalescence $\rho_{11} = 1.53 \times 10^{48} cm^{-5}$
^{b} update of [49] using $\rho_{11} = 1.32 \times 10^{48} cm^{-5}$ [50]
^{c} from Refs. [47,25], with $\lambda_f = \tilde{\lambda}_f - \lambda_e$
^{d} reanalysis of early bubble chamber experiments [42]

ratio R. The surprising deviation of R from unity observed for the two isospin–symmetric reaction channels of L=1 fusion [39] was recently explained by an essentially charge–independent R matrix analysis of the A=4 system [51]. Extensions of the experiment to lower temperature can also determine R for L=0 fusion [26].

pd fusion. After the revised interpretation of the Wolfenstein–Gershtein effect, theory and experiment agree concerning the low–energy pd fusion rates (table 2). Further experimental work should corroborate these results by providing critical tests of this interpretation and by clearing up remaining inconsistencies in the kinetics. Important experiments in this context are precise measurements of the absolute yield of fusion gammas (where theory and most experiments disagree) and of the branching ratio between muon and gamma emission as a function of c_d [11].

5.4 Weak Interactions.

The semileptonic reactions

$$\mu + p \rightarrow n + \nu_\mu$$
$$\mu + p \rightarrow n + \nu_\mu + \gamma$$
$$\mu + d \rightarrow n + n + \nu_\mu \qquad (11)$$

are basic medium–energy processes to determine the weak nucleon form factors [52]. For the interpretation of experimental results an accurate knowledge of the spacial and spin wave function of the system at the moment of the nuclear capture is indispensible since these reactions take place with the μ^- bound in a hydrogen atom or molecule. In addition, scattering of muonic atoms should also be well understood because diffusion to the target wall often constitutes a dangerous source of background. Several important problems in this field have been solved by μCF methods. However, many challenging problems still lie ahead.

$d\mu$ capture from the doublet state. The determination of $\tilde{\lambda}_{3/2,1/2}$ by μCF techniques [30] solved the crucial question of $d\mu$ hyperfine populations, which had impeded the interpretation of earlier experiments. Recent experiments in pure D_2 find capture rates of $409 \pm 40\ s^{-1}$ at $\Phi = 0.04$ [35] and $470 \pm 29\ s^{-1}$ in liquid D_2 [53], consistent within errors. The latest theoretical calculation gives $402\ s^{-1}$ [54] in better agreement with the smaller experimental value.

Normal and radiative muon capture in the $pp\mu$ molecule. Strong interest for muon induced processes in liquid H_2 is stimulated by the first measurement of radiative muon capture [55] and the possibility for a more accurate interpretation of normal capture rates observed in $pp\mu$ molecules (cf.Ref. [56]). In particular, the current knowledge of the transition rate between the $pp\mu$ ortho and para state [56,57] and also of the $pp\mu$ formation rate [58] is unsatisfactory. For precision experiments a careful theoretical evaluation of relativistic effects in the molecular wavefunction is essential [57].

Muon capture from the upper $p\mu$ and $d\mu$ hyperfine states in rarified gases. A prerequisite for such a measurement is the knowledge of the lifetimes of these states, a problem closely related to the above-mentioned studies of the muonic cascade and thermalization [5]. Also the question should be investigated whether elastic scattering in excited states (in particular in the 2s state) can change the statistical population of hyperfine states during the muonic cascade.

Precision measurement of the $p\mu$ singlet capture rate. The detection of muons in a hydrogen ionisation chamber [39,59] is an attractive method for an active hydrogen target where the position of every muon stop can be reconstructed. This method or extensions towards a hydrogen drift chamber may pave the way towards a precision experiment to test $\mu - e$ universality in muon capture.

Acknowledgements

Support by the Austrian Science Foundation and the Paul Scherrer Institute is gratefully acknowledged. It is a pleasure to thank the workshop organizers Dr. C. Petitjean and Prof. L. Schaller for their kind invitation and for creating a stimulating atmosphere at the meeting. Finally, I want to express my sincere gratitude to all members of our international collaboration at PSI for many years of fruitful and enjoyable joint work which forms the basis of the present survey.

References

[1] W.H. Breunlich, P. Kammel, J.S. Cohen and M. Leon, Ann. Rev. Nucl. Part. Sci. **39**, 311 (1989), L.I. Ponomarev, Contemporary Physics **31**, 219 (1991).

[2] H.K. Walter, this conference, S.Monti, ENEA NUC-RIN Bologna, int. report (1991).

[3] L.I. Men'shikov, Muon Catal. Fusion **2**, 173 (1988).

[4] see W. Chaplinski et al., Muon Catal. Fusion **5**, 59 (1990), and references given therein.

[5] R.T. Siegel, this conference, J.B. Kraiman et al., Phys. Rev. Lett. **63**, 1942 (1989), J.B. Kraiman et al., Muon Catal. Fusion **5**, 43. (1990)

[6] J. Crawford et al., Phys. Lett. **213B**, 391 (1988).

[7] D. Bovet et al., PSI proposal R-86-05 (1986).

[8] W. Chaplinski et al., Muon Catal. Fusion **5**, 55 (1990).

[9] see P. Weber et al., Phys. Rev. A41, 1 (1990) and references given therein.

[10] K. Derrick et al., Phys. Rev. **151**, 82 (1966).

[11] P. Ackerbauer et al., PSI–proposal, addendum, R-81-05.2 (1990).

[12] P. Ackerbauer et al., PSI Nucl. Part. Physics Newsletter 1991, 53 (1991), Verhandl.DPG (VI) **27**, C1.4 (1992), and contribution to the International Workshop on Muon Catalyzed Fusion, Uppsala 1992, to be published in Muon Catal. Fusion, J.P. Egger et al., this conference.

[13] G.M. Marshall et al., this conference.

[14] M. Jeitler, this conference, M. Jeitler et al., Muon Catal. Fusion **5**, 217 (1990), V. Markushin, this conference.

[15] L. Bracci et al., Muon Catal. Fusion **4**, 247 (1991).

[16] A. Adamczak and V.S. Melezhik, Muon Catal. Fusion **4**, 303 (1991).

[17] C. Chiccoli et al., Muon Catal. Fusion **7**, 87 (1992).

[18] L. Bracci et al., Muon Catal. Fusion **5**, 21 (1990).

[19] L. Bracci et al., Phys. Lett. **A134**, 453 (1989).

[20] M. Bubak and M.P.Faifman, JINR preprint E4–87–464, Dubna (1987).

[21] J. Cohen, Phys. Rev. **A43**, 4668 (1991).

[22] A. Bertin et al., Nuovo Cim. **72A**, 225 (1982).

[23] V.M. Bystritsky et al., Sov. Phys. JETP **60**, 219 (1984).

[24] A.Adamczak, contribution to the International Workshop on Muon Catalyzed Fusion, Uppsala 1992, to be published in Muon Catal. Fusion, and private communication.

[25] J. Zmeskal et al., Muon Catal. Fusion **1**, 109 (1987), Phys. Rev. **A42**, 1165 (1990), W.H. Breunlich, Nucl. Phys. **A508**, 3c (1990), A. Scrinzi et al., submitted to Phys. Rev. A.

[26] D.V. Balin et al., Muon Catal. Fusion **5**, 163 (1990).

[27] V. M. Bystritsky et al., Muon Catal. Fusion **5**, 141 (1990).

[28] M. Leon, Phys. Rev. **A33**, 4434 (1986).

[29] L.I. Men'shikov et al., Sov. Phys. JETP **65**, 656 (1987), M. P. Faifman et al., Muon Catal. Fusion **4**, 1 (1989).

[30] P. Kammel et al., Phys. Lett. 112B, 319 (1982), Phys. Rev. **A28**, 2611 (1983).

[31] S.A. Alexander and H.J. Monkhorst, Phys. Rev. **A38,** 26 (1988).

[32] D.D. Bakalov et al., Sov.Phys. JETP **67,** 1990 (1988).

[33] M.P. Faifman, Muon Catal. Fusion **4,** 341 (1989).

[34] N. Naegele et al., Nucl. Phys. **439A,** 397 (1989).

[35] M. Cargnelli, Ph.D. thesis, Technical University of Vienna, Vienna, 1986 (unpublished); M. Cargnelli et al., in Proceedings of the 23rd Yamada Conference on Nuclear Weak Processes and Nuclear Structure, Osaka, 1989, edited by M.Morita, E,Ejiri, H.Ohtsubo and T.Sata (World Scientific, Singapore, 1989), p.115.

[36] S.E. Jones et al., Phys. Rev. Lett. **56,** 588 (1986).

[37] pointed out by M.Leon, private communication.

[38] J. van Kranendonk, Solid Hydrogen, Plenum Press, New York, 1983.

[39] D.V. Balin et al. Phys. Lett. **141B,** 173 (1984), JETP Lett. **40,** 112 (1984).

[40] S.E. Gershtein, Sov. Phys. JETP **13,** 488 (1961).

[41] W.H. Bertl et al., Atomkernenerg. Kerntech. **43,** 184 (1983).

[42] L.N. Bogdanova and V.E. Markushin, Muon Catal. Fusion **5,** 189 (1990).

[43] J.L. Friar et al., Phys. Rev. Lett. **66,** 1827 (1991).

[44] C. Petitjean et al., Muon Catal. Fusion **5,** 199 (1990).

[45] G. Aissing et al., Phys. Rev. **A42,** 116 (1990).

[46] A. Scrinzi and K. Szalewicz, Phys. Rev. **A39,** 4983 (1989).

[47] W.H. Breunlich et al., Muon Catal. Fusion **5,** 149 (1990).

[48] K.A. Aniol et al., Muon Catal. Fusion **2,** 63 (1988).

[49] L.N. Bogdanova et al., Phys. Lett. **115B,** 171 (1982); **167B,** 485(E) (1986).

[50] S.A. Alexander et al., Phys. Rev. **A41,** 2854 (1990); Phys. Rev. **43,** 2585(E) (1991).

[51] G. Hale, Muon Catal. Fusion **5,** 227 (1990).

[52] E. Zavattini, in Muon Physics, edited by C.S. Wu and V.W. Hughes (Academic, New York, 1975), Vol.II, p.219.

[53] G. Bardin et al., Nucl. Phys. **A453,** 591 (1986).

[54] M. Doi et al., Nucl. Phys. **A511,** 507 (1990).

[55] M. Hasinoff, this conference.

[56] G. Bardin et al., Phys. Lett. **104B,** 320 (1981).

[57] D. Bakalov, this conference.

[58] L.I. Ponomarev, ibidem.

[59] K. Lou et al., ibidem.

μ-Capture in the Mesic Molecule $pp\mu$

L.I. PONOMAREV
Russian Scientific Center
Kurchatov Institute
Pl. Kurchatova, Moscow 123182, Russia

Abstract. The status of theory and experiment on the capture of muons stopped in hydrogen by protons is presented. The possibilities of refinement of the theoretical description taking into account ortho-para transitions in the $pp\mu$-molecule are discussed.

1. Introduction

There are two kinds of experiments to study weak interactions. The first one is the interaction of a high energy neutrino with a nucleus. It is analogous to Rutherford's scattering experiments with α-particles. The second is the study of nuclear muon capture of mesic atoms and mesic molecules. It is analogous to Millikan's experiment of measuring the electron charge.

The experiments of the second type are sometimes preferable to clear up the weak interaction characteristics. The reaction of μ-capture by the proton

$$\mu^- + p \to n + \nu_\mu \tag{1}$$

takes a special place among these experiments, as its description is free of the uncertainties brought in by the structure of complex nuclei.

In real experiments the reaction (1) takes place from the bound states of mesic atoms and mesic molecules, i.e.

$$p\mu \to n + \nu_\mu \tag{2}$$

$$pp\mu \to p + n + \nu_\mu \tag{3}$$

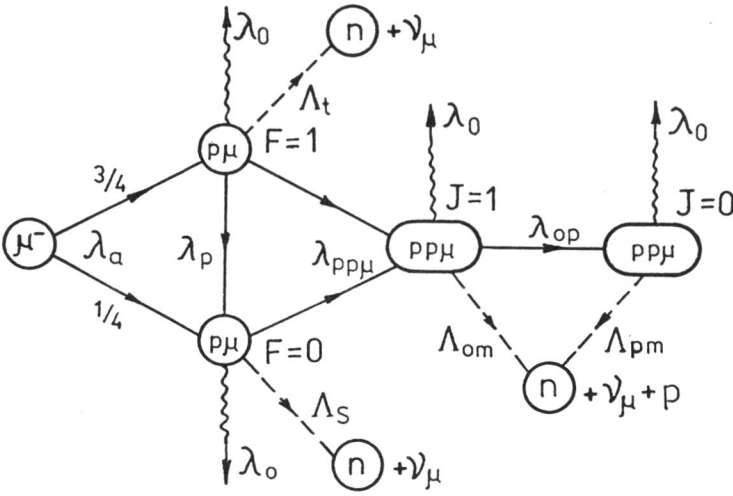

Figure 1: Sequence of mesic atomic and mesic molecular processes accompanied by μ-capture in hydrogen.

The differences of the initial states of the reactions (1) and (2,3) can be estimated rather precisely, as the Coulomb interaction between the particles is well known.

It is more difficult to take into account the influence of the total sequence of mesic atomic and mesic molecular processes, leading to the formation of mesic atoms $p\mu$ and mesic molecules $pp\mu$. But it is necessary, because the neutron time distribution dN_n/dt in reactions (2,3), which is usually measured, depends on this kinetics. In the following I will present the status of theory and experiment on μ-capture in liquid hydrogen at the beginning of the eighties and discuss the necessary research for modern understanding of the weak interaction physics in mesic molecules (see also [1,2,3,4]).

2. The main definitions

The observed rate Λ_c of the reaction (1) depends not only on the weak interaction constants, but also on the kinetics and rates of mesic atomic processes, shown in Fig. 1, where the following notations are used:

λ_0 = $0.455 \cdot 10^6 s^{-1}$ - the decay rate of the free muon
λ_a = $3 \cdot 10^{11} s^{-1}$ - the formation rate of $p\mu$-atoms in the 1s-state
λ_p = $1.7 \cdot 10^{10} s^{-1}$ - the transition rate $(p\mu)F = 1 \rightarrow (p\mu)F = 0$ from the triplet to the singlet state of the $p\mu$-atom
$\lambda_{pp\mu}$ = $2.5 \cdot 10^6 s^{-1}$ - the formation rate of mesic molecules $pp\mu$
(all rates are given for liquid hydrogen density).

In liquid hydrogen the formation rate of mesic molecules is so high ($\lambda_{pp\mu} \gg \lambda_0$), that the μ-capture process (1) comes from the bound state of the $pp\mu$-molecule. This state is formed in the reaction

$$p\mu + H_2 \to [(pp\mu) + pe]^+ + e \tag{4}$$

and becomes the heavy "nucleus" of the mesic molecular complex $[(pp\mu)^+pe]^+$, which is analogous to the molecular ion $(DH)^+$, in which the deuterium nucleus is replaced by a mesic molecular ion $(pp\mu)^+$.

At kinetic energy $\epsilon_{p\mu} < 1\text{eV}$ of the $p\mu$-atom the mesic molecules $pp\mu$ are mainly formed in the orthostate, i.e. in the state with total momentum $J = 1$, vibrational quantum number $v = 0$, binding energy $\epsilon_{Jv} = \epsilon_{10} = 253\text{eV}$ and total spin of nuclei $I = 1$.

In the parastate ($J = 0$, $v = 0$, $I = 0$, $\epsilon_{00} = 107\text{eV}$) only a small fraction ($< 0.3\%$) of mesic molecules $pp\mu$ is formed, as the formation rate of $pp\mu$-molecules in this state is rather low:

$$\lambda_{pp\mu}^{ortho} = 2.2 \cdot 10^6 s^{-1} \gg \lambda_{pp\mu}^{para} = 0.72 \cdot 10^4 s^{-1} [5] \ . \tag{5}$$

The spin-flip rate λ_p of the process $(p\mu)_{F=1} \to (p\mu)_{F=0}$ is much higher than the rate $\lambda_{pp\mu}$ of formation of $pp\mu$-molecules; that is why the mesic atoms $p\mu$ always enter reaction (4) in the singlet state with total spin $F = 0$. Therefore, the mesic molecules $pp\mu$ are practically always formed in the state with total spin $S = 1/2$ (which in the orthostate is split in two fine structure sublevels with total momenta $J = 1/2$ and $J = 3/2$ correspondingly (see Fig. 2).

In the theoretical analysis of 1981 [6,7] the following μ-capture rates from singlet ($F = 0$) and triplet ($F = 1$) states of mesic atom $p\mu$ were adopted [8]:

$$\Lambda_s = (664 \pm 20)s^{-1}$$
$$\Lambda_t = (11.9 \pm 0.7)s^{-1} \ . \tag{6}$$

The following γ-factors for ortho- and para-states of the $pp\mu$-molecule were used [9]

$$2\gamma_o = 1.009 \pm 0.001$$
$$2\gamma_p = 1.143 \pm 0.001 \tag{7}$$

(γ_o and γ_p are equal to the ratios of the density probability to find μ-mesons in the vicinity of a proton in ortho- and parastates of the $pp\mu$-molecule and of the $p\mu$-atom, respectively). With the values (6) we obtain the following μ-capture rates from ortho- and parastates of the $pp\mu$-molecule [6,7]

$$\Lambda_{om} = 0.756\Lambda_s + 0.253\Lambda_t = (505 \pm 15)s^{-1}$$
$$\Lambda_{pm} = 0.286\Lambda_s + 0.857\Lambda_t = (200 \pm 6)s^{-1} \ . \tag{8}$$

In accordance with the kinetic scheme (see Fig. 1) the μ-capture rate in hydrogen is:

$$\Lambda_c = \Lambda_s \frac{\lambda_0}{(\lambda_0 + \lambda_{pp\mu})} + \frac{\lambda_{pp\mu}}{(\lambda_0 + \lambda_{pp\mu})} \frac{\lambda_0}{(\lambda_0 + \lambda_{op})} (\Lambda_{om} + \Lambda_{pm} \frac{\lambda_{op}}{\lambda_o}) \ , \tag{9}$$

where λ_{op} is the rate of ortho-para transition from the excited orthostate ($J = 1, v = 0$) of the $pp\mu$-molecule to the ground (para-)state ($J = 0, v = 0$).

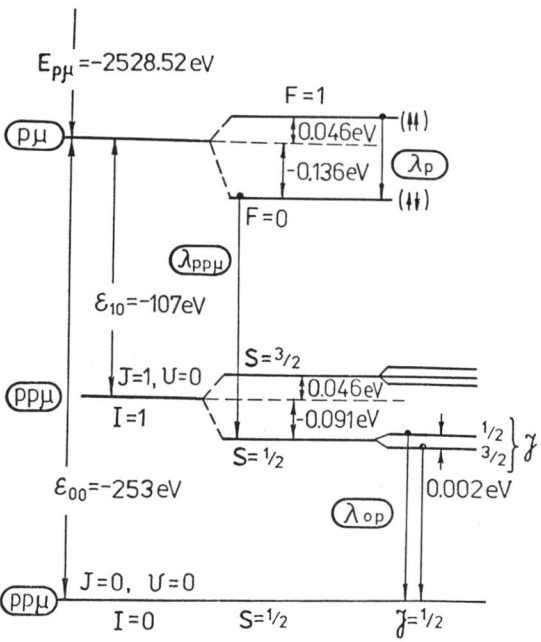

Figure 2: The energy level splitting of the mesic atom $p\mu$ and of mesic molecules $pp\mu$ due to the fine and hyperfine interaction. F is the total spin of the $p\mu$-atom, J, S and $\mathcal{J} = S + J$ are the angular momentum, spin and total momentum of the $pp\mu$-molecule, I is the total spin of the protons.

3. The problem of ortho-para transitions

The problem of ortho-para transitions in the $pp\mu$-molecule has a long history. For the first time it was discussed in [10], where there was the conclusion that the rate of these transitions is negligible, $\lambda_{op} = 0$.

Really, in nonrelativistic approximation the electromagnetic E1- transitions between the states ($J = 1$) and ($J = 0$) of the mesic molecule $pp\mu$ are forbidden, as in such a transition the total spin of nuclei must be changed ($I = 1$) → ($I = 0$), but this contradicts the selection rule $\Delta I = 0$.

However, the interaction of spins of the particles in the $pp\mu$-molecule has the result that the wave function of the $pp\mu$ orthostate, in addition to the "big" component with the quantum numbers ($J = 1, I = 1$), contains admixtures $\sim \alpha$ of "small" components with $(J = 0, I = 0)$. Vice versa, the wave function of the para-state $(J = 0, I = 0)$ contains admixtures $\sim \alpha$ of the components $(J = 1, I = 1)$. Thus, matrix elements of the operator of ortho-para $E1$-transitions differ from zero due to the contributions from cross-overlapping between "big" and "small" components of the wave functions with the selection rules $\Delta J = 1, \Delta I = 0$. In order of magnitude they are $\sim \alpha^2$ in comparison with the matrix elements of ordinary $E1$-transitions in mesic molecules with different nuclei, e.g. in $pd\mu$-molecules. In comparison with these rates $(\lambda_{dex} \sim 10^{12} s^{-1})$ [11] the ortho-para transition rate must be suppressed as $\sim \alpha^4 \cdot \lambda_{dex} \sim 10^4 s^{-1}$. The comparatively high rate of ortho-para transition in the $pp\mu$-molecule:

$$(pp\mu)^{J=1} \to (pp\mu)^{J=0} \tag{10}$$

is stimulated by the fact that the deexcitation of $pp\mu$-molecules occurs in mesic molecular complexes $[(pp\mu)pe]^+$, $[(pp\mu)pee]$ etc. rather than in the isolated $pp\mu$-molecule, and it is accompanied by Auger electron ejection from the "host molecule", for example:

$$[(pp\mu)^{J=1}pee] \to [(pp\mu)^{J=0}pe]^+ + e, \tag{11}$$

which carries away the transition energy $\Delta\epsilon = \epsilon_{10} - \epsilon_{00} = 146$eV. At such a low transition energy the conversion coefficient of the $E1-$ transition is rather high ($\sim 10^5$); that is why the ortho-para transition rate λ_{op} is in order of magnitude comparable with the muon decay rate λ_0 and should be taken into account in evaluating the experimental data.

The calculated rate of ortho-para transition [5] is

$$\lambda_{op}^{(a)} = (9.5 \pm 0.9) \cdot 10^4 s^{-1} . \tag{12}$$

In reality, following the $pp\mu$ formation in the complex $[(pp\mu)pe]^+$, in accordance to the reaction (3), a complicated sequence of ion-molecular processes takes place [12]. As a result, $pp\mu$-molecules become with probabilities W_a W_b and W_c "the nuclei" of the mesic molecular complexes [1]

$$(MH)^+ + H_2 \to \begin{cases} \xrightarrow{W_a} MH + H_2^+ \\ \xrightarrow{W_b} MH_2^+ + H \\ \xrightarrow{W_c} H_3^+ + M \end{cases} \tag{13}$$

with the probabilities [12] $W_a \approx 0$, $W_b \approx 0.75$, $W_c \approx 0.25$. (Here $M = (pp\mu)^+ e$ is the "atom" with the "heavy nucleus" $(pp\mu)^+$, MH is the "hydrogen isotope molecule" with the same nucleus, etc.)

The conversion coefficients of $E1$-transitions in complexes (13a/b/c) are proportional to the electronic densities in the $(pp\mu)^+$ vicinity and are in relations (with relative error 10%) [12]

$$K_a : K_b : K_c = 1 : 0.78 : 0.66 . \tag{14}$$

---

[1] These reactions are analogous to the well known ion-molecular reactions $DH^+ + H_2 \to DH + H_2^+$, $DH^+ + H_2 \to DH_2^+ + H$, etc. the rates of which are $10^{13} s^{-1}$ [11].

The observable rate of ortho-para transitions is

$$\lambda_{op} = (W_a + W_b K_b + W_c K_c)\lambda_{op}^{(a)} = (7.1 \pm 1.2)10^4 s^{-1} . \tag{15}$$

Using this value of λ_{op} and the value of $\lambda_{pp\mu} = (2.50 \pm 0.11)10^6 s^{-1}$ which is averaged over all experimental data [6] we obtain by means of formula (9) the following theoretical rate of μ-capture in liquid hydrogen:

$$\Lambda_c = (494 \pm 16)s^{-1} . \tag{16}$$

It is necessary to compare the calculated values λ_{op} and Λ_c with the experimentally measured ones [13,14]

$$\lambda_{op} = (4.1 \pm 1.4) \cdot 10^4 s^{-1} \tag{17}$$

$$\Lambda_c = (460 \pm 20)s^{-1} . \tag{18}$$

Fig. 3 shows the theoretical dependence (9) of Λ_c versus λ_{op} at the values Λ_s and Λ_t from (6) and $\lambda_{pp\mu} = (2.50 \pm 0.11) \cdot 10^6 s^{-1}$.

The experimental error bars include the uncertainties of the values Λ_s, Λ_t, $\lambda_{pp\mu}$ and λ_{op}. The experimental values are taken from the papers [13,14].

4. Discussion

The visible discrepancy between the theoretical and experimental values Λ_c and λ_{op}, presented in Fig. 3, should not be considered as serious.

First of all, in the calculations of 1981 rates were used which have changed nowadays (see table 1). Besides, it becomes also clear that in the wave functions of the $pp\mu$-molecule it is necessary to take into account additional relativistic corrections [15].

At last, the kinetics of ion-molecular processes should be considered more carefully. Thus, it is reasonable to undertake a new theoretical analysis of the problem only with regard to the whole complex of new data.

At the same time it is necessary to perform new experiments to measure the rate of ortho-para transitions in the mesic molecule $pp\mu$, and the rate $\lambda_{pp\mu}$ of $pp\mu$-molecule formation.

In combination with the thorough theoretical analysis this will not only allow to determine the values of Λ_s and Λ_t, but also to get informations about the other constants of weak interaction [2,4,16,17,18].

Acknowledgments

The author is thankful to Profs. A. Bertin, D. Bakalov, A. Vitale and E. Zavattini for numerous and useful discussions.

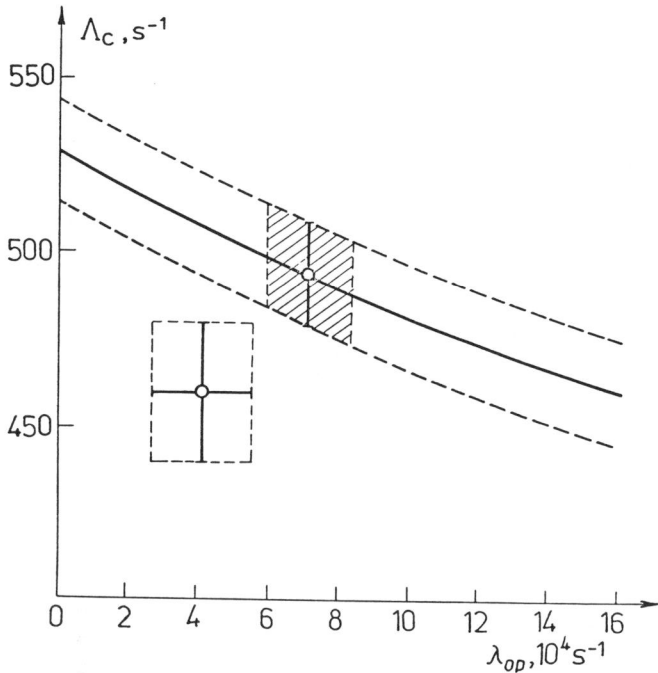

Figure 3: The dependence (6) of the μ-capture rate Λ_c on the value λ_{op} of ortho-para transitions. The region of the theoretically compatible values of Λ_c and λ_{op} is shaded, the experimental values are presented on the lower left.

Table 1: The characteristics used for the description of μ-capture in hydrogen.

Value	1981	1991
Λ_s	$664 \pm 20 s^{-1}$	
Λ_t	$11.9 \pm 0.7 s^{-1}$	
λ_a	$10^{11} s^{-1}$	$4 \cdot 10^{12} s^{-1}$
λ_p	$1.7 \cdot 10^{10} s^{-1}$	$1.69 \cdot 10^{10} s^{-1}$
$\lambda_{pp\mu}$	$(2.50 \pm 0.11) \cdot 10^6 s^{-1}$	
$\lambda_{pp\mu}^{ortho}$	$2.2 \cdot 10^6 s^{-1}$	$1.8 \cdot 10^6 s^{-1}$
$\lambda_{pp\mu}^{para}$	$7.2 \cdot 10^3 s^{-1}$	$7.5 \cdot 10^3 s^{-1}$
ϵ_{10}	$107 eV$	107.266 eV
ϵ_{00}	$253 eV$	253.152 eV
$2\gamma_o$	1.009	
$2\gamma_p$	1.143	

References

[1] E. Zavattini, in "Muon Physics", Eds. V. Hughes and C.S. Wu, Academic Press, New York, v. II, p. 219, 1975.

[2] A. Bertin and A. Vitale, Riv. Nuovo Cimento, 7, 1 (1984).

[3] J. Martino, in Fundamental Interactions in Low-Energy Systems, Eds. P. Dalpiaz, G. Fiorentini and G. Torelli, Plenum Press, New York, p. 43 (1984).

[4] A. Bertin et al., The talk at 2-nd Int. Symposium on Muon and Pion Interaction with Matter, Dubna, USSR, June 30 - July 4, 1987.

[5] L.I. Ponomarev and M.P. Faifman, Zh. Exp. Teor. Fiz. 71 (1976) 1689; Sov. Phys. JETP, 44, 886 (1976).

[6] D. Bakalov, M.P. Faifman, L.I. Ponomarev and S.I. Vinitsky, Nucl. Phys. 384, 302 (1982).

[7] D. Bakalov, S.I. Vinitsky, L.I. Menshikov, L.I. Ponomarev and M.P. Faifman, Preprint JINR P4-82-633, Dubna, 1982.

[8] H. Primakoff, In Nuclear and Particle Physics at Intermadiate Energy, Plenum Press, New York, 1975.

[9] S.I. Vinitsky et al., Zh.Exp.Teor.Fiz. 79, 698; (1980); Sov. Phys. - JETP 52, 353 (1980).

[10] Ya.B. Zeldovich and S.S. Gerstein, Zh. Exp. Teor. Fiz. 35, 821 (1958);
S. Winberg, Phys. Rev. Lett. 4, 575 (1960).

[11] W.A. Chupka, M.E. Russel and K.J. Rafaev, J. Chem. Phys., 48, 1518 (1968).

[12] L.I. Menshikov, Preprint IAE-3810/12, Moscow, 1983.

[13] G. Bardin, J. Duclos, A. Magnon, J. Martino and A. Richter, Phys. Lett. 104B, 320 (1981).

[14] G. Bardin, J. Duclos, A. Magnon, J. Martino and A. Richter, Nucl. Phys. A352, 365 (1981).

[15] D. Bakalov, Private communication.

[16] A. Bertin, M. Capponi, I. Massa, M. Piccinini and G. Vannini, Nuovo Cimento, 86A, 123 (1985).

[17] A. Bertin and A. Vitale, in Hadronic Physics at Intermadiate Energy, Eds. T. Bressani and R.A. Ricci, P. 189, Elsevir Science Publishers, 1986.

[18] A. Bertin and A. Vitale, the talk at Alma Mater Studiorum Saecularia Nona, Bologna, 1988.

Monte-Carlo Modeling of Epithermal Effects in Muon-Catalyzed dt Fusion

M. JEITLER*
Institute for Medium Energy Physics
Austrian Academy of Sciences
Boltzmanngasse 3
A-1090 Vienna
Austria

*) Representing the IMEP (Austrian Academy of Sciences) - Paul Scherrer Institute - Technical University of Munich - Lawrence Berkeley Laboratory - Los Alamos National Laboratory - ETH Zurich collaboration

Abstract. Theory and experiment agree that $dt\mu$ molecular formation rates for non-thermalized ("epithermal") muonic tritium atoms are much higher than at low-temperature thermal equilibrium. The detailed comparison of experimental data and theoretical predictions shows, however, that there are still major quantitative discrepancies.

In the resonant process of $dt\mu$ molecular formation [1]

$$t\mu + DA_{\nu_i=0, K_i} \to [(dt\mu)_{11}aee]_{\nu_f, K_f} \qquad (a = d \text{ or } t, A = D \text{ or } T) \qquad (1)$$

the strongest resonances correspond to collision energies of the muonic atom and the hydrogen isotope molecule that lie high above the thermal equilibrium for room temperature [2,3]. This is explained by the fact that the strongest contributions to the transition ($\nu_i = 0$) $\to$ ($\nu_f = 2$) are forbidden because of energy conservation - to fulfill the resonance condition, the $t\mu$ atom would have to have a negative energy ("sub-threshold resonances") - while the transition to the strong resonance of the next vibrational state ($\nu_f = 3$) is only possible if the $t\mu$ atom has a kinetic energy of several hundred meV.

Experimentally, this fact was first established during experiments in low-density deuterium-tritium gas mixtures [4,5] carried out at PSI (Paul Scherrer Institute, Switzerland). The time spectra of the $dt\mu$ fusion neutrons showed a strong, quickly decaying component (fig. 1), which was subsequently interpreted [2,6] as being due to $t\mu$ atoms that had not yet been slowed down to thermal energies ("non-thermalized" or "epithermal" muonic atoms) and could therefore form muonic molecular complexes $[(dt\mu)_{11}aee]_{\nu_f,K_f}$ ($a = d$ or t) via the strong resonances with $\nu_f \geq 3$.

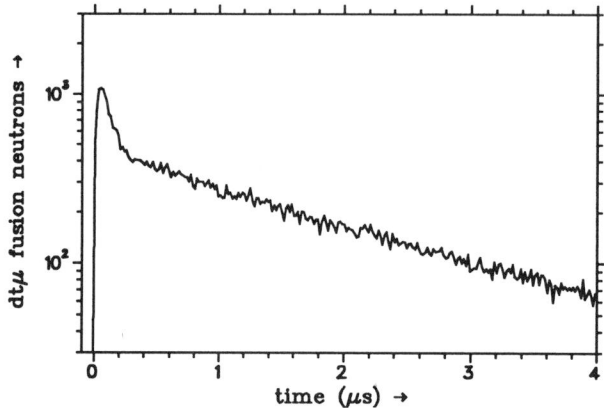

Figure 1: Experimental time distribution of $dt\mu$ fusion neutrons.

The low density of the target gas ($\Phi = 0.01$ of liquid hydrogen density) is an important experimental feature because the thermalization rate of muonic atoms scales with density. In dt mixtures with $\Phi \sim 1$ the epithermal component in the time spectra is so short-lived that it is hard to carry out an accurate analysis. Another approach is to do measurements in triple mixtures of all three hydrogen isotopes (H, D and T) and to make use of the very low scattering cross sections of $t\mu$ on protons [7,8]. $t\mu + p$ scattering can practically be neglected and the thermalization rate scales with the absolute deuterium and tritium concentration only. As in $dt\mu$ formation on D_2 or DT molecules, the strongest resonance for the process $t\mu + HD_{\nu_i=0,K_i} \to [(dt\mu)_{11}pee]_{\nu_f,K_f}$ lies at a high $t\mu$ energy [9]. Such experiments have also been carried out at PSI [10,11].

An increase in molecular formation rates at higher collision energies of the $t\mu + DA$ system has also been found in experiments carried out at LAMPF. Measurements of the $dt\mu$ cycling rate showed a steady increase with temperature, which did not yet level off at 800 K, the highest temperature measured [12].

In order to compare the experimental neutron time distributions with the theoretical molecular formation rates one has to consider the competing processes of epithermal molecular formation and thermalization of muonic atoms. For a deuterium-tritium mixture a calculation of the fusion neutron time distribution predicted by the theory must include [13]:

1) the $dt\mu$ molecular formation rates dependent on
- the center-of-mass kinetic energy of the $t\mu - DA$ ($A = D$ or T) system

- the kind of molecule (D_2 or DT) that forms a compund molecule with the $t\mu$ atom
- the initial rotational state K_i of the molecule (depending on the target temperature)
- the hyperfine spin state of the $t\mu$ atom ($F = 0$ or $F = 1$);

2) the differential scattering cross sections of $t\mu$ and $d\mu$ dependent on
- the center-of-mass kinetic energy of the collision partners
- the collision partner (d or t nuclei bound in molecules D_2, DT or T_2)
- the scattering angle θ
- the final spin state of the $t\mu$ atom (conservation of spin or spin-flip by charge exchange on a t nucleus)
- the change of the total energy available in the center of mass due to spin-flip or because of rotational or vibrational excitation of the colliding molecule (depending on its initial rotational state K_i and thus on the target temperature);

3) the energy-dependent rate of the transfer process $d\mu + t \rightarrow t\mu + d$;

4) the initial kinetic energy distribution of $t\mu$ and $d\mu$ atoms.

The resonant and non-resonant $dt\mu$ molecular formation rates for a wide range of $t\mu$ kinetic energies averaged for several different target temperatures have been calculated by Faifman and co-workers [14,15].

Cross sections for the scattering of $t\mu$ and $d\mu$ atoms on hydrogen isotope nuclei and for $d\mu \rightarrow t\mu$ transfer were first calculated and published by Bubak and Faifman [7]. Refined calculations were carried out by Melezhik et al. [16,8]. Total molecular cross sections have been calculated by Adamczak et al. [17,18].

The thermalization of muonic atoms strongly depends on the angular distribution of the scattering processes. Therefore differential scattering cross sections $d\sigma/d\Omega$ have to be used. The angular anisotropy is especially strong in the process $t\mu + d$ because of an important nuclear p-wave contribution (this is due to the existence of the weakly bound muonic molecular state $(dt\mu)_{v=1, J=1}$ - the very state which makes resonant molecular $dt\mu$ formation possible). Due to the high total $t\mu+d$ cross section this is the dominant scattering process (except at very high tritium concentrations of $c_t > 0.9$).

The final energy E_f (measured in the laboratory system) of a particle after an elastic scattering process with an angular distribution that is isotropic in the center-of-mass system (pure s-wave scattering, e.g. $t\mu_{F=1} + t \rightarrow t\mu_{F=1} + t$ in the $t\mu$ energy region shown in fig. 2) is distributed (almost) evenly between zero and the initial energy E_i if the two colliding particles have (approximately) equal masses ($0 \leq E_f \leq E_i$). If small scattering angles are favoured as in the process $t\mu + d$, the scattered particle tends to lose less energy per collision (fig. 3).

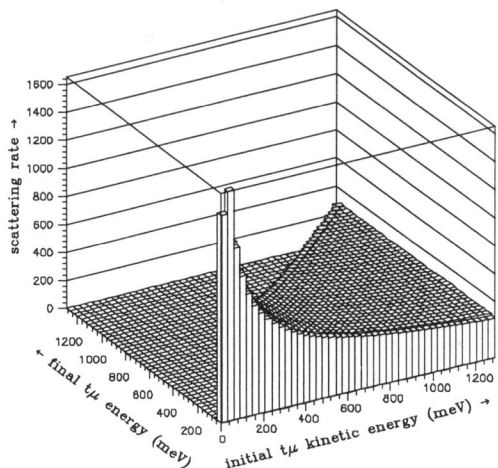

Figure 2: Differential elastic scattering rates (in arbitrary units) for given $t\mu$ kinetic energies before and after the scattering process $t\mu_{F=1} + t \to t\mu_{F=1} + t$.

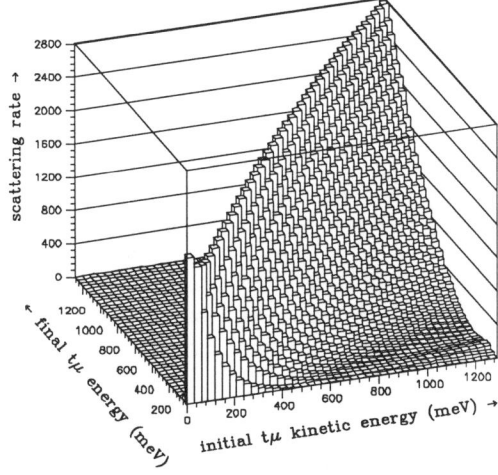

Figure 3: Differential elastic scattering rates (in arbitrary units) for given $t\mu$ kinetic energies before and after the scattering process $t\mu + d \to t\mu + d$.

Differential molecular cross sections are being calculated at the moment. However, they do not include the effect of the nuclear p wave. A preliminary analysis has shown that the effect of the nuclear p wave on the $t\mu$ thermalization process is far more pronounced than molecular effects. Therefore the present analysis has been based only on the nuclear scattering cross sections.

For the $d\mu + t \to t\mu + d$ transfer rate the angular distribution of the reaction products is of no importance for the $t\mu$ thermalization process due to the high energy ($\sim 19 eV$) gained by the $t\mu$ atom in the process. Therefore, the total rate including electronic effects calculated in [19] could be used.

There are no reliable calculations for the inital kinetic energy of muonic atoms if they were not formed by transfer processes but by direct muon capture. Measurements have yielded an energy of $\sim 1 eV$ for $d\mu$ atoms [20]. For the $t\mu$ case no measurements have been carried out so far.

For a calculation of theoretical $dt\mu$ fusion neutron time distributions one has to compare the rates for all processes that influence the $t\mu$ population and the resulting fusion yield (the rates for molecular formation, backdecay, elastic scattering and spin-flip are shown in fig. 4 and 5). Because of the complexity of the system this is most easily done by a Monte Carlo program.

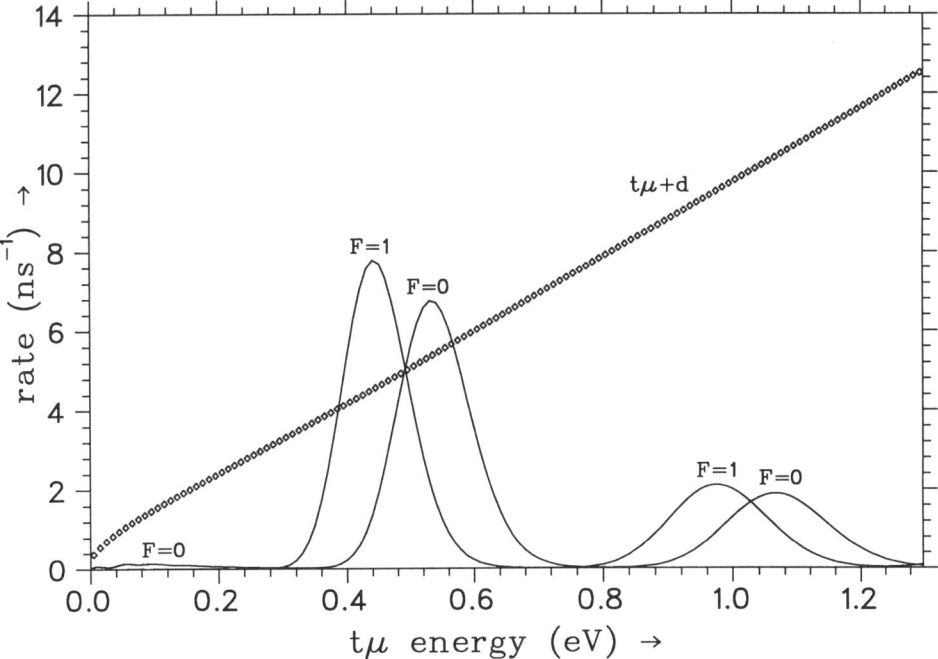

Figure 4: Total interaction rates as functions of the $t\mu$ kinetic energy: processes important at low tritium concentrations - the molecular formation rates $\lambda_{dt\mu-d}^{F=0}$ and $\lambda_{dt\mu-d}^{F=1}$ and the elastic scattering rate for $t\mu + d \to t\mu + d$. The probability of backdecay after molecular formation is not shown.

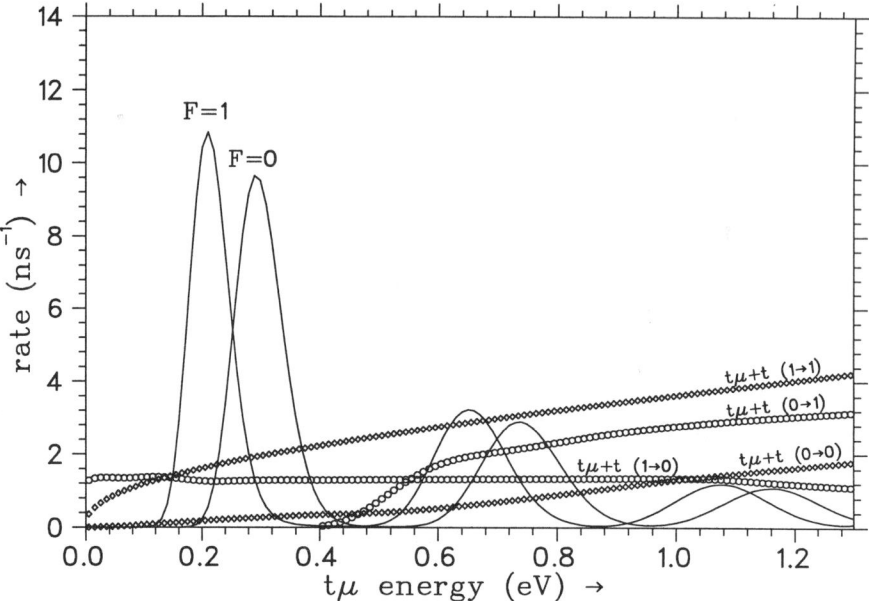

Figure 5: Total interaction rates as functions of the $t\mu$ kinetic energy: processes that become important at high tritium concentrations - the molecular formation rates $\lambda_{dt\mu-t}^{F=0}$ and $\lambda_{dt\mu-t}^{F=1}$, the elastic scattering rates for $t\mu_{F=0} + t \to t\mu_{F=0} + t$ and $t\mu_{F=1} + t \to t\mu_{F=1} + t$, and the spin-flip rates for $t\mu_{F=1}+t \to t\mu_{F=0}+t$ and $t\mu_{F=0}+t \to t\mu_{F=1}+t$. The probability of backdecay after molecular formation is not shown.

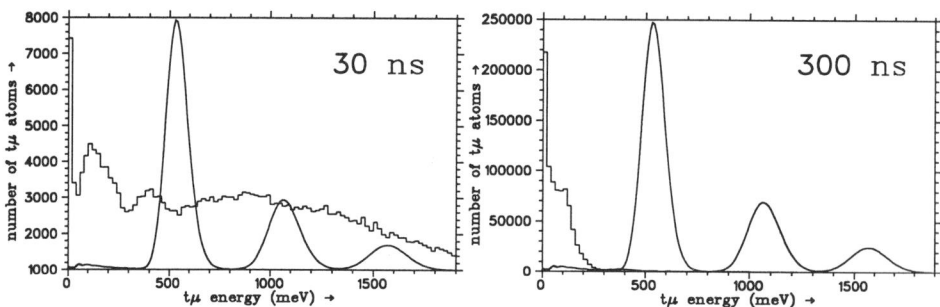

Figure 6: The development of the $t\mu$ kinetic energy distribution in time (histogram; target conditions: $c_t = 0.49$, $\Phi = 0.01$). The decrease of the overlap with the molecular formation rate $\lambda_{dt\mu-d}^{F=0}$ (solid line) illustrates the importance of epithermal effects at early times. The scale on the ordinate refers to the number of $t\mu$ atoms. The scale for the molecular formation rate is not given.

A comparison of the resulting energy distributions at various times with the theoretical molecular formation rates (fig. 6) shows that at early times there is a large population of $t\mu$ atoms in the region of the strongest $dt\mu$ molecular formation resonances.

Because of the large amount of physical parameters for which no accurate independent measurements exist it is essential to compare theory and experiment at different physical conditions (tritium concentration and temperature). At high tritium concentrations most deuterons are bound in DT molecules and epithermal molecular formation is mostly due to the process $t\mu + DT \rightarrow [(dt\mu)tee]$ (molecular formation by thermalized $t\mu$ atoms at low temperatures of $\sim 30K$ proceeds mostly via $t\mu_{F=0} + D_2 \rightarrow [(dt\mu)dee]$ even at high c_t because of the very low rate for the other processes). At intermediate tritium concentrations epithermal molecular formation occurs both on D_2 and on DT molecules. At low tritium concentrations the sensitivity to the molecular formation process is small and the fusion yield is mostly determined by the $d \rightarrow t$ transfer process (λ_{dt} and q_{1s}). The shape of the molecular formation resonances and the width of the distribution of thermalized muonic atoms strongly depend on the temperature. To check the consistency between experiment and theory it is therefore essential to make comparisons at various temperatures.

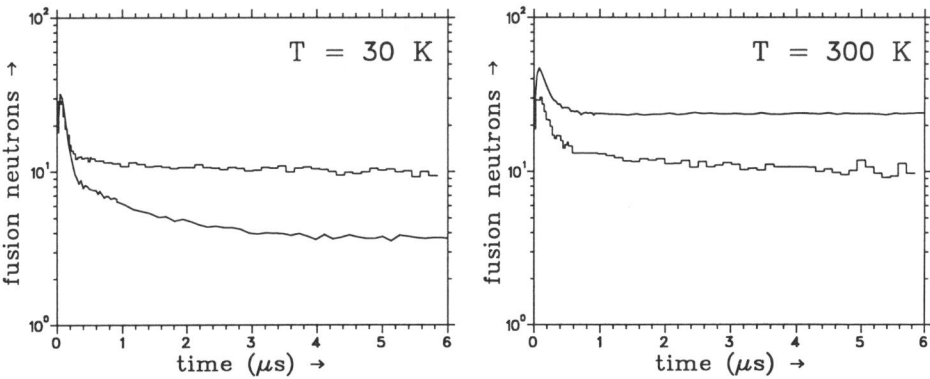

Figure 7: Comparison of experimental (histograms) and calculated fusion neutron time distributions in a deuterium-tritium target ($c_t = 0.49$).

The fusion neutron time distribution at various tritium concentrations and temperatures is the only result that can be compared with experiment. The available experimental data do not yield direct information on the energy distribution of the muonic atoms that produce fusions, on molecular formations resulting in backdecay etc. The agreement of the unmodified theoretical model with the experimental data is rather poor (see fig. 7).

While at low temperatures ($30K$) and intermediate tritium concentrations ($c_t \sim 0.5$) the fusion yield drops below the experimental value it is far too big for room temperature. For low temperatures a satisfactory fit can be achieved by increasing the rate for $dt\mu$ molecular formation on D_2 molecules by thermal $t\mu$ atoms in the lower hyperfine state to $\lambda_{dt\mu-d}^{F=0,th} = 130\mu s^{-1}$ while for high temperatures the molecular formation rates must be reduced by a factor of 2 to achieve agreement with the experimental data (see fig. 8).

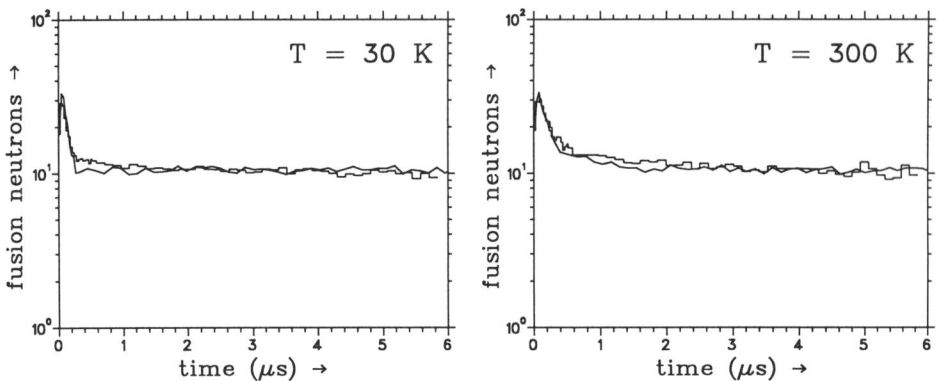

Figure 8: Comparison of experimental (histograms) and calculated fusion neutron time distributions in a deuterium-tritium target ($c_t = 0.49$) with modified theoretical parameters (30 K: $\lambda_{dt\mu-d}^{F=0,th} = 130\mu s^{-1}$; 300 K: all molecular formation rates reduced by a factor of 2).

This agreement does not mean, however, that the exact values for epithermal molecular formation rates can be obtained by fitting the experimental data. In particular, the time distributions are not very sensitive to the exact shape of the epithermal molecular formation resonances, and changes in their size could be offset by modifying the initial energy distribution of muonic atoms and the assumptions on $d \to t$ transfer in excited states (the q_{1s} parameter).

The sensitivity of the fusion neutron time distributions to elastic scattering rates is small. Due to the very low elastic nuclear cross section for $t\mu_{F=0} + t$ and the fast depopulation of the $t\mu_{F=1}$ state the scattering process is dominated by the (spin-independent) cross section for $t\mu + d$ except at very high tritium concentrations $c_t > 0.9$. It is not expected that the inclusion of differential molecular effects will have a major effect on the calculated time distributions (except possibly at such high tritium concentrations). The experimental data were fitted without modifying the cross sections published in [16,8]. However, due to the low sensitivity of the fusion yield to the scattering process this cannot be regarded as an experimental verification of the scattering cross sections.

The initial energy distribution of $t\mu$ atoms formed directly (not by muon transfer from another hydrogen isotope) was fitted by introducing a phenomenological parameter p_{th} that describes the percentage of $t\mu$ atoms with thermal initial energy while the rest was considered to be formed with an initial energy of $\sim 2eV$. Considering the lack of theoretical information this crude parametrization is sufficient (what counts most for the fusion process is whether the initial energy lies above or below the important molecular formation resonances with $\nu_f = 3$). At intermediate and high tritium concentrations ($0.49 \leq c_t \leq 0.95$) the fit of the experimental data yielded $0.5 \leq p_{th} \leq 0.7$. At low tritium concentrations an increase in p_{th} leads to a decrease in q_{1s} and vice versa. As neither theory nor other experiments have yielded conclusive evidence on either of these parameters it is not yet possible to disentangle these two effects. The present analysis has shown, however, that at least a certain percentage of all $t\mu$ atoms formed directly (not by transfer) must have an initial energy of $\geq 2eV$ to account for the initial build-up in the fusion neutron time distribution (see fig. 1; if the maximum initial $t\mu$ energy is lower the epithermal resonances are reached more quickly and the initial build-up becomes too short or disappears altogether).

The basic features of epithermal effects in muon-catalyzed fusion are now well understood and in many details there is remarkable agreement between theory and experiment. Still, to obtain a complete understanding of epithermal effects some more theoretical and experimental investigations should be carried out. Concerning the $dt\mu$ molecular formation rates refined calculations that go beyond the dipole approximation and include subthreshold and ortho-para effects will certainly be needed. Reliable calculations of the initial $t\mu$ kinetic energy after atomic muon capture and of the probability of $d \to t$ transfer in excited states (the q_{1s} parameter) are indispensable for a correct interpretation of experimental data. On the experimental side, direct measurements of q_{1s} based on the detection of muonic X rays are being prepared at PSI. Another experiment that is planned at PSI aims at reaching the high-energy molecular formation resonances by $t\mu$ atoms thermalized in a very hot target of up to 2000 K.

Acknowledgments

I would like to thank all the members of our collaboration - W.H.Breunlich, M.Cargnelli, P.Kammel, J.Marton, N.Naegele, P.Pawlek, A.Scrinzi, J.Werner, J.Zmeskal, H.Daniel, F.J.Hartmann, G.Schmidt, T. von Egidy, C.Petitjean, J.Bistirlich, H.Bossy, K.M.Crowe, M.Justice, J.Kurck, R.H.Sherman and W.Neumann - for their permanent support and many fruitful discussions. I would also like to thank A.Adamczak, J.S.Cohen, M.P.Faifman, M.Leon and V.S.Melezhik for making the results of their calculations available to me. Financial support by the Austrian Science Foundation is gratefully acknowledged.

References

[1] W.H.Breunlich, P.Kammel, J.S.Cohen and M.Leon, Ann.Rev.Nucl.Part.Sci. **39**, 311 (1989).

[2] P.Kammel, Nuovo Cimento Lett. **43**, 349 (1985).

[3] J.S.Cohen and M.Leon, Phys.Rev.Lett. **55**, 52 (1985).

[4] W.H.Breunlich et al., Phys.Rev.Lett. **53**, 12 (1984).

[5] W.H.Breunlich et al., Phys.Rev.Lett. **58**, 4 (1987).

[6] W.H.Breunlich et al., Recent results of μCF experiments at SIN, Muon Catalyzed Fusion **1**, 67 (1987).

[7] M.Bubak and M.P.Faifman, Cross sections for hydrogen muonic atomic processes in two-level approximation of the adiabatic framework, JINR Communication E4-87-464, Dubna (1987).

[8] V.S.Melezhik et al., The atlas of the cross sections of mesic atomic processes III. The processes $p\mu + (d,t)$, $d\mu + (p,t)$ and $t\mu + (p,d)$, to be published in Muon Catalyzed Fusion.

[9] M.P.Faifman and L.I.Ponomarev, Resonant formation of $DT\mu$ mesic molecules in the triple $H_2 + D_2 + T_2$ mixture, Preprint IAE-5329/12, Moscow 1991.

[10] K.Lou, contribution to this conference.

[11] V.Markushin, contribution to this conference.

[12] S.E.Jones et al., Phys.Rev.Lett. **51**, 1757 (1983).

[13] M.Jeitler et al., Epithermal Effects in muon catalyzed dt fusion, Muon Catalyzed Fusion **5/6**, 217 (1990).

[14] M.P.Faifman, L.I.Menshikov and T.A.Strizh, Calculation of the mesic molecular formation rates, in Muon Catalyzed Fusion **4**, 1 (1989).

[15] M.P.Faifman, Nonresonant formation of hydrogen isotope mesic molecules, in Muon Catalyzed Fusion **4**, 341 (1989).

[16] V.S.Melezhik et al., The atlas of the cross sections of mesic atomic processes I. The processes $p\mu + p$, $d\mu + d$ and $t\mu + t$, in Muon Catalyzed Fusion **4**, 247 (1989).

[17] A.Adamczak et al., Cross sections of processes $p\mu + H_2$, $d\mu + D_2$ and $t\mu + T_2$, in Muon Catalyzed Fusion **5/6**, 65 (1990/91).

[18] A.Adamczak, V.Korobov and V.S.Melezhik, Atlas of cross sections for muonic hydrogen scattering on hydrogen molecules, to be published in Muon Catalyzed Fusion.

[19] A.Adamczak et al., Muon transfer rates in hydrogen isotope mesic atom collisions, JINR preprint E4-92-140, Dubna 1992; submitted to Phys. Lett. B.

[20] R.Siegel, Contribution to this conference.

Direct Measurement of Sticking in Muon Catalyzed DT Fusion and Physics of "Hot" μt Atoms

K. LOU, C. PETITJEAN
Paul Scherrer Institute (PSI)
CH-5232 Villigen, Switzerland
T. CASE, K.M. CROWE
Lawrence Berkeley Laboratory (LBL) and University of California, Berkeley (UCB)
Berkeley CA 94720, USA
W.H. BREUNLICH, M. JEITLER, P. KAMMEL, B. LAUSS, J. MARTON,
W. PRYMAS, J. ZMESKAL
Austrian Academy of Sciences, Institute for Medium Energy Physics (IMEP)
A-1090 Vienna, Austria
D.V. BALIN, V.N. BATURIN, YU.S. GRIGORIEV, A.I. ILYIN, E.M. MAEV,
G.E. PETROV, G.G. SEMENCHUK, YU.V. SMIRENIN, A.A. VOROBYOV,
N.I. VOROPAEV
Petersburg Nuclear Physics Institute (PNPI)
Gatchina 188350, Russia
P. BAUMANN, H. DANIEL, F.J. HARTMANN, M. MÜHLBAUER, W. SCHOTT,
P. WOJCIECHOWSKI
Physics Department, Technical University of Munich (TUM)
D-8046 Garching, Germany

Abstract. New results are reported from a recent PSI experiment about the fusion reactions $d\mu t \to \mu + \alpha + n$ and $d\mu t \to \mu\alpha$ (sticking) + n. The apparatus consisted of a high pressure ionization chamber to detect charged particles directly and of an array of neutron counters to measure the fusion neutrons. The principle and performance of the experiment are described as well as a detailed study of its systematics by using a new Monte Carlo code. The preliminary results for the probability ω_s of final dt sticking are $(0.50 \pm 0.06)\%$ from a reanalysis of a 1989-run and $(0.47 \pm 0.06)\%$ for the recent 1991-run, nearly 3 standard deviations lower than the theoretical calculations. The initial time distributions of the neutrons at various low deuterium and tritium concentrations of the H/D/T mixture are presented providing new insights of the fast non-thermalized transfer of the muon from the μd to the μt atom.

1. Introduction

Sticking in $d\mu t$ fusion can be described in the frame of the dt cycle shown in fig.1: a negative muon in a mixture of hydrogen isotopes forms muonic atoms with protium, deuterium or tritium and subsequently, a muonic molecule $d\mu t$. This molecule undergoes fusion producing a charged particle α of 3.5 MeV and a neutron of 14.1 MeV kinetic energy. In most cases, the muon is freed and catalyzes further fusions. But, occasionally, the muon gets trapped to the α after the fusion. This phenomenon is called "initial sticking" and the probability of this process is theoretically predicted to be 0.92% (see ref. [1] and refs. therein).

One has to distinguish between "initial sticking" and "final sticking": about 1/3 of the muons, originally stuck to the α are stripped off due to the recoil motion of the $\mu\alpha$. Such stripped muons are then free to return to the fusion cycle. The remaining part of unstripped muons composes the fraction "final sticking" ω_s which is the probability of a muon finally stuck to an α. This parameter sets an ultimate limit to the number of fusions which one muon can catalyze.

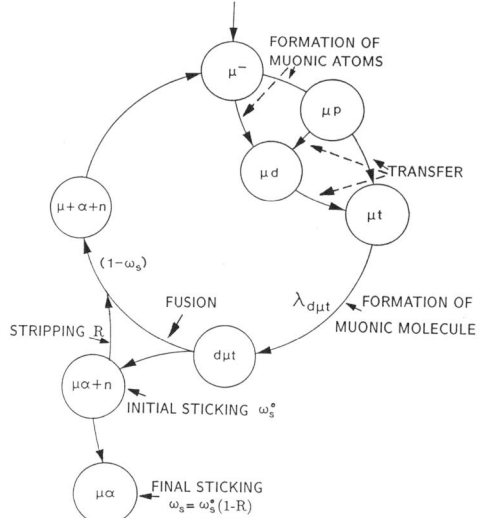

Figure 1:

Scheme of the dt fusion cycle, explaining the difference between initial sticking ω_s^o and final sticking ω_s.

2. The experiment

2.1 General description. The apparatus uses a high pressure ionization chamber (IC) as central detector of the incoming muons and of charged products from fusion reactions. The history and development of the IC as well as many experimental aspects have been given in previous reports, see e.g. refs. [2-8]; here they are only briefly reviewed.

The first runs at PSI were done in 1989 and improved runs were started at PSI in the fall of 1991 and are presently continued in 1992.

The sensitive volume of the IC is ~ 3 cm^3, 1 cm in height and 2 cm in diameter. It is filled with a gaseous mixture of hydrogen isotopes at a pressure of 160 atm at room temperature. The density ϕ of the mixture is 0.17 normalized to that of liquid hydrogen. The IC is surrounded by an array of 19 thick plastic neutron detectors and one liquid NE213 neutron counter; between the IC and the neutron counter array are thin plastic counters to identify charged particles, e.g. electrons from muon decay.

After a muon passes through the trigger counters and stops in the sensitive volume of the IC, a dt fusion may occur, producing an ionized track from the α particle. With an efficiency of about 30 percent the neutron is detected by the neutron counter array in coincidence to the α signal.

The negative charges (electrons) from the track are collected by at least one of the anodes of the IC; this signal is then preamplified, and its form is analyzed by 6-bit flash ADC's (FADC) of 1024 channels each, covering an overall period of 10 μs. In a first (online) analysis the signals are parameterized into start-time, width and charge integral; their shape characteristics are studied in offline analysis to distinguish among different event types.

2.2 Principle of the experimental method. A dμt molecule can fuse via two reaction channels, in most cases producing an α^{++} particle, which has charge two, and occasionally a $\mu\alpha^+$, which has charge one due to the μ^- sticking to the α^{++}. Both α and $\mu\alpha$ have about the same kinetic energy, 3.5 MeV. But, the ionization density of the α is approximately 4 times greater than that of the $\mu\alpha$, which results in a $\mu\alpha$ track that is about 4 times longer than the α track.

In the IC there is a non-linear charge recombination effect which results in only about 1/3 of the α charge being collected at the anodes whereas approximately 1/2 of the charge is collected in the case of $\mu\alpha$ due to the lower charge density along the track. This recombination effect allows us to distinguish on the energy scale "sticking" $\mu\alpha$ events from normal α's.

A second way of $\mu\alpha/\alpha$ separation is the analysis of the width distribution of the signals: α's (track length 0.64 mm) have baseline widths up to 0.25 μs, while $\mu\alpha$ widths (track length 2.13 mm) range up to 0.5 μs.

In order to reduce the noise from tritium β decays, the anode part of the IC was segmented into 19 separate anodes, and the concentration of tritium was kept low. A main component of the background comes from dμd fusions going into a proton and a triton (dμd $\rightarrow$ p + t). It was kept low by using a small deuterium concentration and could be further suppressed by requesting a neutron signal in the neutron counter array. A smaller part of the background comes from the pμd fusion chain, since at the high protium concentration ($C_p > 90\%$) most muons form pμd molecules. On the one hand, this incidentally allows us to determine the pμd formation rate $\lambda_{p\mu d}$ from the dt fusion time slope with high accuracy [5]. On the other hand, about 20% of the pμd molecules produce a μ^3He atom by the fusion reaction pμd $\rightarrow$ μ^3He + γ, and finally the muon capture reactions μ^3He $\rightarrow$ t + ν_μ, μ^3He $\rightarrow$ d + n + ν_μ or μ^3He $\rightarrow$ p + p + n + n + ν_μ take place. All these reactions produce charged particles detected in the IC, but they do not have electrons. Therefore the muon capture reactions can be rejected by requesting an electron from the array of the electron counters (detection efficiency $\approx$ 15%) in delayed coincidence. At the current level of accuracy, this background does not play a big role in the determination of dt sticking because the detected energy of the charged particles (≈ 0.9 MeV for tritons) is mostly below that of the $\mu\alpha$'s (≈ 1.8 MeV) and because the probability of the reactions is low ($\approx 5\%$).

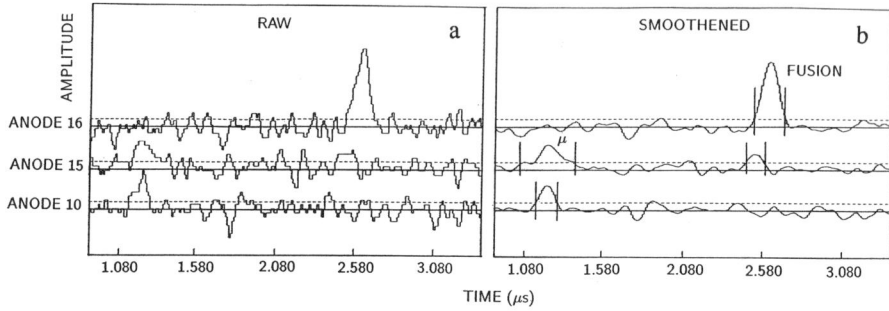

Figure 2: FADC display of a dt fusion event, amplitude versus time (10 ns per channel): a) raw data; b) the same event after offline filtering procedure.

A real event of dt fusion recorded by FADC's of 3 neighboring anodes is shown in fig.2. An incident μ was detected by the two neighboring anodes 10 and 15, and subsequently a dt fusion occurred, firing the two neighboring anodes 15 and 16. The tritium noise seen in the figure contributed about 80 percent to the energy resolution in the 1989-run. By reducing the amount of tritium ($C_t = 0.036\%$ instead of 0.047%), we have successfully improved the energy resolution in the 1991-run from 11.8% to 9.7% (σ_E/E).

3. Systematics

We have studied the systematics of the IC and the surrounding array of counters by creating a complete Monte Carlo program. It generates the tracks of all charged particles, simulates the recombination effect and drifts the electrons towards the anode region to produce the electronic signals. Then the signals are added onto an anode spectrum of tritium noise extracted from the experimental data. The Monte Carlo simulation is in very good agreement with the experimental data, see comparison in fig.3. The events shown are signals collected from single anodes. For the p + t events shown in fig.3b a special window of large width from 0.64 μs to 1.2 μs is applied; widths of up to 1.2 μs are expected since the p + t tracks are 5.23 mm long. The minimum width of 0.64 μs was required to keep the p + t events clean from any dt fusions which produce shorter signals only.

4. Data analysis and results on sticking

4.1 Background reduction. Fig.4 shows the energy spectra of fusion events. In the 1991-

run the sticking events are better resolved than in the 1989-run due to smaller tritium noise.

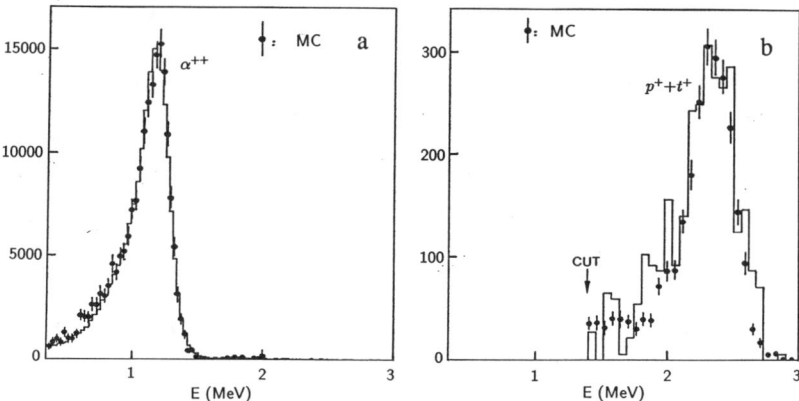

Figure 3: Comparison between the experimental data of the 1991-run and the Monte Carlo simulations (MC) of α's (charge-two) [a] and p + t 's (charge-one) [b]. The signals are taken from the inner anode region where we have better understanding of the systematic errors.

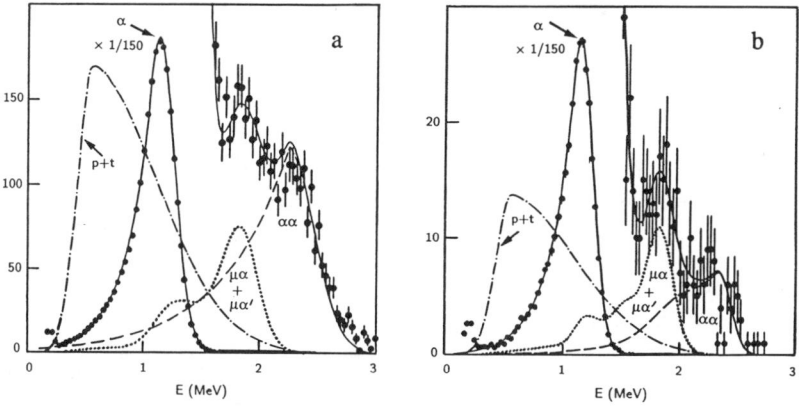

Figure 4: Energy distribution of the fusion products from: a) the 1989-run; b) the 1991-run. The solid line represents the fit to the experimental data obtained with a maximum likelihood analysis in an energy region from about 1.4 MeV to 2.8 MeV. The fit functions for the sum of sticking and stripping ($\mu\alpha + \mu\alpha'$, in dotted line, where $\mu\alpha'$ stands for the stripping events), for p + t (in dot-dashed line) and $\alpha\alpha$ pileup (in dashed line) resulted from the Monte Carlo simulation. The function for the α was taken from the experimental data and its upper tail was adjusted by the MC. There are only 2 free parameters in the fit: the intensities of the ($\mu\alpha + \mu\alpha'$)'s and of the $\alpha\alpha$'s, while the intensity of the α's was determined by a separate fit in a low energy region (from about 0.6 MeV to 1.3 MeV) and then fixed. Only data from the inner chamber region were analyzed.

The events in fig.4 are signals collected from single anodes, whereas all events giving also signals in neighboring anodes are rejected. A window of small width from 0.12 μs to 0.56 μs is made to keep the p + t background low. The intensity of the remaining p + t events in fig.4 is inferred accurately from the p + t events observed at large widths (see fig.3b) using the Monte Carlo program.

In order to obtain better separation between α and sticking $\mu\alpha$, a special shape analysis was adopted to cut the noise pileups at the ends of the signals. The $\alpha\alpha$ pileups are reduced by elimination of signals with double peaks; the possible loss of $\mu\alpha$ events ($\approx$ 10%) is controlled by applying the same requirements to the Monte Carlo data.

In the 1991-run, a significant additional background underneath the $\mu\alpha$ line was found which is due to a small amount of nitrogen impurity in the target ($C_N \approx$ 3-10 ppm). In this case the muon can be transferred to N, and subsequently a muon capture process on nitrogen could happen, which may produce 3 alphas detected by the IC, i.e. $\mu N \rightarrow \alpha + \alpha + \alpha + n + n + \nu_\mu$, and prevents of course electrons from muon decay. By requiring that an electron from muon decay is detected after a reaction has occurred, this background is eliminated. Due to the detection efficiency of the electron-counter array, the events in fig.4b where this condition is applied are only one seventh of the whole statistics. The nitrogen impurity in the 1989-run was at least one order of magnitude less, and in fact no such background underneath the $\mu\alpha$ line was found.

The background to the sticking events from the muon capture on ^{3}He was found to be negligible. It was omitted in the analysis of the 1989-run and incidentally rejected by requiring a delayed electron coincidence in the 1991-run.

With respect to the statistics of the 1991-run, we will improve the analysis by using the full statistics and by properly taking into account the μN capture background. Additional (almost pure) μN capture events were recorded in a background H/D run in which the concentrations of protium and deuterium were kept the same as in the main run.

4.2 Results. Table 1 lists the preliminary results on the sticking parameter ω_s, obtained by a maximum likelihood analysis in which we assumed the reactivation coefficient R (stripping) to be 0.31 [9]. The corresponding fit curves are shown in fig.4.

The results on sticking from the 1989-run and 1991-run are consistent with each other but somewhat lower than the previously reported value (0.59 $\pm$ 0.07)% [5]. The main reason is that our understanding of the background components, namely the p + t and $\alpha\alpha$ pileup events, has been improved due to the new Monte Carlo code. The errors of the new results are combined from statistical and systematic uncertainties. The main part is still the systematic error.

The current results on the final dt sticking parameter ω_s are by nearly 3 standard deviations lower than the theoretical calculations. This raises again the suggestion that there is a systematic discrepancy between experiments and theory, as the indirect measurements from PSI on final sticking have previously indicated [10].

Table 1: Preliminary results on the sticking parameter ω_s

$\phi = 0.17$	1989-run	1991-run	theory [1]
C_p (%)	90.5	96.0	
C_d (%)	9.5	4.0	
C_t (%)	0.047	0.036	
N_α	361,220	51,315	
$N_{\mu\alpha+\mu\alpha'}$	1,474 ± 68	204 ± 25	
$N_{\alpha\alpha}$	3,145 ± 82	167 ± 18	
N_{p+t}	4,624	411	
ω_s (%)	0.50 ± 0.06 (stat.+syst.)	0.47 ± 0.06 (stat.+syst.)	0.65 ± 0.03

5. The physics of "hot" μt atoms

Surprisingly, we have found a peak in the initial time distribution of the dt-fusion neutrons detected with the neutron counter array. This is shown in fig.5, where the "spike" is from the initial processes of "hot" μt atoms while the main peak (slow transfer of thermal μd to μt) is due to the disappearance of the μd atoms by muon decay and $p\mu d$ formation. The IC does not see directly this peak since there is no separation between the μ stop and an α signal in the first 200 ns [6].

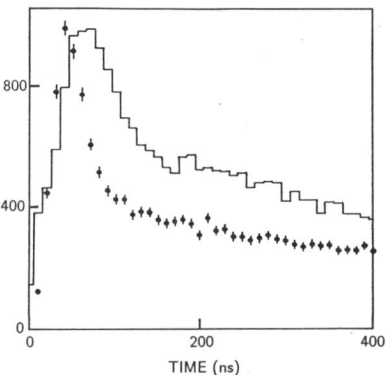

Figure 5:

Initial time distribution of neutrons from dt fusions. Error-bars: the 1989-run; histogram: the 1991-run. The time is measured between μ stop and neutron event.

This spike originates from two processes. One is the direct transfer of the muon from the μp to the μt atom and the other is the fast non-thermalized transfer from the μd to the μt atom. The μd and μt atoms formed in the transfer reactions have an initial kinetic energy of about 20 to 50 eV, and are called "hot" atoms since they are energetic. This phenomenon is now well understood by theory and known as the Ramsauer-Townsend effect, see refs. [11,12].

The 1991-run has reduced concentration of deuterium, thus the rate of elastic collisions in $\mu d + d$ and $\mu t + d$, which is the major process to slow down (thermalize) the μd and μt atoms, is reduced; therefore the μd as well as the μt atoms stay hot for a longer period after the transfer, causing the spike in the time distribution to be widened.

6. Outlook and summary

6.1 The Survived muon method. To make a direct measurement of the final sticking independent of any assumption about $\mu\alpha$ stripping, we will rely on the survived muon method which is described here in an example of a "double-fusion" event, shown in fig.6. One sees that the incident muon made 2 fusions: the first fusion signal definitely was not an event of final $\mu\alpha$ sticking, since by the same muon, a second fusion was produced later. The difference between the normalized spectrum of the single fusion events and that of first fusions from the double fusion events, gives just the $\mu\alpha$ energy distribution from final sticking. Stripping, $d\mu d \rightarrow p + t$, etc. get automatically deducted.

This method was first successfully used for the measurement of μ^3He sticking in muon catalyzed dd fusion ($\omega_{dd} \approx 12\%$) [4]. However, under our experimental condition of the H/D/T triple mixture, the rate of the double fusion events is only about 1-2% of the single fusion events. Therefore this method suffered so far from insufficient statistics to reach high precision [5]. A measurement with better statistics will be made in a forthcoming 1992-run.

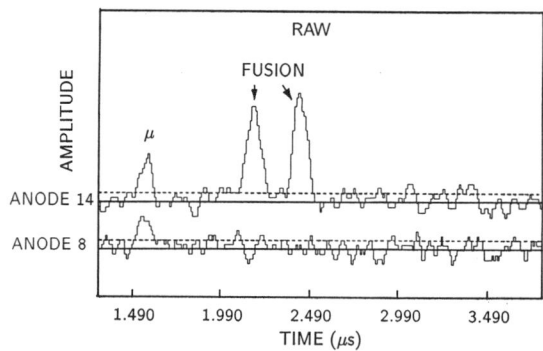

Figure 6:

A FADC display of an event with two successive fusions by the same μ, so as to explain the "survived muon method". This is from the 1991-run.

6.2 Summary.

1. The preliminary results for the final dt sticking probability ω_s are $(0.50 \pm 0.06)\%$ from a renewed analysis of the 1989-run and $(0.47 \pm 0.06)\%$ for the recent 1991-run. In further work we plan to improve the analysis and reduce the systematic errors.

2. The physics of "hot μt atoms" reveals the kinetics of the energetic μd and μt atoms, showing very large fusion rates at the initial time when the μt atoms are hot. This phenomenon can be understood by the calculations in ref. [13] predicting large resonances of $d\mu t$ formation in the eV region, and by the kinetics interpretation given in ref. [11].

Acknowledgements

We would like to thank the German Bundesministerium für Forschung und Technologie, the Austrian Academy of Sciences, the Russian Academy of Sciences, the U.S. DOE and the Paul Scherrer Institute and its technical staff for the continuous support of this experiment.

References

[1] M. Kamimura, AIP Conf. Proc **181**(1989) 330. L.N. Bogdanova et al., Sov. J. Nucl. Phys. **50**(1989) 848. B. Jeziorski et al., Phys. Rev. **A42**(1990) 3768.
[2] D.V. Balin et al., Phys. Lett. **B141**(1984) 173.
[3] A.A. Vorobyov, Muon Cat. Fusion **2**(1988) 17.
[4] D.V. Balin et al., Muon Cat. Fusion **7**(1992) 1.
[5] C. Petitjean et al., Muon Cat. Fusion **5/6**(1990) 261.
[6] T. Case et al., Muon Cat. Fusion **5/6**(1990) 327.
[7] K. Lou et al., Muon Cat. Fusion **5/6**(1990) 525.
[8] D.V. Balin et al., Muon Cat. Fusion **5/6**(1990) 481.
[9] V.E. Markushin, Muon Cat. Fusion **3**(1988) 395.
[10] C. Petitjean et al., Muon Cat. Fusion **1**(1987) 89 and **2**(1988) 37.
[11] V.E. Markushin, contribution to this conference.
[12] V.E. Markushin et al., Muon Cat. Fusion **7**(1992) 155.
[13] M.P. Faifman and L.I. Ponomarev, Phys. Lett. **B265**(1991) 201.

The Exotic-Atom Cascade

F.J. HARTMANN
Physik-Department, E18, Technische Universität München
James-Franck-Straße, W-8046 Garching, Germany

Abstract. The atomic cascade of exotic particles is described from the start after Coulomb capture to the end by particle decay or nuclear capture. A new way of taking electron depletion and refilling during the exotic-particle quantal cascade into account is proposed.

1. Introduction

At least since the first detailed measurements of x rays from muonic atoms with Ge detectors (in the early sixties [1]) the exotic-atom cascade has found interest among experimental and theoretical physicists: On the one hand, the apparently simple problem of calculating the electromagnetic processes during the cascade (only radiative and Auger transitions if one restricts oneself to exotic atoms with $Z > 2$) contains some traps, connected, e.g., with the electron balance in the host atom. On the other hand, a detailed knowledge of the exotic-particle cascade *and* on the population of the electron shells during this cascade is mandatory for many experiments dealing with exotic atoms: Knowledge of the number of K (and to a smaller extent also of L) electrons is of great interest in precision measurements of mesic transition energies, since the electron screening correction, which can be calculated with high precision if the number of electrons is known, is uncertain due to missing knowledge on the number of K (and L) electrons in the exotic atom. In a recent experiment the pion rest mass was determined from the energy of the 4f → 3d transition in pionic Mg [2]. Corrections depend strongly on the filling of the electronic K shell which had to be determined separately by a dedicated experiment on the cascade in π-Mg. Another example: the population of the 2s level in muonic atoms determines if an experiment on the parity- nonconserving effects in muonic atoms is feasible or not [3].

The aim of this contribution is to briefly sketch the present knowledge of the exotic-atom cascade as it emerged from all kinds of measurements on muonic, pionic, kaonic and antiprotonic atoms during the last decades. First the question of the first levels populated after capture of the exotic particle shall be raised. Then the "missing link" between capture and quantal cascade shall be examined and finally the quantal cascade shall be treated. In this context a new attempt to keep track of the electron-shell population in exotic atoms shall be shortly described. The cascade in *muonic* atoms will be in the centre of attention. As the Stark effect and the possibility of collisional deexcitation makes the exotic-atom cascade in hydrogen unique I will not deal with this special case.

2. The first bound orbit in exotic atoms

What do we know about the first bound orbit of the exotic particle ? Not too much, neither from theory nor from experiment. As far as theory is concerned, semiclassical calculations [4] can say nothing about the first *bound* orbit as they are characterized by a *smooth* transition from unbound to bound state. Quantum-mechanical calculations for elements with atomic number $Z \geq 2$, on the other hand, are rare [5,6,7]. As a rule of thumb one can assume that the first bound state of the exotic particle populated with maximum probability has a wave function with maximum overlap with the wave function of the ejected electron. This leads to values of the principal quantum number n shown in Table 1 if one assumes the most loosely bound electron to be ejected in the Auger process. This assumption is supported by theory at least for muonic He and Li [6,7].

Table 1: Rough estimate for the principal quantum number of the first bound orbit in exotic atoms.

Exotic atom	n of the first bound orbit	Outmost electron	Exotic atom	n of the first bound orbit	Outmost electron
Muonic He	14	1s	Muonic Pb	80	6s
Muonic Fe	50	3d	Antiprotonic He	38	1s

The initial *angular-momentum* distribution in the exotic atom has been studied in many theoretical papers (see, e.g., [8,9,10] and [11] with the references quoted therein). The simplest assumption is that states with different magnetic quantum number but the same angular-momentum quantum number ℓ are populated with equal probability. This leads to the well known statistical initial distribution

$$p(\ell) \propto 2\ell + 1. \tag{1}$$

For large n this finding is practically equivalent to a naive classical picture: All exotic particles (mass m) are assumed to have the same energy T_0 when hitting the atom and all impact parameters up to $b_{\max} = r_0$ (with r_0 the atomic radius) are equally likely. With

$$p(\ell)d\ell \propto q(b)db \propto b\,db \tag{2}$$

and
$$\hbar \ell \approx \sqrt{2 \cdot m \cdot T_0} \cdot b \quad (3)$$
we get
$$p(\ell) \propto \ell. \quad (4)$$

If, less naively, we assume all impact parameters up to $b_{\max} = r_0$ to be equally probable, but this time the energy spectrum of the exotic particles hitting the atom to be white up to an energy T_0,
$$p(\ell) \propto \ell \cdot \ln(\ell_{\max}/\ell) \quad (5)$$
follows [12], with $\ell_{\max}$ given by $\hbar \ell_{\max} \approx \sqrt{2 \cdot m \cdot T_0} \cdot r_0$.

More elaborate semiclassical calculations [13] lead to angular-momentum distributions which may deviate strongly from statistical ones.

The *experiment* suffers from the fact that the first transitions after Coulomb capture of the exotic particle into states with large n proceed by Auger effect (if this is energetically possible) with the most loosely bound electrons. The energy of the Auger electrons is so small that no one has even tried up to now to detect them in a counter experiment. Only Auger electrons from the final stage of the cascade have been detected with *counters* [14]. Perhaps measurements of Auger electrons emitted during the initial stage of the cascade will become possible when intense low-energy muon beams will be available at the meson factories [15]. Altogether it seems justified to say that only indirect experimental information is available about the first bound orbits in exotic atoms. One source is the observation of x rays from high-lying levels of the exotic atom which compete with the deexcitation by Auger effect and favor transitions with large Δn to low-lying states of the exotic atom. Transitions from orbits with quantum number $n \approx 20$ have been seen in muonic atoms of low to medium Z [16,17]. The fraction of muons captured into large mesomolecular orbits [18] was shown to be small for P and Se [17].

Recently new information about the first bound state in exotic atoms was derived, in this case for hadronic He. After forming hadronic atoms, negative hadrons like π^-, K^- or $\bar{p}$ are expected to vanish immediately by nuclear absorption. It was observed [19], however, that when negative kaons are stopped in liquid He a few percent show free decay (lifetime 10 ns). Very recently, in an experiment with antiprotons [20] it was found that the time distribution of the annihilation products after $\bar{p}$ stop in *liquid* He contains, with a fraction of roughly 3%, a delayed component with a trapping time of 3 μs. No such delayed component was observed for $\bar{p}$ in liquid nitrogen or in liquid argon. Similar observations were also made for $\bar{p}$ in gaseous He at pressures up to 5 bar. A convincing explanation for this "trapping" seems to be that the neutral $\bar{p}$-He atom can only slowly deexcite from states with $n \geq 35$ and $\ell = n - 1$ or $\ell = n - 2$ by radiative transitions or by Auger transitions with $\Delta \ell < -3$: Auger transitions with smaller angular-momentum transfer are energetically forbidden and Stark mixing is improbable as the remaining electron shields the $\alpha \bar{p}$ system in the antiprotonic He atom. From variational calculations with the hadron and the remainig electron described by hydrogen-like wave functions Russell [21] found the transition rates for hadronic He atoms shown in Fig. 1(left). Recently even lower rates for $\bar{p}$-He were calculated [12]. This means that, e.g., for $\bar{p}$-He a fraction of the antiprotons are Coulomb-captured in states with principal quantum number higher than $n = 35$, in agreement with the rough estimate (cf. Table 1). If we assume an initial distribution which is Gaussian in n with mean $n_0 = 35$ and standard deviation $\sigma_n = 2$ and which is statistical in ℓ, roughly 16% of the antiprotonic He atoms are formed in levels were only deexcitation by radiation and Auger effect with $\Delta \ell < -3$ is possible. As the observed fraction of trapped antiprotons is only around 3% one may speculate that the initial ℓ distribution in $\bar{p}$-He must be flatter than statistical.

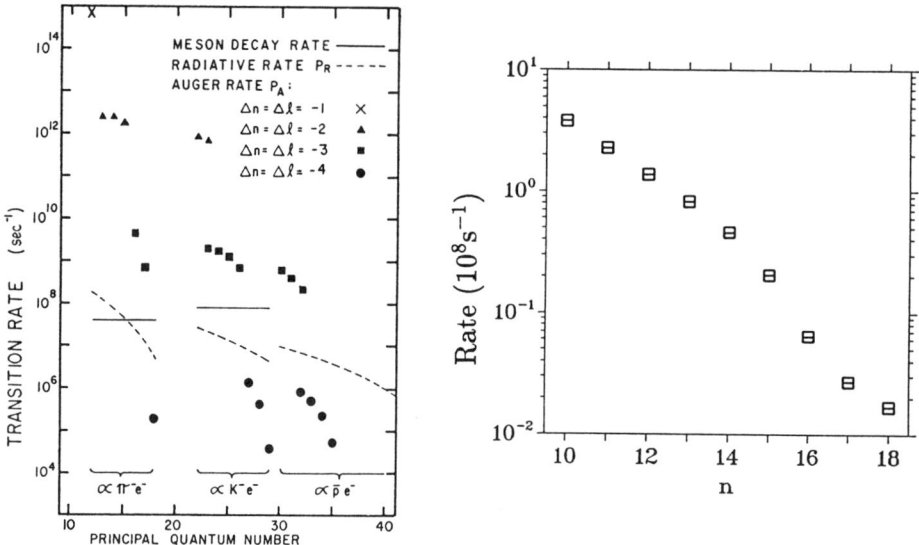

Figure 1: Transition rates in hadronic and muonic He. Left: transition rates in hadronic He; right: rates in muonic He for transitions from circular orbits.

One can immediately apply the techniques used to calculate energy levels and life times in $\bar{p}$-He to μ-He. The results are shown in Fig. 1(right). According to these calculations trapping times of around 100 ns are to be expected and should be seen in time spectra of the x rays from muonic He [22].

3. Cascade through the electron cloud

After capture the exotic particle finds itself in the middle of the electron cloud. In this stage it may be viewed as moving on a classical trajectory loosing energy by interaction

The Exotic-Atom Cascade

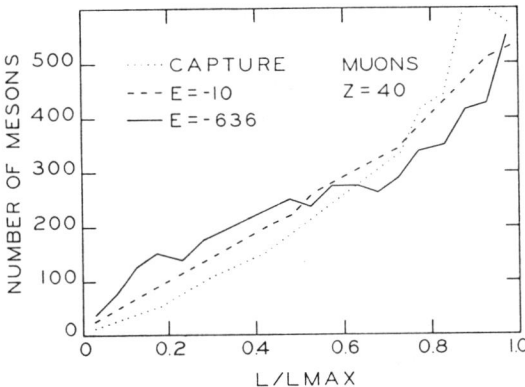

Figure 2: Angular momentum distribution [9] for muonic Zr (Z=40) after capture (dotted line), at an energy of $E = -270$ eV (dashed line) and at $n \approx 16$.

with the atomic electrons and by emitting bremsstrahlung. Rook [23] was the first to solve this problem. He employed an 1/r potential which is only a poor approximation. As for the interaction with the atomic electrons he assumed, as had already been done in the work by Fermi and Teller [24], the exotic particle to continuously loose energy by interaction with a Fermi gas of electrons. This electron gas exerts a frictional force on the exotic particle, which hereby looses energy *and* angular momentum. Rook derived the relation

$$(\ell + \frac{1}{2})/n = \text{const} \tag{6}$$

between ℓ and n. Thus the *shape* of the angular momentum distribution is *not* changed by Auger transitions.

Corresponding expressions for the losses in energy E and angular-momentum $\vec{L}$ by bremsstrahlung can be derived [23]. An elegant method has been pointed out by S.S. Gershtein [25]. According to his calculations angular-momentum and principal quantum numbers are related by

$$\frac{1}{\ell + \frac{1}{2}}[\frac{1}{(\ell + \frac{1}{2})^2} - \frac{1}{n^2}] = \text{const}, \tag{7}$$

if one assumes $E \Rightarrow m \cdot (Ze^2)^2/(2\hbar n^2)$ and $L^2 \Rightarrow \hbar^2(\ell + \frac{1}{2})^2$. From this equation one may see that circular orbits ($\ell = n - 1$) are reached quickly.

If one compares the energy losses for Auger effect and for bremsstrahlung calculated in this simple manner one finds out [23] that for high n the exotic atom is deexcited nearly exclusively by Auger effect and hence the shape of the angular momentum distribution should be unchanged during the cascade through the electron cloud.

Calculations with a more realistic potential roughly along the lines outlined above [9,10] corroborated the assumption that the shape of the angular-momentum distribution remains essentially unchanged during the cascade of the exotic particle from energy zero to a bound orbit of the dimensions of the electronic K shell (cf. Fig. 2).

4. The quantal cascade

When the exotic particle has reached orbits around $n = \sqrt{m/m_e}$ (with m_e the electron mass) the shielding of the nucleus by the electron cloud becomes less and less important. Already in the early fifties the first formulae were given for the rates of Auger and radiative transitions [26,27] for the hydrogen-like problem. Later on handier formulae were produced [28], the movement of nucleus and exotic particle around a common center of gravity was taken into account [29] and in the late seventies also quadrupole and octupole contributions to the Auger rates were calculated [30]. All kinds of corrections like penetration of the electron cloud between exotic particle and nucleus and retardation [30] were evaluated. It turned out that, by accident, the corrections mostly cancel and therefore the application of the simpler formulae [28] remains justified. For fast computations Ferrells formula [31], which connects Auger rates, radiative rates and the cross-section for photoeffect, is very useful.

I think it is legitimate to say that the problem of rate calculations is solved. One peculiarity of the cascade, however, is up to now still treated in a rather crude way in all the cascade programs although it has a major influence on the population of the different levels and hence also on the x-ray intensity pattern: The electron balance in the exotic atom during the cascade, in other words, the competition between electron depletion by Auger effect and electron refilling from inside and outside the exotic atom. Even in the more elaborate cascade codes the refilling of K electrons is assumed to proceed with a constant rate irrespective of all changes in higher electron shells; the depletion of L electrons or the refilling (or even both) are neglected at all.

Electron holes produced by Auger transitions of the exotic particle can be refilled by (usually fast) Coster-Kronig transitions from higher subshells of the same electron shell [32], by electronic Auger or radiative transitions from higher shells, by transfer of electrons from other atoms during collisions (in the case of gases), or by transition of electrons from the conduction band (in the case of metals) or from the valence band (in the case of nonmetals). Here the highly ionized exotic atom strongly attracts electrons in the vicinity and even may deform the crystal potential to such an extent that transitions from the valence shells of neighboring atoms to the depleted atom become possible [33]. Hence refilling of electrons from the valence band may be important even in insulators. To improve the description of the exotic-atom cascade a model with the following features was developed [34]:

1. The K refilling rate was adjusted according to the depletion of the L electron shell.

2. L-shell depletion by K-hole refilling was taken into account.

3. Electron refilling from outside was also taken into consideration.

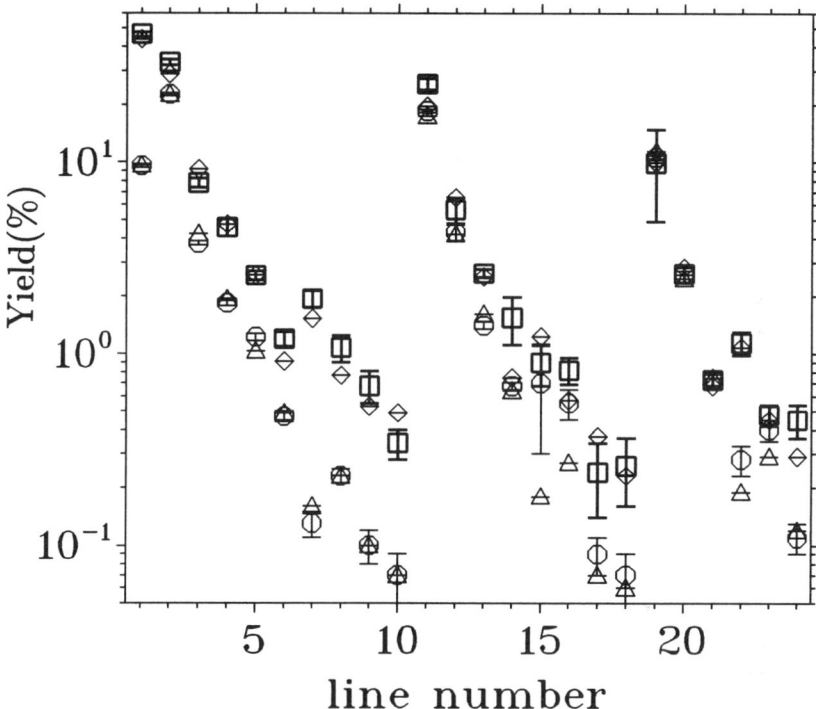

Figure 3: Experimental and calculated x-ray intensities from muonic and pionic Fe. Squares: μ experiment; diamonds: μ calculation. Circles: π experiment; triangles: π calculation. Line 1: L_α; lines 2 to 10: Paschen series; lines 11 to 18: Brackett series; lines 19 to 24: Pfund series

Exotic x-ray intensities from calculations on the muonic and pionic cascade in Fe starting at n= 17 are shown in Fig. 3 together with experimental values [35,36]. The initial ℓ distribution chosen for the muon case was

$$p_\mu(\ell) \propto 1 + a \cdot \ell + b \cdot \ell^2. \qquad (8)$$

Agreement between experiment and calculation is good, although the only parameters adjusted for the muon case were the parameters for the initial population of the ℓ sublevels (all quantities governing depletion and refilling of electrons are unambiguously fixed by the electronic properties of the elements). In the case of pionic atoms the pion-nucleus interaction plays an important role already at high n. To take it into account, the distribution

$p_\mu(\ell)$ found for muons was multiplied by $c_1 \cdot [1 - \exp(-c_2 * \ell^2)]$, and nuclear absorption was introduced into the calculations as a process competing with radiative and Auger transitions. The constants c_1 and c_2 were adjusted to give best agreement with the pionic x-ray data measured.

References

[1] D. Kessler, H.L. Anderson, M.S. Dixit, H.J. Evans, R.J. McKee, C.K. Hargrove, R.D. Barton, E.P. Hincks, and J.D. McAndrew, Phys. Rev. Lett. **18**, 1179 (1967).

[2] S. Thomann, Diploma thesis, Eidgenössische Technische Hochschule, Zürich, 1990.

[3] R. Bacher, P. Blüm, D. Gotta, K. Heitlinger, M. Schneider, J. Missimer, and L.M. Simons, Phys. Rev. A **39**, 1610 (1989).

[4] H. Daniel, Ann. Phys. (NY) **129**, 303 (1980) and references quoted therein.

[5] H. Daniel, Radiat. Effects **28**, 189 (1976).

[6] N.A. Cherepkov and L.V. Chernysheva, Yad. Fiz. **32**, 709 (1980) - Sov. J. Nucl. Phys. **32**, 366 (1981).

[7] P.K. Haff and T.A. Tombrello, Ann. Phys. **86**, 178 (1974).

[8] M. Leon and R. Seki, Nucl. Phys. A **282**, 445 (1977).

[9] M. Leon and J.H. Miller, Nucl. Phys. A **282**, 461 (1977).

[10] P. Vogel, A. Winther, and V. Akylas, Phys. Lett. **70B**, 39 (1977).

[11] H. Daniel, Progr. Theor. Physics **65**, 1481 (1981).

[12] T. Yamazaki and K. Ohtsuki, Phys. Rev. A **45**, 7782 (1992).

[13] H. Daniel, Z. Phys. A - Atoms and Nuclei **302**, 195 (1981).

[14] R. Callies, H. Daniel, F.J. Hartmann, and W. Neumann, Phys. Lett. **91A**, 441 (1982).

[15] R. Abela, F. Foroughi, C. Petitjean, D. Renker, and E. Steiner, Muon Cat. Fusion **6**, 459 (1991).

[16] F.J. Hartmann, H.J. Pfeiffer, K. Springer, and H. Daniel, Z. Physik **271**, 353 (1974).

[17] K. Kaeser, T. Dubler, B. Robert-Tissot, L.A. Schaller, L. Schellenberg, and H. Schneuwly, Helv. Physica Acta **52**, 238 (1979).

[18] L. Ponomarev, Ann. Rev. Nucl. Sci. **23**, 395 (1973).

[19] T. Yamazaki, M. Aoki, M. Iwasaki, R.S. Hayano, T. Ishikawa, H. Outa, E. Takada, and H. Tamura, Phys. Rev. Letters **63**, 1590 (1989).

[20] M. Iwasaki, S.N. Nakamura, K. Shigaki, Y. Shimizu, H. Tamura, T. Ishikawa, R.S. Hayano, E. Takada, E. Widmann, H. Outa, M. Aoki, P. Kitching, and T. Yamazaki, Phys. Rev. Letters **67**, 1246 (1991).

[21] J.E. Russell, Phys. Rev. A **1**, 721 (1970).

[22] A. Blaer, J. French, A. M. Sachs, M. May, and E. Zavattini, Phys. Rev. A **40**, 158 (1989).

[23] J.R. Rook, Nucl. Phys. B **20**, 14 (1970).

[24] E. Fermi and E. Teller, Phys. Rev. **72**, 399 (1947).

[25] S.S. Gershtein, International Workshop on the Electromagnetic Cascade and Chemistry of Exotic Atoms, Erice, 1989.

[26] G.R. Burbidge and A.H. deBorde, Phys. Rev. **89**, 189 (1953).

[27] A.H. deBorde, Proc. Phys. Soc. London **67**, 57 (1954).

[28] Y. Eisenberg and D. Kessler, Nuovo Cimento **19**, 1195 (1961).

[29] Z. Fried and A.D. Martin, Nuovo Cimento **29**, 574 (1963).

[30] V.R. Akylas and P. Vogel, Comp. Phys. Commun. **15**, 291 (1978); V.R. Akylas, Ph. D. thesis, California Institute of Technology, Pasadena, Calif., 1978.

[31] R.A. Ferrell, Phys. Rev. Letters **4**, 425 (1960).

[32] W. Bambynek, B. Craseman, R.W. Fink, H.U Freund, H. Mark, C.D. Swift, R.E. Price, and P. Venugopala Rao, Rev. Mod. Physics **44**, 716 (1972).

[33] G.T. Condo, Phys. Rev. Lett. **37**, 1649 (1976).

[34] F.J.Hartmann, to be published.

[35] F.J. Hartmann, T. von Egidy, R. Bergmann, M. Kleber, H.-J. Pfeiffer, K. Springer, and H. Daniel, Phys. Rev. Lett. **37**, 331 (1976).

[36] F.J.Hartmann, H. Daniel, W. Neumann, G. Schmidt, and T. von Egidy, Z.Phys. A **341**, 101 (1991).

Muon Transfer

Muon Transfer Processes. Old and New Problems

S.S. GERSHTEIN
Institute for High Energy Physics,
142284, Protvino, Moscow region,
Russian Federation

Abstract. It is shown that the observed transfer rates of muons from mesic hydrogen atoms to the nuclei of heavier elements as well as the isotopic exchange in the excited states of mesic hydrogen atoms and their Coulomb deexcitation, are explained satisfactorily by the crossings or quasicrossings of molecular terms, discovered earlier. But still, the precision measurements done over recent years at meson facilities require that a natural development of the quasiclassical theory, related to quantum oscillations and retardation processes of mesic hydrogen atoms in medium, should be made.

Introduction

The transition of $\mu^-(\pi^-)$ - mesons from hydrogen isotope nuclei to those of other elements, often present in the form of slight admixtures, has a great importance for the study of such phenomena as

1) capture of μ^- in hydrogen and deuterium for investigation of weak interactions;

2) muon-catalyzed fusion (μCF);

3) meson chemistry.

In a number of cases this transition is unwanted and therefore, to avoid it, a high degree of cleaning hydrogen from admixtures is required. Yet, in some cases, when this process is controlled, it is very helpful as it allows one to estimate experimentally the probabilities of a number of important mesomolecular processes, e.g. the rate of mesic molecule production, of nuclear processes in mesic molecules, etc.

In addition to that, the experimental measurement of the transfer probabilities of μ^- mesons from isotopes to different mesic atom levels of admixed atoms makes it possible to visualize theoretically the process mechanisms. And this is a very important point as similar processes in atomic physics proved essential for the study of controlled thermonuclear fusion.

1. Background of the problem

After the universal $(V-A)$ interaction was discovered in 1958, its verification for μ-capture studied by that time rather insufficiently became an outstanding problem. Therefore a few groups of experimentalists started planning experiments on μ-capture by the simplest nuclei, e.g. those of protons, deuterons and ^3He. In particular, the Dubna group headed by B. Pontecorvo was planning an experiment on μ-capture by ^3He and the other headed by V.P. Dzhelepov was going to study mesomolecular processes in hydrogen and deuterium because the latter determined essentially the probabilities of muon capture in these media.

Since B. Pontecorvo did not have a sufficient number of ^3He he planned to study μ-capture in a mixture of hydrogen and ^3He assuming that muons stopped in hydrogen transfer to ^3He fairly fast. His assumption was corroborated by the data obtained in propane chambers, indicating that the muons, stopped during their lifetime, have enough time to transfer to carbon nuclei. In late 1959 B. Pontecorvo asked me to estimate the probabilities of muon transfer from protons to ^3He nuclei. Proceeding from the low-order terms of the $peHe$-system calculated by that time I noticed that these transfers could be nonadiabatic only and, hence, their collision probabilities at low energies must be small. But at that time it seemed unclear why the probability of muon transfer from protons to carbon nuclei was high, as it followed from the experimental data. In our attempts to understand that, V.D. Krivchenkov and I decided first to establish the qualitative behaviour of the terms in the $Z_1 e Z_2$ or $Z_1 \mu Z_2$ system versus the distance between the nuclei. For this to be accomplished, it was necessary to introduce the so-called correlation diagrams connecting the quantum numbers of ions spaced at long distances with those of the united atom. Such diagrams were known only in the systems with the same charges of nuclei (P. Morse and E.C. Stueckelberg, 1929 [1]).

2. Correlation diagrams. Crossings and quasicrossings of terms

It is not difficult to establish correlation diagrams [2] by taking advantage of the fact that an elliptical (or spheroidal) coordinate system, in which the variables for the problem on muon motion in the field of two motionless nuclei separate, for $R \to 0$, transforms into a spherical one for the united atom $(Z_1 + Z_2)$, whereas for $R \to \infty$ it transforms into a parabolic one, in which the Stark split levels of isolated atoms are described. In this case the number of nodes (zeros), n_ξ, of the wave function of the elliptic variable $\xi = (r_1 + r_2)/R$ for $R \to 0$ transforms into the radial quantum number of the united atom

n_r:
$$n_r = n_\xi, \qquad (1)$$
whereas for $R \to \infty$ it transforms into a parabolic quantum number n_1 (or n_1') of atoms Z_1 (or Z_2) spaced at long distance, i.e.
$$n_\xi = \begin{cases} n_1 & (Z_1) \\ n_2 & (Z_2) \end{cases}. \qquad (2)$$

As to the number of nodes of the wave function versus the hyperbolic variable $\eta = (r_1 - r_2)/R$, for $R \to 0$ it is related to the orbital quantum number of the united atom l as
$$l = n_\eta + |m|, \qquad (3)$$
whereas for $R \to \infty$ it is expressed through the parabolic quantum numbers of atoms at long distances, or, putting it more precisely, through the number of the nodes with respect to the relevant variable
$$n_\eta = n_2 + n_2'. \qquad (4)$$
The exception is the case when for $R \to \infty$ the muon situated close to the nucleus possessing a larger charge, $Z_2 > Z_1$, is in the $n'(n_1', n_2', m)$ state with the quantum numbers $Z_1/n = Z_2/n'$, $n_1' = n_1$. In the latter case we have
$$n_\eta = n_2 + n_2' + 1. \qquad (4')$$

If we put $Z_1 = 1$, $Z_2 = Z$, then the above reasonings suggest that for $R \to 0$ the occupation level $n(n_1, n_2, m)$ of the mesic atom $p\mu$ transfers into the state (N, l, m) of the united atom having the leading quantum numbers N, l equal to
$$N = nZ + n_2, \quad l = N - n_1 - 1, \qquad (5)$$
whereas the state of the mesic atom $Z\mu$ possessing the quantum numbers $n'(n_1', n_2', m)$ transforms into the one of the united atom (N, l, m):
$$N = \begin{cases} n' & \text{for } n_2' < n'(1 - \frac{1}{Z}) \\ n' + 1 + \text{Ent}[n_2' - n'(1 - \frac{1}{Z})] & \text{for } n_2' \geq n'(1 - \frac{1}{Z}) \end{cases}$$

$$l = N - n_1' - 1. \qquad (6)$$

Hence, the ground state of $p\mu$ corresponds to the occupation level of the united atom,
$$N = nZ, \quad l = N - 1, \quad m = 0, \qquad (5')$$
whereas those of the mesic atom $Z\mu$ $n'(n_1', n_2', m)$ lying below the ground state of $p\mu$ correspond to the states of the united atom
$$N = n'; \quad l + N - n_1' - 1 \quad (n' < Z). \qquad (6')$$

With Coulomb repulsion of nuclei taken into account, the terms corresponding to the interaction of the mesic atom $Z\mu$ with the proton behave at large distances R as

$$W^Z \simeq -\frac{Z^2}{2n'^2} + \frac{Z-1}{R} + \frac{3}{2}\frac{(n'_1 - n'_2)n'}{Z}\frac{1}{R^2} + O(\frac{1}{R^3}) + \ldots \, . \tag{7}$$

The 2nd term in this expression corresponds to the Coulomb repulsion of the $(Z-1)$-charged mesic atom $Z\mu$ from the proton and the 3rd one corresponds to the interaction with the proton charge of the dipole moment of the mesic atom $Z\mu$, occurring due to the $Z\mu$ polarization during the Stark splitting of its occupation levels in the electric field of the proton. It is worth noting that for $n'_2 - n'_1 > 0$, when the muonic cloud is shifted to the proton, this term corresponds to attraction.

At the same time, the term corresponding to the neutral mesic atom $p\mu$ (in the ground state) in the field of the nucleus Z at large distances R has but just a weak Van der Waals shift:

$$W^p \simeq -\frac{1}{2} - \frac{9}{4}\frac{Z^2}{R^4}. \tag{8}$$

Therefore the terms corresponding to the lower occupation levels of the mesic atom $Z\mu$ ($n' < Z$) and growing with a decrease of the distance R due to Coulomb repulsion, $(Z-1)/R$, can cross term (8)[1]. **The discovered crossing of terms revealed the possibility of adiabatic transfer of muons from protons to heavier nuclei in the processes of slow collisions of mesic atoms with nuclei [2].** This became the basis for further consideration of not only mesic hydrogen atoms - nuclei charge exchange but also of that between conventional hydrogen atoms and ions. This also explained the fact that the cross section for the transfer of muons from mesic hydrogen atoms to helium nuclei is a few orders of magnitude smaller than that of the transfer to the nuclei of heavier elements[2]. It also followed from the consideration of the correlation diagrams that for the excited states of hydrogen mesic atoms there exists a large number of crossings of the relevant terms with those of the states lying below the energy occupation level. As it was noted in ref. [5], this should not only lead to a large cross section of the transfer of mesons from the excited states of mesic hydrogen atoms to the nuclei of other elements but also to the processes of Coulomb deexcitation of mesic hydrogen atoms during their collision with protons and to those of isotopic exchange from excited states. As it was indicated in ref. [5], the latter point has a paramount importance for interpreting the phenomenon of the capture of negative pions in a mixture of hydrogen and deuterium containing a large amount of the latter. These phenomena are presently treated in a large number of publications.

---

[1]Further estimates made by L.I. Ponomarev et al. showed that this crossing of terms takes place at $Z > 4$ (for the details and refs. see [3]).

[2]Later Yu. Aristov et al. [4] found an effective mechanism of the transfer of muons from mesic hydrogen atoms to helium nuclei, caused by the production of a quasimolecule of $(p\mu He^{++})$. But still, this mechanism corroborated experimentally that the probability of the transfer to helium is 2-3 orders of magnitude less than the one of the transfer to the nuclei of heavier elements.

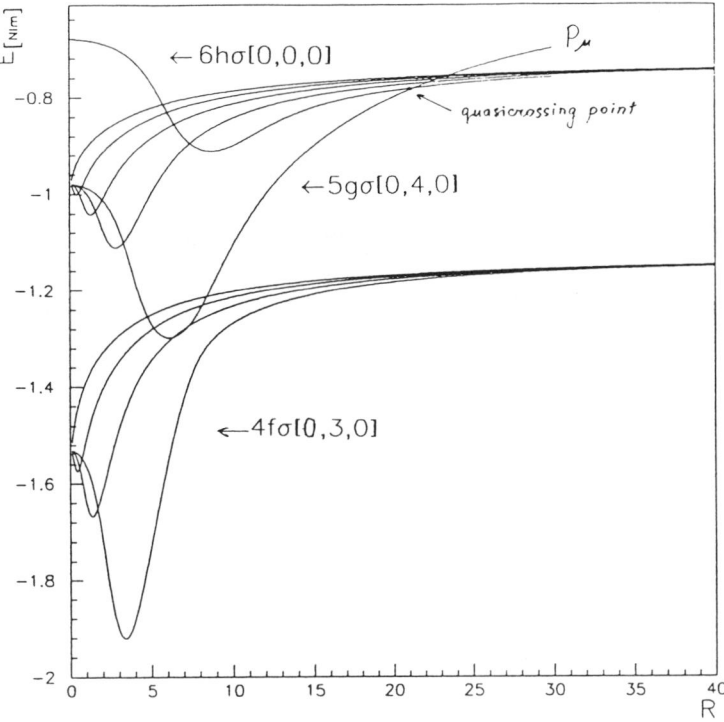

Figue 1: σ-terms in the system $Z_1=1$, $Z-2=6$. The notations correspond to the quantum numbers of the United Atom. The values in the brackets are parabolic quantum numbers $[n-1, n_2, m]$ of the isolated atoms at $R \to \infty$. The term $6h\sigma$ $[0,0,0]$ corresponds to the ground state of isolated hydrogen atoms. It's quasicrossing point with the $5g\sigma$ $[0,4,0]$ at $R \simeq 21.5$, which is invisible in the scale of figure.

The question of the crossing of terms in the problem on two Coulomb centres requires additional remarks.

1. As it might seem at first sight, the crossing of the terms of the same geometric symmetry, i.e. of the same projection of the moment on the symmetry axis, contradicts the well-known theorem of Neumann-Wigner. However, this is not the case. The problem on two point-like Coulomb centres has an additional intrinsic symmetry manifesting itself in the existence of additional conserving integrals of motion and leading to the separation of the variables ξ and η. Within the frames of this intrinsic symmetry the terms having different values of the elliptic quantum numbers, n_ξ and n_η, possess different symmetries and, hence, their crossing does not contradict the Neumann-Wigner symmetry (see S.P. Allilyev, A.V. Matveenko [6]).

2. As discovered by L.I. Ponomarev et al [3], some of the crossing points of the terms specified in [2] are actually quasicrossing ones. This refers to the terms having the same quantum numbers, $n_1 = n'_1$. As for the terms corresponding to the ground state of the mesic hydrogen atom $(n_1 = n_\xi = 0)$, for $R \to \infty$ quasicrossings rather than crossings lying below the σ-terms of the mesic atoms $Z\mu$ with $n'_1 = n'_\xi = 0$ should take place. The states with the maximum quantum numbers n'_2 possible, whose wave functions are most "extended" to the direction of the nucleus $Z_1 = 1$, correspond to these terms. As it was established, the energy term reaching the vertex of the potential barrier in the equation with respect to the variable η corresponds to the quasicrossing point and to the muon motion in the joint field of the two nuclei. (For more details, see the book by I.V. Komarov, L.I. Ponomarev, S.Yu. Slavyanov [3]).

3. As noted in ref. [2], the analysis of the correlation diagrams suggests that if the terms of $Z\mu$ and $p\mu$ do cross, this crossing should occur twice, at a long and a short distance (see Fig. 1). The same is also the case with quasicrossings. Short-distance quasicrossings correspond to a series of branching poins in the complex plane of the variable R, discovered by E. Soloviev [7].

3. The mechanism of adiabatic transfer $p\mu \to Z\mu$. Advances and new problems.

The theory of adiabatic transfers $p\mu \to Z\mu$, based on the existence of crossings of terms was developed 30 years ago (see S.S. Gershtein [5]). G. Fiorentini and G. Torelli [8] analyzed and generalized it further. The major qualitative result of the theory [5], though the specific estimates were made for the transfers to the nuclei of carbon, $Z = 6$, and oxygen, $Z = 8$, were as follows:

1. Muons transfer primarily to the excited states of mesic atoms $Z\mu$, whose terms cross the hydrogen term at distances of about $R_{int} = 7 - 10$ of mesic units and have the quantum numbers $n_\xi = n'_1 = 0$ and $n_\xi = n'_1 = 1$. These terms correspond to the most pronounced overlapping of the functions $p\mu$ and $Z\mu$, the first of them really experiencing a quasicrossing.

2. The probability of the transfer to the above terms is high, amounting to several tenths of unity.

The consideration was carried out under the following approximations:

a) the approximate wave functions of isolated mesic atoms were used and the energy of the terms was calculated from formulas (7-8) without taking into account the exchange terms $exp(-R)$;

b) since the collision energy was small, the transfers with respect to nuclear motion in the S-wave were considered. In virtue of this, the transfers with a variation of the moment projection onto the axis ($|\Delta m| = 1$) were not taken into account either;

c) electron screening was not considered;

d) the consideration was made under the quasiclassical approximation with averaging the oscillations, and for $v \to 0$ was matched with a quantum-mechanical one.

We obtained under the above approximations:

1) the approximate linear Z-dependence for the transfer probability;

2) the ratio between the transfers $p\mu \to Z\mu$ and $d\mu \to Z\mu$,

$$\frac{\Lambda^Z_{d\mu}}{\Lambda^Z_{p\mu}} \simeq \sqrt{\frac{M_p}{M_d}}$$

at sufficiently low collision energies.

The estimates carried out yielded the following results:

1) the probabilities of the transfer $p\mu \to Z\mu$ for a number of elements, which agree with the experiment by one order of magnitude;

2) the prediction for the variation in the structure of a series of cascade γ-transitions, which is observed in the mesic atom $Z\mu$ after muon transfer in the process $p\mu \to Z\mu$ to certain excited levels of $Z\mu$; this may be compared with the series observed for direct muon capture into the mesic atom $Z\mu$ [9].

This study was made possible due to new experimental feasibilies, i.e. some meson factories put into operation. Its results are in a good agreement with the theoretical predictions (see Fig. 2). As compared with the direct production of the mesic atoms $Z\mu$, a noticeable enhancement of higher-order components of the Lyman series, caused by the transfer $p\mu \to Z\mu$, is observed. In my opinion, **this is the best demonstration of the validity of the theoretical predictions on the mechanism of the transfer $p\mu \to Z\mu$.** However, the study also revealed a good deal of discrepancy between the experiments and theoretical estimates.

1. As it was established, the ratio between the probabilities of the transfers $p\mu \to Z\mu$ and $d\mu \to Z\mu$ is in strong disagreement with the simple theoretical estimates [5,8]. This discrepancy was noticed long ago, though various experimental data on the probability of such transfers were also in discordance for a long time. They have been measured quite reliably only recently [10,11]. Surprising is a strong discrepancy between the probabilities of the transfers to neon atoms, $p\mu \to (Ne)\mu$ and $d\mu \to (Ne)\mu$: the latter turned out to be much larger than the former. The same was also observed for the transfers to nitrogen atoms, $p\mu \to (N)\mu$ and $d\mu \to (N)\mu$.

2. A strong disagreement with the simplest linear Z-dependence is observed: the Z-

dependence is not monotonous.

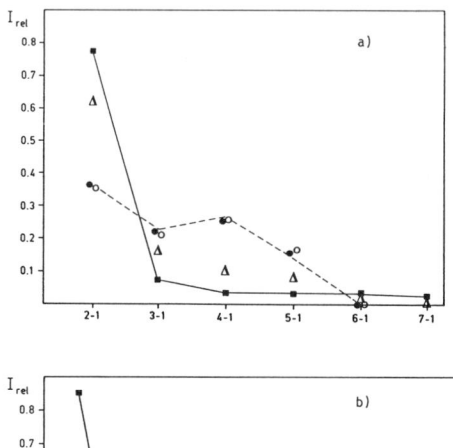

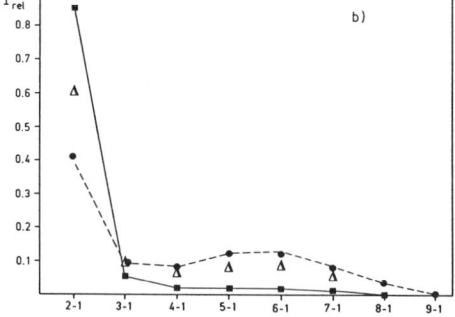

Figure 2: a) Relative muonic X-ray intensities of the Lyman series measured in oxygen of SO_2. The squares correspond to the intensities measured in pure SO_2 at 2.4 bar. The intensities of the higher members of the Lyman series measured in a $H_2 + (0.4\% \, SO_2)$ gas mixture at 14.9 bar, resulting from muon transfer from μp atoms in the ground state (black disks) are much larger and agree well with theoretical predictions (broken line). The intensities of the unexpected second time component of the oxygen X-rays (open circles) are identical to those of muon transfer from the μp ground state. The triangles represent the prompt intensities in the $H_2 + (0.4\% \, SO_2)$ gas mixture at 14.9 bar. The structure of these intensities somewhat differs from those observed in direct capture in pure SO_2. b) Relative muonic X-ray intensities of the Lyman series measured in sulphur of SO_2. The structure of these intensities does not agree with those observed in direct capture in pure SO_2.

3. A new enigmatic phenomenon was discovered after precision measurements made at PSI in a gas mixture of H_2 with small admixtures of SO_2, SF_6, N_2, Ne [10-18]. As it turned out, the time dependence of the muon transfer to the nuclei of oxygen, fluorine and neon is not described by a simple exponential dependence. In addition to the specific "slow"

time for the transfer $p\mu \to Z\mu$ from the ground state of $p\mu$, coinciding with the theoretical estimates, "fast" times of unknown nature are observed. The muons corresponding to this "fast" transfer time were given a special name, "black" muons.

This phenomenon remained unexplained during a number of years. However, recently there have appeared papers by Yu.S. Sayasov, pointing to ways of possible interpretation [19,20].

4. Nonmonotonous Z-dependence and muon transfer from different hydrogen isotopes

When calculating the probability of transfer between the crossing terms which, according to the theory of Landau-Zenner-Stueckelberg, has the form

$$w = 4e^{-\delta}(1 - e^{-\delta})cos^2(S_0 + \frac{\pi}{4}), \tag{9}$$

the authors of [5,8] put for the value of the oscillating factor, as it was the case in the quasiclassical approach, the mean value of $\overline{cos^2(S_0 + \pi/4)} = 1/2$. Here the quantity S_0 is the difference in the quasiclassical phases corresponding to the initial U_i and final U_f terms,

$$S_0 = \int_{r_i}^{r_c} p_i dr - \int_{r_f}^{r_c} p_f dr; \quad p_{i,f} = \sqrt{2M(E - U_{i,f})}, \tag{10}$$

where r_c corresponds to the crossing of the terms, $U_i(r_c) = U_f(r_c)$, and $r_{i,f}$ corresponds to the stopping points, $p_i(r_i) = 0$. Basically, as noted by Yu. Sayasov, consideration for the oscillations proves very essential in the conditions for the transfer $p\mu \to Z\mu$. As it follows from his calculation [19], the probability of the transfer $p\mu \to Z\mu$ differs from the one obtained in [5] by the oscillating factor $F(\xi)$ expressed through the Airy functions of the parameter ξ:

$$\xi = (\frac{3}{8}\frac{Z}{r_c}\sqrt{2M})^{2/3}. \tag{11}$$

For $\xi >> 1$

$$F(\xi) \simeq sin^2 4\xi^{3/2}, \tag{12}$$

whereas the value taken in [5] was $F = 1/2$.

As is seen from fig. 3, the consideration of the oscillating factor explains well the observed nonmonotonous Z-dependence of the transfer rate and its difference for the transfers $p\mu \to Z\mu$ and $d\mu \to Z\mu$, which remained unclear for a long time. Particularly impressing is the explanation of a very small transfer rate λ_{pNe} because the function $F(\xi)$ has the minimum in the vicinity of the value of ξ, corresponding to the transfer $p\mu \to Ne\mu$. However, one should bear in mind that the agreement between the theoretical results presented in ref. [19] and the experimental data is qualitative rather than quantitative because some earlier experiments did not draw the distinction between "fast" and "slow" transfers (see the section below).

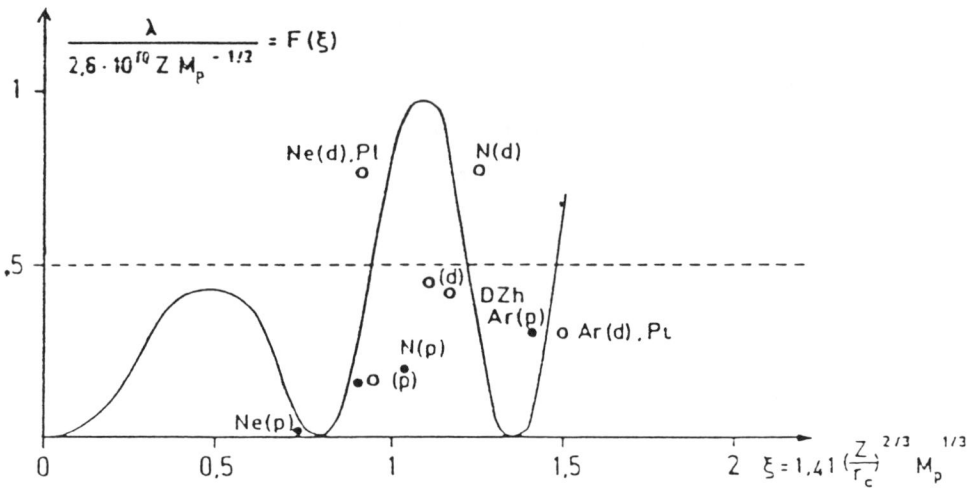

Figure 3:

5. Ephemeral Muonic Hydrogen Atoms

The interpretation of the precision measurements of the time dependence of the transfer $p\mu \to Z\mu$ [10-18] requires that the following outstanding questions should be answered:

1. Why for a number of elements (O, Ne, F) the time dependence differs from the simple exponential observed for other elements, e.g. N_2, S, Ar? (see Fig. 4,5).

2. Why is this difference observed for "short" times even in cases when these elements are found simultaneously in compounds SO_2, SF_6 (see Fig. 4a,4b)?

3. What is the origin of the build-up time observed most distinctly in the transfer of muons to fluorine and carbon nuclei[3] (see Fig. 5a)?

This seems rather intriguing because it means a certain short-time "delay" of muon transfer after the mesic atoms $p\mu$ are produced. This was the reason why we started speaking about ephemeral mesic hydrogen atoms [18]. Particularly enigmatic is the fact that the phenomenological analysis of the "fast" transfer rate revealed the existence of a part independent of the admixture concentration [18]. Very essential for understanding the nature of the phenomenon is the transfer of muons from "black" (ephemeral) mesic hydrogen atoms

---
[3]F. Mulhauser, private communication

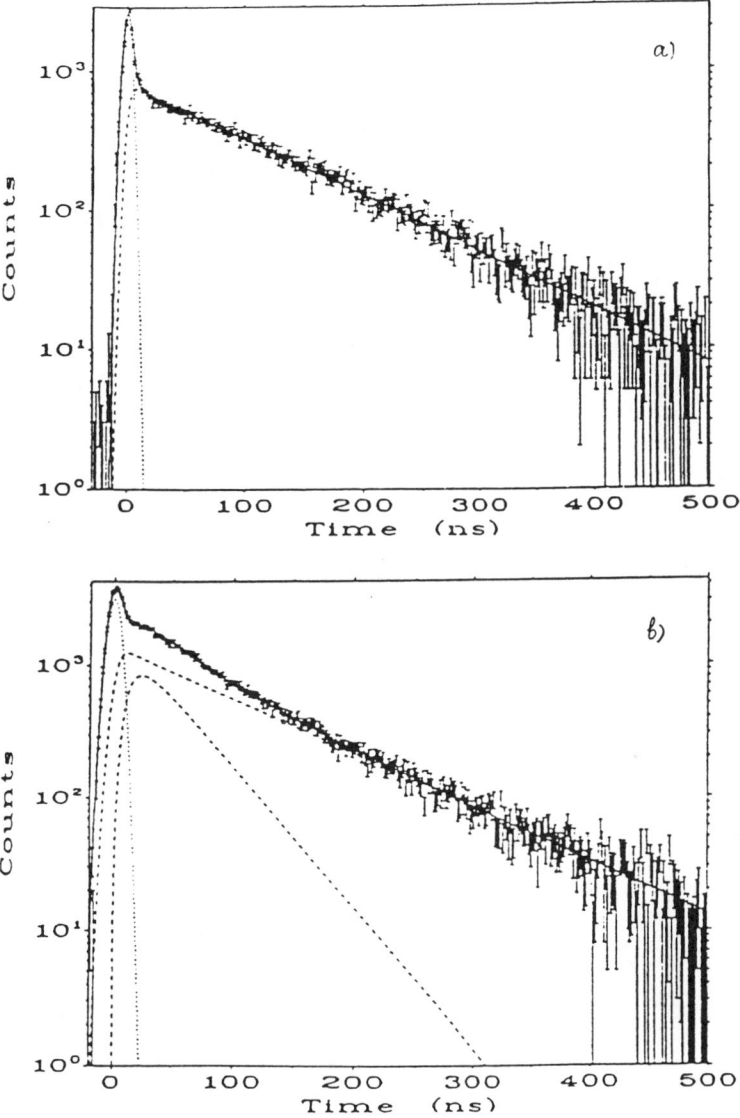

Figure 4: a) Measured time spectrum for the sulphur 2p-1s X-ray transition in an $H_2 + 0.4\% \, SO_2$ gas mixture. b) Measured time spectrum for the oxygen 2p-1s X-ray transition in an $H_2 + 0.4\% \, SO_2$ gas mixture.

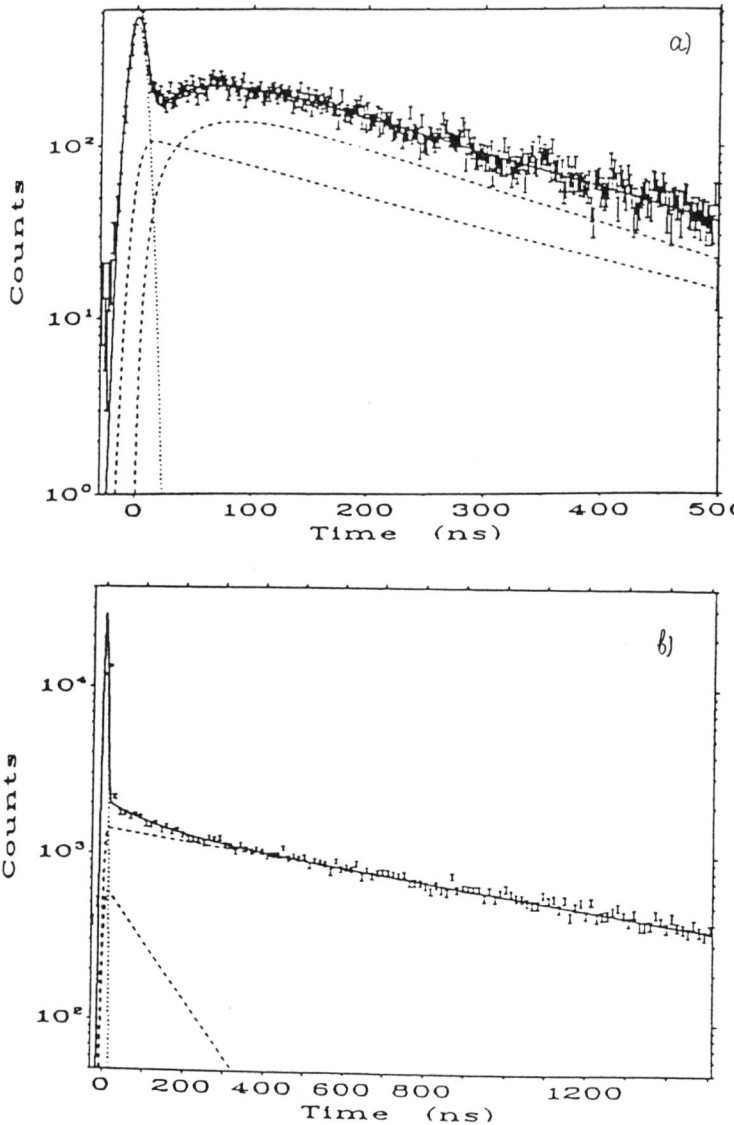

Figure 5: a) Measured time spectrum for the fluorine 2p-1s X-ray transition in an $H_2 + 0.1\% \, SF_6$ gas mixture. b) Measured time spectrum for the neon 2p-1s X-ray transition in an $H_2 + 0.69\% \, Ne$ gas mixture.

into the same states of the mesic atoms $Z\mu$ as for normal "white" mesic atoms. This is confirmed by the observed structure of the Lyman series in mesic atoms $Z\mu$ (see Figs. 2a,b). Another important point noticed in [20] and pointing to a possible interpretation of the phenomenon of ephemeral mesic atoms is the fact that the "short" times observed turn out to be close to the retardation times in mesic atoms $p\mu$ having after cascade transitions into the ground state an energy of 3-4 eV[4]. This value was obtained from direct experimental data [21,22]. Taking into account that at energies of several eV the transfer $p\mu \to Z\mu$ should take place not only from the S-wave with respect to the nuclear motion, as it was assumed in ref.[7] but also from the P-wave, whose contribution decreases in the course of retardation of atoms $p\mu$, the author of paper [20] show that the time dependence of the transfer at short periods of time differs from a simple exponential shape. This was the way he explained the effect of "black" muons during the transfer to hydrogen nuclei in compounds of SO_2. Apparently, this approach cannot explain the origin of build-up time observed in the transfers to hydrogen nuclei, though not so distinctly as in those to fluorine ones. Neither can it explain the transfers to sulphur nuclei in compounds of SO_2 having a simple exponential shape. From the theoretical viewpoint this approach is also unsatisfactory since at $p\mu$ energies of $\sim 1eV$ the cross section of the transfer $p\mu \to Z\mu$ in the S-wave depends on the velocity as

$$\sigma_{pZ} \sim \frac{1}{v^2}, \tag{13}$$

rather than as $\sigma_{pZ} \sim \frac{1}{v}$ at thermal velocities. This was noted in ref. [5]. But still, in my opinion the idea [20] to relate the emergence of ephemeral muonic atoms with retardation of mesic atoms in a medium and to the dependence of the probability for the transfer $p\mu \to Z\mu$ on the energy of the mesic atoms $p\mu$ seems quite interesting. In this connection the general expressions for the process kinematics might be helpful.

Consider the retardation of atoms $p\mu$ in hydrogen accompanied by the muon transfer to elements Z_1 and Z_2 contained in hydrogen having a low concentration of c_1 and c_2. Let $f(E,t)$ be the function of the distribution of mesic atoms $p\mu$ in energy at the time instant t, that has a specified value at the initial time instant:

$$f(E,0) = f_0(E); \quad \int f_0(E)dE = 1. \tag{14}$$

The energy lost by the mesic atoms $p\mu$ at velocities exceeding essentially the thermal velocities in a medium can be presented as

$$-\frac{dE}{dt} = \Psi(E). \tag{15}$$

At energies of $p\mu$ exceeding 1 eV the function $\Psi(E)$ is determined to a good accuracy by elastic scattering of mesic atoms on free protons [20]:

$$\Psi(E) = \frac{1}{2}v\overline{\sigma}EN, \tag{15'}$$

---

[4]The mechanisms of acceleration of mesic atoms $p\mu(d\mu)$ in cascade transfers are described in [23] and references therein.

where $\bar{\sigma}$ is the transport cross section

$$\bar{\sigma} = \int (1 - \cos\Theta) d\sigma, \tag{16}$$

and N is the number of hydrogen nuclei in cm^{-3}. In a more general case one should have considered the scattering of $p\mu$ on the molecules of H_2, the transfers between the occupation levels of the superfine structure of $p\mu$ and elastic scattering on the nuclei of admixtures.

With formula (15) taken into account, the distribution functions $f(E,t)$ obeys the kinematic equation

$$\frac{\partial f}{\partial t} = \frac{\partial (f\Psi)}{\partial E} - \lambda f, \tag{17}$$

where

$$\lambda = Nc_1\sigma_{pZ_1}v + Nc_2\sigma_{pZ_2}v + \lambda_0 + \lambda_{pp\mu}\frac{N}{N_0}, \tag{18}$$

and $\sigma_{pZ_1}(v)$, $\sigma_{pZ_2}(v)$ are the cross sections of the muon transfer from $p\mu$ to the nuclei of Z_1, Z_2, λ_0 is the probability of muon decay, $\lambda_{pp\mu}$ is the rate for the production of mesic molecule normalized for the liquid hydrogen density N_0.

Introducing the function

$$\Phi(E) = \int_{E_0}^{E} \frac{dE'}{\Psi(E')}, \tag{19}$$

and variable

$$\xi = t + \Phi(E), \tag{20}$$

we can find the general solution to the kinematic equation (17) in the form

$$f(E,t) = \frac{\phi(\xi)}{\Psi(E)} \exp\left(\int_{E_0}^{E} \frac{\lambda(E')}{\Psi(E')} dE'\right), \tag{21}$$

where $\phi(\xi)$ is an arbitrary function that can be found from the initial condition (14):

$$f(E,0) = f_0(E) = \frac{\phi(\Phi(E))}{\Psi(E)} \exp\left(\int_{E_0}^{E} \frac{\lambda(E')}{\Psi(E')} dE'\right). \tag{22}$$

From this formula it follows that

$$\phi(\xi) = f_0(\Phi^{-1}(\xi))\Psi(\Phi^{-1}(\xi)) \exp\left(-\int_{E_0}^{\Phi^{-1}(\xi)} \frac{\lambda(E')}{\Psi(E')} dE'\right), \tag{23}$$

where $\Phi^{-1}(\xi)$ is a function inverse to $\Phi(E)$:

$$\omega(E,t) = \Phi^{-1}(\xi) = \Phi^{-1}[t + \Phi(E)], \tag{24}$$

or

$$\Phi(\omega(E,t)) = \Phi(E) + t; \tag{25}$$

$$\int_E^{\omega(E,t)} \frac{d\varepsilon}{\Psi(\varepsilon)} = t; \omega(E,0) = E. \qquad (26)$$

This formula suggests that $\omega = \omega(E,t)$ has the meaning of energy E the particle had at the initial time instant t. Hence, we obtain the formula

$$f(E,t) = \frac{f_0(\omega(E,t))\Psi(\omega(E,t))}{\Psi(E)} \exp(\int_E^{\omega(E,t)} \frac{\lambda(E')}{\Psi(E')} dE'). \qquad (27)$$

Noting from formulas (19), (25) that $\frac{dE}{\Psi(E)} = \frac{d\omega}{\Psi(\omega)}$ one may express the distribution functions $f(E,t)$ with respect to the variable related to the initial energy ω:

$$f(E,t)dE = f_0(\omega)d\omega \exp(-\int_{E(\omega,t)}^{\omega} \frac{\lambda(E')}{\Psi(E')} dE'), \qquad (28)$$

where the quantity $E(\omega,t)$,

$$E(\omega,t) = \Phi^{-1}[\Phi(\omega) - t], \qquad (29)$$

$$\Phi(\omega) - \Phi(E(\omega,t)) = t, \qquad (30)$$

has the meaning of the energy E the particle had at the initial instant of time ω. Applying the distribution function $f(E,t)$ one can express the function for the distribution of X-ray quanta in mesic atoms Z_i after the transfer $p\mu \to Z_i\mu$ in the form

$$\frac{dn_\gamma^{(i)}}{dt} = Nc_i \int v\sigma_{pZ_i}(v)f(E,t)dE, \qquad (31)$$

or, with account of formula (28), as

$$\frac{dn_\gamma^{(i)}}{dt} = Nc_i \int f_0(\omega)\chi_i(E(\omega,t)) \exp(-\int_{E(\omega,t)}^{\omega} \frac{\lambda(E')}{\Psi(E')} dE')d\omega, \qquad (32)$$

where

$$\chi_i(E) = \sigma_{pZ_i}(v)v. \qquad (33)$$

In a particular case, when the initial distribution of atoms $p\mu$ in energy is described by a δ-function,

$$f_0(\omega) = \delta(\omega - E_0), \qquad (34)$$

expression (32) is reduced to the form

$$\frac{dn_\gamma^{(i)}}{dt} = Nc_i\chi_i(E(E_0,t)) \exp(-\int_{E(E_0,t)}^{E_0} \frac{\lambda(E')}{\Psi(E')} dE'). \qquad (35)$$

In the limit of low velocities, $v \to 0$,

$$\sigma_{pZ_i} \sim \frac{1}{v}, \qquad (36)$$

and the quantities $\lambda(E)$ and χ_i become constants, in which case, as it follows from expressions (14), (26), the time distribution (32) is described by a simple exponential dependence, common for all nuclei of admixtures. The same should be the case when the mesic atoms $p\mu$ reach the mean thermal velocity even if the law (36) is not fulfilled for them yet. However, in this case the rates of transfers $p\mu \to Z\mu$ should depend on the medium temperature.

Qualitatively the dependence (13) of the cross section for the transfer $p\mu \to Z\mu$ on the rate could actually provide the build-up time effect. An essential point in this case is the fact that the time-dependent factor $\chi_i(t)$ in expression (32) is independent of the mixture concentration and therefore, regarding the phenomenological treatment of time distribution [18], it could contribute to the concentration independent term in the rate of "fast" transfer.

However, it is impossible to describe qualitatively with the help of this simple model the time distribution observed experimentally. The model is applied just to point to the way of understanding the development of the build-up time effect. The realistic description of the phenomena observed, perhaps, should take into account the existence in the transfer cross section of the factor oscillating with energy (see expressions (9), (10)), varying irregularly from nucleus to nucleus and, possibly, having the minimum in the energy range of retardation of the mesic atoms $p\mu$. If the energy of $p\mu \to Z\mu$ is several eV one should bear in mind that the transfer $p\mu \to Z\mu$ can take place from several partial waves with respect to nuclear motion. This means that one should have a more comprehensive understanding of the kinetics of retardation of the mesic atoms $p\mu$ in a medium. Difficult for the interpretation of the experimental data is the question why the time dependence of the muon transfer to sulphur nuclei in a mixture of $H_2 + SO_2$ is a simple exponent though despite a possible difference of the preexponential factors in expression (32) for sulphur and oxygen the muon transfer to these nuclei has a common exponential dependence. In this connection the transfers $p\mu \to Z\mu$ to sulphur nuclei observed even for very short instants of time, the so-called prompt transfers, deserve attention. The evidence of such transfers is a noticeable variation of the Lyman series structure in a mixture of $H_2 + SO_2$ as compared with the one observed in pure SO_2 medium. This is shown by triangles in fig. 2b. For oxygen this effect manifests itself much weaker (see fig. 2a).

Conclusions

1. The precision measurements of the transfer process $p\mu(d\mu) \to Z\mu$ made at PSI revealed some new phenomena whose interpretation allowed a deeper insight into the details of the transfer mechanism.

2. The observed structure of the K-series in the mesic atoms $Z\mu$ after muon transfer from $p\mu$ is in conformity with the prediction that the transfers take place mainly on the terms of the quasicrossing with the hydrogen term $p\mu(d\mu)$.

3. The nonmonotonous Z-dependence and the difference in the rates of the transfers

$p\mu \to Z$, $d\mu \to Z$ can apparently find their natural explanation by quantum oscillations.

4. It may well be that the effects of "ephemeral" muons discovered experimentally are related to retardation of hydrogen mesic atoms in a medium and the deviation of the cross section for the transfer $p\mu \to Z\mu$ from the simple dependence $\sigma \sim \frac{1}{v}$, that must definitely take place in the retardation energy range. The structure of the Lyman series for "fast" transfers and the fact that the time of fast transfers coincides by order of magnitude with that of retardation of the mesic atoms $p\mu$ and depends on the medium density favour this hypothesis. However, this hypothesis encounters certain difficulties in explaining the data on the muon transfer to different nuclei of Z_1, Z_2 of the same admixtures.

5. It is necessary to accomplish the programme of accurate quantum and mechanical calculations of the cross sections for the transfers of muons from the ground and excited states of mesic hydrogen atoms to the nuclei of other elements, simultaneously taking into account long-ranging forces, electron screening and other factors, and to determine the dependence of the cross sections on the collision energy.

Acknowledgements

I find it a pleasant duty to thank my noteworthy colleagues F. Mulhauser, L.A. Schaller, L. Schellenberg and H. Schneuwly for helpful friendly discussions and valuable remarks facilitating my better understanding of the problem. I am also grateful to the Directorate of PSI and the Institute de Physique de l'Université de Fribourg for their hospitality.

References

[1] P.M. Morse, E.C. Stueckelberg, Phys. Rev. **33**, 932 (1929).
[2] S.S. Gershtein, V.D. Krivchenkov, Zh. Exsp. Teor. Fiz. **40**, 491 (1961).
[3] L.I. Ponomarev, T.P. Puzynina, Zh. Exsp. Teor. Fiz. **52**, 1273 (1967);
     V.I. Komarov, L.I. Ponomarev, S.Yu. Slavyanov, in the book "Spheroidal and Coulomb Spheroidal Functions" (in Russian), "Nauka" Publishers, Moscow, 1976.
[4] Yu.A. Aristov, A.V. Kravtsov, N.P. Popov, G.E. Solyakin, N.F. Truskova and M.P. Faifman, Yad. Fiz. **33**, 1066 (1981); Sov. J. Nucl. Phys. **33**, 564 (1981).
[5] S.S. Gershtein, Zh. Exp. Teor. Fiz. **43**, 706 (1962); Sov. Phys. JETP 16, 50 (1963).
[6] S.P.Alliluev, A.V. Matveenko. Zh. Exsp. Teor. Fiz. **51**, 1873 (1966).
[7] E.A. Soloviev, Usp. Fiz. Nauk. **147**, 437(1989), and refs. therein.
[8] G. Fiorentini, G. Torelli, Nuovo Cimento, A **36**, 317 (1976).
[9] G. Holzwarth and H.-J. Pfeifer, Zh. Phys. A **272**, 311 (1976);
     V.R. Akulas and P. Vogel, Comp. Phys. Comm. **15**, 291 (1978).
[10] H. Schneuwly, R.Jacot-Guillarmod, F. Mulhauser, P. Oberson, C. Piller and L. Schellenberg, Phys. Lett. A **132**, 335 (1988).

[11] H. Schneuwly, Muon-Catalyzed Fusion, **4**, 87 (1989).

[12] H. Schneuwly, in the book "Electromagnetic Cascade and Chemistry of Exotic Atoms", eds. L.M. Simons, D. Horwarth and G. Torelli (Ettore Majorana Int. Sc. Series), p. 205.

[13] E. Mulhauser, R. Jacot-Guillarmod, C. Piller, L. Schellenberg and Schneuwly, ibid, p. 217.

[14] R. Jacot-Guilalrmod, F. Mulhauser, C. Piller, L.A. Schaller, L. Schellenberg and H. Schneuwly, ibid, p. 223.

[15] E. Mulhauser, R. Jacot-Guillarmod, C. Piller, L.A. Schaller, L. Schellenberg and Schneuwly, Muon-Catalyzed Fusion **5/6**, 101 (1190/91).

[16] F. Mulhauser, H. Schneuwly, R. Jacot-Guillarmod, C. Piller, L.A. Schaller and L.Schellenberg, Muon-Catalyzed Fusion, **4**, 365 (1989).

[17] R. Jacot-Guillarmod, F. Mulhauser, C. Piller and H. Schneuwly, Phys. Rev. Lett. **65**, 709 (1990).

[18] H. Schneuwly and F. Mulhauser, Phys. Lett. A, **160**, 71 (1991).

[19] Yu. S. Sayasov, Helv. Phys. Acta **63**, 517 (1990).

[20] Yu.S. Sayasov, Preprint "Moderation of Mesic Atoms $p\mu$ in Hydrogen-Containing Gas Mixtures and its Influence on the Kinetics of Muon Transfer Reactions", Phys. Dep., Univ. of Fribourg.

[21] J.B. Kraiman et al. Proc. Int. Symp. on Muon-Catalyzed Fusion μCF-89, ed. J.D. Davis, 47 (1990).

[22] J.B. Kraiman, G. Chen et al, Phys. Rev. Lett. **63**, 1942 (1989).

[23] V.E. Markushin, in the book "Electromagnetic Cascade and Chemistry of Exotic Atoms", P.73.

Muon Transfer with a New High Pressure Gastarget

L.SCHELLENBERG, P.BAERISWYL, R.JACOT-GUILLARMOD, B.MISCHLER,
F.MULHAUSER, C.PILLER and L.A.SCHALLER

Institut de Physique de l'Université,Pérolles
CH1700 Fribourg, Switzerland

Abstract:: The present situation of muon transfer to helium is summarized. A new gas target for a maximal pressure of 40 b with four thin beryllium windows for the detection of low energy photons is described. First results of an analysis of the time spectra of muonic carbon X-rays in a H_2+CH_4 gas mixture taken with the new target show a pressure dependent structure.

1. Motivation

In recent years the study of the muon transfer from muonic hydrogen atoms to heavier elements has met with renewed interest. New experimental data show a much more complex structure than expected from earlier theoretical calculations in the framework of the quasiclassical approximation of the Coulomb three body interactions. The status of the present theoretical situation is given in another contribution to this conference [1]. The transfer to helium is of special interest for the muon catalyzed fusion cycle and has therefore been studied under very different experimental conditions in gaseous and liquid hydrogen and helium mixtures.

Muonic hydrogen atoms are formed in liquid and gaseous targets in high quantum states $10 \leq n \leq 12$. They deexcite to the 1s ground state within a time $< 10^{-10}$ s, depending on the hydrogen density of the target. Transfer from excited states must have rates comparable to the deexcitation rate for a pure hydrogen target in order to compete. This is the case in H_2-Z mixtures with comparable concentrations. The quoted normalized transfer rates Λ_Z refer to the transfer from the μp_{1s} ground state. They are generally obtained by measuring the time distribution of one of the particles or photons of the different decay channels of the μp_{1s} state. In a target of hydrogen with one or more admixed elements or a molecule the main processes contributing to the disappearance rate λ are free muon decay (λ_0), $p\mu p$ formation (λ_{pp}), transfer to deuterium (λ_d) and to the admixed elements or molecules ($\lambda_{Z1}, \lambda_{Z2}$):

$$\lambda = \lambda_0 + \lambda_{pp} + \lambda_d + \lambda_{Z1} + \lambda_{Z2}$$

Any contamination of the gas or transfer to the walls of the target vessel also contribute to the total rate and must be minimized. In our case the muonic X-rays emitted after the muon

Table 1. Experimental and theoretical transfer rates from muonic hydrogen isotopes to ^{4}He.

Isotope	ref.	pressure (bars)	C(He/H2) (%)	Λ(exp) (10^8 s^{-1})	Λ(theor.) (10^8 s^{-1})
p,^{4}He	(8)	16-34	32-136	0.36 (.10)	
p,^{4}He	(9,17)	15.0	34-50	0.88 (.09)	0.32
p,^{4}He	(10)	5.0	100	0.032 (.013)	
d,^{4}He	(11)	6.6	10	<0.2	
d,^{4}He	(12,17)	88.2	3-6	3.68 (.18)	3.22
d,^{4}He	(13,17)	1350.0	0.1-2	2.75 (.22)	2.96
d,^{4}He	(14,17)	liquid	0.04	13.1 (1.2)	11.8
d,^{4}He	(14,17)	liquid	0.04	6.7 (1.2)	11.8

transfers to one of the admixed elements Z is measured. Their time distribution should show a single exponential structure with the total disappearance rate λ as the characteristic constant. The other rates being known, one can calculate from λ the transfer rate λ_{Z1} or for molecules (λ_{mZ1nZ2}), as well as the corresponding normalized rates Λ_Z.

The reduced transfer rates Λ_Z are of the order of $10^{11} - 10^{12}$ s^{-1}. In contrast to the expected monotonous Z dependence the transfer rates vary considerably between different elements [2]. An especially striking example is the transfer μp to Ne with a rate as low as Λ_{pNe}=0.082 10^{11} s^{-1} [3,4] compared to Λ_{pN}=0.34 10^{11} s^{-1} for transfer to nitrogen [5].

The transfer to helium is about three orders of magnitude smaller. In the helium case direct transfer from the μp1s ground state is strongly suppressed, due to the absence of level crossing [6]. The muon transfer proceeds predominantly via the formation of a muonic molecular ion with the emission of a molecular photon [7]:

$$\mu p_{1s} + \text{He} \rightarrow (p\mu\text{He})^* \rightarrow p + (\mu\text{He})_{1s} + h\nu \ (6.9 \text{ keV})$$

In view of the importance of this transfer rate in the muon catalyzed fusion cycle, a number of experiments employing different methods have been performed for μp and μd transfer to helium, as listed in table 1. The reduced transfer rates given in table 1 have been obtained using different methods, namely the measurement of the life time of the muonic X-rays, of the molecular photon transition hν, of neutrons from the fusion reactions or of the absolute yield of the molecular photon transition in liquid and gaseous targets. The agreement between different experiments is generally not very good. Especially striking is the difference of an order of magnitude between the life time and yield measurements for the transfer from μp to ^{4}He, for which no explanation could be found.

In a recent experiment at KEK [15,16], a search was done for the molecular (dμ^4He) and (dμ^3He) photon transitions using a liquid deuterium target with slight admixtures of ^{4}He and ^{3}He A suppression by a factor four to five of the molecular photon emission was found in the (dμ^3He) system compared with the (dμ^4He) molecule.

With the same experimental set up and a liquid protium target with a small admixture of ^{4}He no molecular photon transition has been found. The photon intensity is estimated to be at least five times lower than calculated for the (pμ^4He) molecule [7].

The surprising differences have been explained by the presence of a hitherto unconsidered decay channel. Kamimuro [18] calculated the rate of the particle decay of the muonic (dμHe) molecules, where the decay energy is directly transmitted to the kinetic energy of the emitted particles, namely d and μHe. The particle decay rate λ_d or λ_p depend strongly on the reduced mass of the system [19]. For instance, λ_{dd} is three times larger than the radiative decay rate in (dμ^3He), but almost the same in (dμ^4He). For the case of (pμ^4He), one gets for the radiative decay a branching ratio $\lambda_\gamma/(\lambda_\gamma+\lambda_p)$ of only 0.13 [19]. This makes the observation of the 6.9 keV photon extremely difficult and probably explains the apparent small transfer rate deduced in ref.[10].

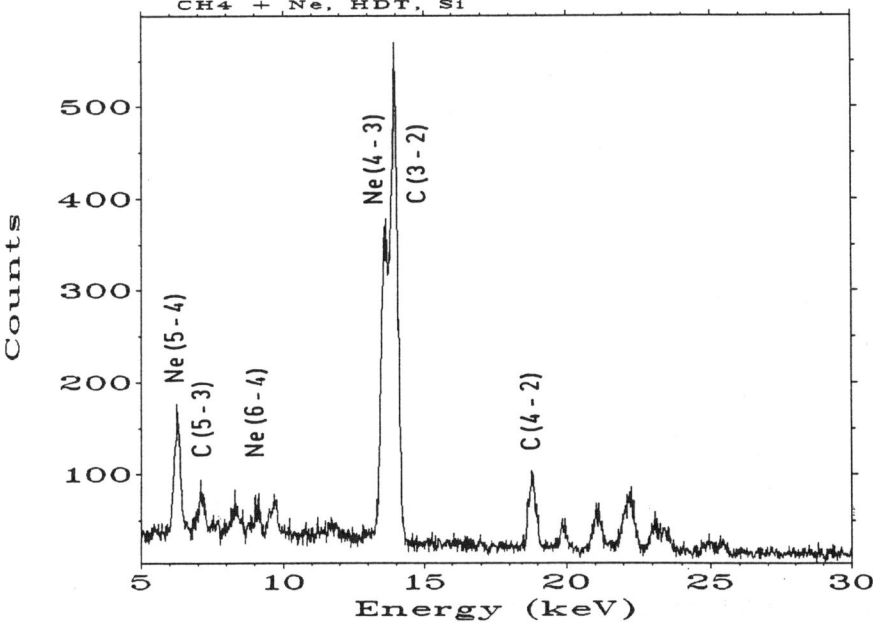

Fig. 1. Photon energy spectrum of a (60% CH$_4$ +40% Ne)gas mixture at 1.7 b pressure, taken wit a 1 cm3 silicon detector. The (5-4) muonic neon X-ray transition at 6.3 keV is clearly seen.

2. Gas target

For the determination of the transfer rates to helium and other light atoms with low energy X-rays (< 10 keV), a special gas target system is necessary. Such a target system for a pressure up to 40 b has been constructed. The cylindrical stainless steel target has a length and a diameter of 20 cm. The inside is silver-plated to minimize the life time of muonic X-rays from the walls.

The target is mounted inside a cylindrical vessel, which can be evacuated. In addition, the target is surrounded by tubes for liquid nitrogen circulation and electrical heating wires. This allows a temperature range from liquid nitrogen temperatures up to 200° C. A standard turbomolecular pumping system assures a low impurity level before filling the target with previously prepared gas mixtures. The muon beam enters through a thin tantalum window of 40 mm diameter sustained by a grid. Two beryllium windows 20 mm diameter each, sustained by stainless steel grids, on each side of the target, allow the detection of the low energy photons at different positions with silicon or other appropriate detectors. As an example, fig.1 shows a muonic X-ray spectrum of a Ne and CH_4 gas mixture taken with a 1 cm^3 silicon detector and using a 0.5 mm thick beryllium window. The (5 - 4) µNe X-ray transition at 6.3 keV is clearly seen.

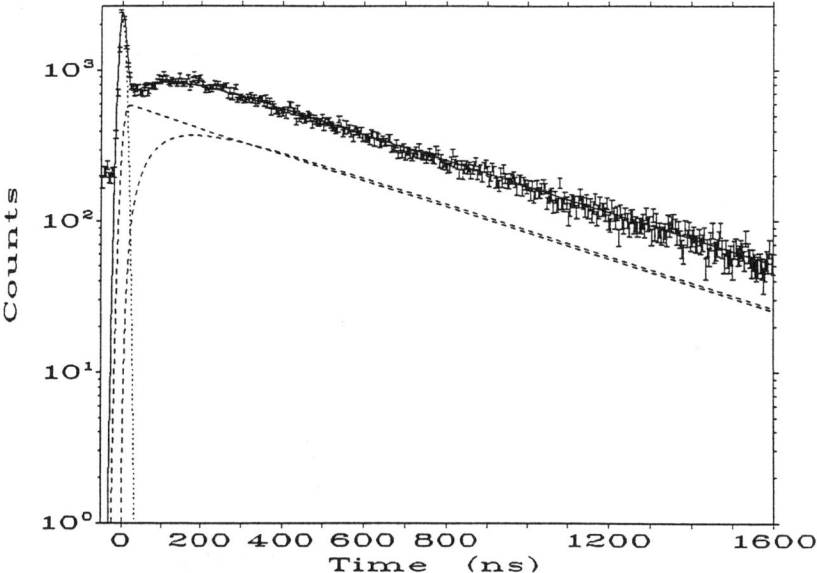

Fig. 2. Time distribution of the muonic carbon (2p-1s) X-ray transition in a H_2+0.17% CH_4 gas mixture at a total pressure of 15 b. The two dashed lines with equal decay times but one in addition with an 80 ns rise time make up the delayed spectrum.

3. Muon transfer to CH_4

The new target was used in a measurement of the muon transfer from µp to CH_4 in a (H_2+0.17% CH_4) gas mixture. This experiment was planed in view of a future triple gas measurement (H_2+He+CH_4) for the simultaneous determination of the transfer rate to helium from the molecular X-ray and the life time of the muonic carbon X-rays. For the triple gas method, the reduced rate for carbon has to be known beforehand. The time spectra of the Balmer and Lyman series in muonic carbon have been measured at three different pressures of

10, 15 and 40 b. The time spectra show a pronounced rise time followed by an apparent single exponential decay as can be seen in fig.2 for the 15 b case. From a preliminary analysis, which depends to some extent on the fitting philosophy used, an inverse pressure dependence of the rise time is observed (see full circles in fig.3). In a triple gas mixture (H_2+50%He+ 0.085% CH_4) no such pressure dependence is observed (see full diamonds in fig.4). The rise time at 15 b is similar to the one at 40 b. This may point to an effect of µp thermalisation, since scattering cross sections for µp on H_2 and He are quite different.

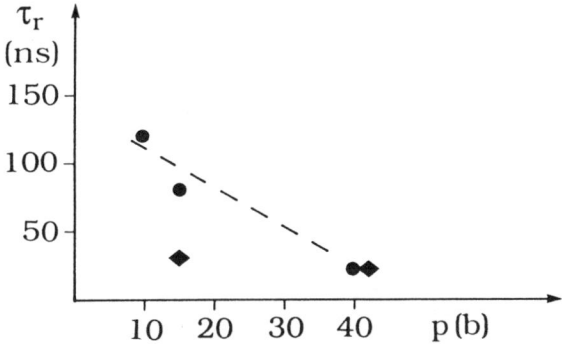

Fig. 3. The full circles show the pressure dependence of the rise time of the muonic carbon X-ray transitions for a gas mixture of (H_2+0.17% CH_4). The full diamonds represent the values obtained in a (H_2+ 50% He+ 0.085% CH_4) triple gas mixture.

A similar pressure dependence of the rise time has been observed in the oxygen X-ray spectra of SO_2, namely 8 ns at 15 b and 12 ns at 10 b [20]. For SF_6, the corresponding rise time in the fluorine spectra was 40 ns at 15 b. Therefore, the rise time seems to depend not only on the thermalization time but also on the specific element.

As already mentioned, the delayed part of the carbon time spectra show approximately 200 ns after the muon stop a single exponential structure. A one exponential fit to this part of the spectrum only gives very consistent results for the reduced transfer rate obtained with the three different pressures. The deviations from the mean value $\Lambda_C = 0.95 \cdot 10^{11}$ s^{-1} are less then five percent.

References

[1] S.S.Gershtein, contribution to this volume
[2] L.Schellenberg, Muon Cat. Fusion **5** (1989) 73
[3] P.Baeriswyl, R.Jacot-Guillarmod, M.Mallinger, B.Mischler, F.Mulhauser, C.Piller, W.Reichart, B.Riedo, L.A.Schaller, L.Schellenberg, H.Schneuwly, A.Werthmüller, PSI Nuclear and Particle Physics Newsletter (1991) 57
[4] R.Jacot-Guillarmod, F.Mulhauser, C.Piller, and H.Schneuwly, Phys. Rev. Lett. **65** (1990) 709.

[5]   R.Jacot-Guillarmod, F.Mulhauser, C.Piller, L.A.Schaller, L.Schellenberg, H.Schneuwly, in E.M. Cascade and Chemistry of exotic atoms, edited by L.M.Simons, D.Horvat and G.Torelli (Plenum Press 1990) 223
[6]   S.S.Gershtein and L.I.Ponomarev, Muon Physics III, eds. V.W.Hughes and C.S.Wu, Academic press, New York (1975) p. 41 and references cited therein.
[7]   Y.A.Aristov, A.V.Kravtsov, N.P.Popov, G.E.Solyakin, N.F. Truskova and M.P. Faifmann, [Sov. J. Nucl. Phys. 33 (1981) 564] Yad. Fiz. **33** (1981) 1066.
[8]   V.M Bystritsky, V.P Dzhelepov, V.I.Petrukhin et al. Zh. Eksp. Teor. **84** (1983) 7257 [JETP **57** (1983) 728]
[9]   R.Jacot-Guillarmod, F. Bienz, M. Boschung, C. Piller, L.A.Schaller, L.Schellenberg, H.Schneuwly, W.Reichart and G.Torelli, Phys. Rev. **A38** (1988) 6151.
[10] H.P.von Arb, F.Dittus, H.Hofer, F.Kottmann and R.Schaeren, Muon catalyzed Fusion **4** (1989) 61.
[11] A.Bertin, M.Bruschi, M.Capponi et al., Muon catalyzed fusion, AIP Conf. Proc. **181**(1989) 161.
[12] D.B.Balin, A.A.Vorobyov, An.A.Vorobyov et al., Pisma Zh. Eksp.Teor.Fiz. **42** (85) 236 [JETP Lett.42(85)293(1985)236 [JETP Lett. **42** (1985) 293].
[13] V.M. Bystritsky, V.P. Dzhelepov, V.G. Zinov et al., preprint μCF90
[14] Y.Watanabe, S.Sakamoto, K.Ishida, T.Matsuzaki,, P.Strasser, M.Iwasaki, K.Nagamine, Muon Cat.Fusion **5** (1990) 93
[15] K.Ishida, S.Sakamoto, Y.Watanabe, T.Matsuzaki, K.Nagamine, *in Int.Workshop on Muon Catalyzed Fusion, μCF-92*, June 92 Uppsala, to be published in Muon Cat. Fusion
[16] K. Nagamine in Proc. 3rd. Inter. Conf. on Particle and Nucl. Physics Intersections, American Institute of physics 1992
[17] V.K.Ivanov, A.V.Kravtsov, A.I.Mikhailov, N.P. Popov and V.I.Fomichev, Zh. Eksp. Teor. Fiz. **91** (1986) 358 [JETP 64 (1986) 210].
[18] Y. Kino and M.Kamimura, submitted to Phys.Rev A
[19] S.S.Gershtein, in μCF-92, Uppsala 92, to be published in Muon Cat. Fusion
[20] H.Schneuwly and F.Mulhauser, Phys. Lett. **A160** (1991) 71

On the Time Spectra of Muonic X-Rays in H_2+SO_2

F. MULHAUSER, H. SCHNEUWLY, R. JACOT-GUILLARMOD,
C. PILLER, L. A. SCHALLER and L. SCHELLENBERG
Institut de Physique de l'Université de Fribourg
Pérolles, CH-1700 Fribourg, Switzerland

Abstract. By analysing the time spectra of the muonic sulphur X-rays in seven different H_2+SO_2 gas mixtures, the muon transfer rates to the SO_2 molecule, deduced from the lifetimes of the μp atoms, agree all well with each other. The muonic oxygen time spectra show an additional structure as if μp atoms of another kind were present. Reduced transfer rates Λ'_O are reproducible if one uses the model of "ephemeral" μp atoms[1].

1. Introduction

In a gas mixture of natural hydrogen with a small amount of Z gas, the muonic hydrogen atom in its ground state disappears by muon decay, with an associated rate λ_0, by formation of a pμp molecule (λ_{pp}) or by transferring its muon either to deuterium (λ_d) or to the Z element (λ_Z). The total disappearance rate, λ, of the μp atom in the ground state is then [2] :

$$\lambda = \lambda_0 + \lambda_{pp} + \lambda_d + \lambda_Z \tag{1}$$

$1/\lambda = \tau$ is then the lifetime of the μp atom in the ground state under the corresponding experimental conditions of pressure, temperature and concentration of element Z. The muonic X-rays, emitted after the muon transfer to the Z atom, are delayed relative to a muon stopping in the gas, and their time distribution is expected to be an exponential function with the disappearance rate λ as characteristic time constant.

$$\frac{dN_{\gamma Z}(t)}{dt} \propto e^{-\lambda t} \tag{2}$$

By analyzing the delayed part of the time spectra of the muonic X-rays of the Z atom, one determines λ. The rates λ_0, λ_d, λ_{pp} beeing known, one can calculate using equation (1), the transfer rate to Z, λ_Z, which depends on the experimental conditions. In order to compare measurements performed under different experimental conditions, one normalizes the rate to the atomic density of liquid hydrogen, ρ_0. The normalized transfer rate Λ_Z, is then :

$$\Lambda_Z = \frac{\rho_0}{\rho_Z}\lambda_Z = \frac{\rho_0}{\rho_Z}(\lambda - \lambda_0 - \lambda_{pp} - \lambda_d) \qquad (3)$$

The μZ atom formed this way in an excited state deexcites immediately by emitting characteristic muonic X-rays. The initial (n,l) population in the μZ atom, after muon transfer from the μp atom in the ground state, can be predicted [3] and the muonic X-ray intensity pattern is then characteristic for this charge transfer process [4,5,6].

2. Transfer to sulphur dioxide

In experiments, where muon transfer to sulphur dioxide was investigated, unexpected time distributions have been observed [6,7,8]. The time spectrum of the delayed muonic sulphur $2p-1s$ transition (Fig. 1a) and those of the other sulphur Lyman and Balmer series transitions have, as expected, the shape of a single exponential function (eq. 2). The fitted time constants, called τ_S, are compatible for all these transitions. However, the time spectra of all four observed muonic oxygen X-ray Lyman series transitions, $2p-1s$ to $5p-1s$, have a more complex structure (Fig. 1b), which could be reproduced by using a function with three exponentials (for details see ref [6]) :

$$\frac{dN_{\gamma O}(t)}{dt} = Ae^{-\lambda t} + \frac{C}{\tau_2 - \tau_r}(e^{-\frac{t}{\tau_2}} - e^{-\frac{t}{\tau_r}}) \qquad (4)$$

The time τ_r corresponds to a rise time, used to fit the bump (cf. Fig. 1b). The time $\tau_1 = 1/\lambda$, the longer of the two time constants, is compatible with the one obtained from the sulphur X-ray transitions, τ_S. These times, τ_1 and τ_S, are interpreted as the lifetime of the μp atoms in the ground state under the given experimental conditions. The time τ_2 corresponds to a second decay time.

The relative muonic X-ray intensities of the Lyman series, resulting from muons transferred from μp atoms in the ground state to oxygen or to sulphur, can be predicted [2,3]. The intensity patterns of the delayed muonic sulphur X-rays of the Lyman and Balmer series agree well with the estimates [6,7]. The same is true for the oxygen Lyman series, for the normal time component as well as for the unexpected one [6]. It is important to notice that the relative intensities of the unexpected time component of the oxygen transitions have the same structure as they would have if they resulted from muon transfer from μp atoms in the ground state.

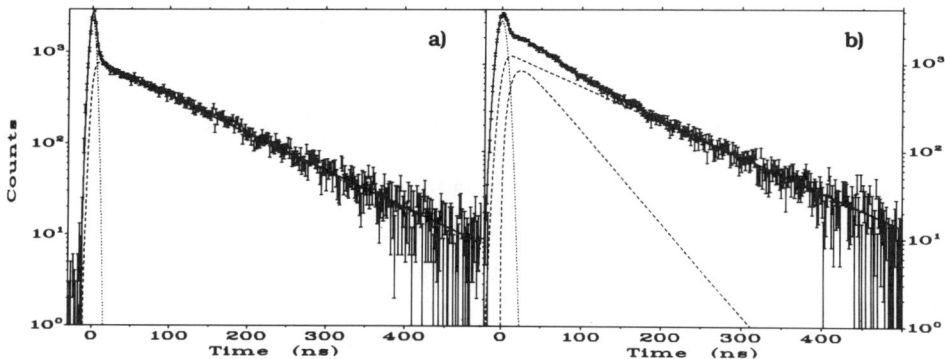

Figure 1: a) Measured time spectrum for the muonic sulphur $2p-1s$ X-ray transition in a H_2 + 0.4% SO_2 gas mixture. The delayed part is fitted with function (2). b) Measured time spectrum for the muonic oxygen $2p-1s$ X-ray transition in the same gas mixture. The delayed part is fitted with function (4).

The experiments of muon transfer to SO_2 have been performed several times, under different experimental conditions. We used different pressures, between 10 and 15 bar, and different concentrations of sulphur dioxide, from 0.1% to 0.6%. In every measurement, the time distribution of the muonic X-rays had the same structure : a single exponential structure for the sulphur X-rays and a triple exponential structure for the oxygen ones. The results of these experiments are given in the Table 1.

The transfer rate to sulphur dioxide, Λ_{SO_2}, is calculated from the lifetime $\tau_1 = \tau_S$ (using equation (3) with $\lambda = 1/\tau_S$), which is common to the time distributions of both the sulphur and the oxygen X-rays (see Fig. 1a and 1b). From Table 1, one observes that all seven measurements, performed under various conditions of total pressure and sulphur dioxide concentrations, yield the same reduced transfer rate Λ_{SO_2}. The reproducibility of the reduced transfer rate shows that the pressure and concentration dependences in the present limited ranges are well understood and that the background problems, like muon transfer to the target walls or nitrogen, oxygen or other gas impurities contained in the target vessel remained at a negligible level.

The rise time, τ_r, which appears only in the oxygen time spectra, seems to depend only on the hydrogen pressure. It might be inversely proportional to the total pressure of the mixture, but independent on the SO_2 concentration. In Table 1, τ_r is given without uncertainties, because the number of parameters, sensitive to the same small fit region, is too high to determine accurate errors.

Table 1: Time constants and reduced transfer rates Λ_{SO_2} and Λ_O^* for H_2+SO_2 gas mixtures.

Conc. SO_2[%]	Pres. [bar]	$\tau_S = \tau_1$ [ns]	Λ_{SO_2} [$\cdot 10^{11} s^{-1}$]	τ_r [ns]	τ_2 [ns]	Λ_O^* [$\cdot 10^{11} s^{-1}$]
0.1	15	356.3(100)	2.56(11)	8	79.5(129)	6.82(139)
0.2	15	197.0(42)	2.54(9)	8	61.1(40)	4.44(37)
0.2	10	270.4(61)	2.60(10)	12	92.1(48)	4.22(30)
0.2	15	198.4(40)	2.55(9)	8	60.1(41)	4.56(41)
0.4	15	107.0(19)	2.51(7)	8	44.5(11)	3.12(10)
0.6	15	68.6(20)	2.65(11)	8	31.7(16)	2.93(15)
0.6	10	102.6(36)	2.56(12)	12	44.0(22)	3.08(15)

The time τ_2, which appears only in the oxygen time spectra, is considered as the lifetime of an unexpected μp atom, which transfers its muon only to oxygen (Fig. 2a). These unexpected μp atoms, called "black" μp atoms [9], give, after transfer of the muon to oxygen, an intensity structure which is not different from the one we obtained from μp atoms in the ground state. From this time constant τ_2, by using equation (3) with $\lambda = 1/\tau_2$, we obtained reduced transfer rates only to oxygen, which we called Λ_O^*. One observes, from Table 1, that these rates Λ_O^* are only partially reproduced, contrary to Λ_{SO_2}. The rates Λ_O^* are the same only for mixtures with the sames SO_2 concentrations, independent of the total pressure. If there was only one kind of "black" μp atoms (Fig. 2a), these Λ_O^* should have been independent of the experimental conditions. This nonreproducibility of Λ_O^* leads us to test another model for the unexpected part of the time distribution of the oxygen X-rays.

The model of "ephemeral" μp atoms [1] is schematically displayed in Fig. 2 b. μInt is still an intermediate state, which disappears with a time τ_r (Fig. 1b). The former "black" μp atom disappears now by two different ways, the first is a transfer to oxygen, dependent on its density ρ_O, with a rate λ'_O and the second, λ_ρ, is dependent only on the hydrogen density, ρ_H. With "ephemeral" μp atoms, equation (1) becomes, for the unexpected time τ_2:

$$\frac{1}{\tau_2} = \lambda_0 + \lambda_{pp} + \lambda_d + \lambda'_O + \lambda_\rho \quad with \quad \Lambda'_O = \frac{\rho_0}{\rho_O}\lambda'_O \quad and \quad \Lambda_\rho = \frac{\rho_0}{\rho_H}\lambda_\rho \quad (5)$$

From the measurements with SO_2 concentrations between 0.2% to 0.6%, we calculated for these two reduced transfer rates mean values [1]:

$$\Lambda'_O = 2.3 \cdot 10^{11} s^{-1} \quad and \quad \Lambda_\rho = 4.5 \cdot 10^8 s^{-1} \tag{6}$$

This non trivial decomposition of the disappearance rate $(1/\tau_2)$ gives us a reduced transfer rate to oxygen Λ'_O which is now independent of the experimental conditions, in particular, independent of the SO_2 concentration. If the rate Λ_ρ were related to a deceleration process, a characteristic time of about 2 ns should appear in liquid hydrogen and of about 200 ns in hydrogen gas at 10 atm. Although there is no obvious explanation of the rates Λ'_O and Λ_ρ, the hypothesis of "ephemeral" μp atoms allows to convert non reproductible rates Λ^*_O to reproducible ones [1], or to avoid in $H_2 + SO_2$ gas mixtures the introduction of "coloured" μp atoms.

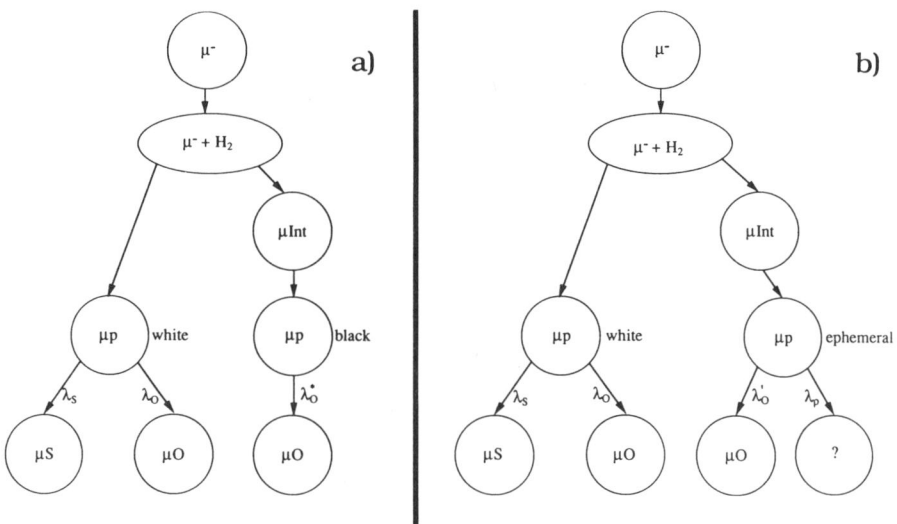

Figure 2: a) Model of "black" and "white" muonic atoms [9]. b) Model of "ephemeral" muonic atoms [1]

The results of the measurement with 0.1% SO_2 can be used to test the hypothesis of "ephemeral" μp atoms. The time spectra of the muonic oxygen X-rays of the $H_2 + 0.1\%$ SO_2 mixture were, indeed, not yet evaluated when the "ephemeral" μp atoms were proposed [1]. Using the $H_2 + SO_2$ mixtures with SO_2 concentration from 0.2% to 0.6%, the value of the time constant τ_2 of the unexpected component of the muonic oxygen time spectra in $H_2 + 0.1\%$ SO_2 should be about $\tau_2 = 81$ ns. As can be seen from Table 1, this time corresponds nicely to the measured time.

Although the model of "ephemeral" μp atoms seems successful in the case of H_2 + SO_2 gas mixtures, it does not solve the problem of the time structures of the muonic X-rays observed in these mixtures nor in others like $H_2 + SF_6$, $H_2 + CO_2$, $H_2 + CH_4$ and H_2 + Ne [10].

References

[1] H. Schneuwly and F. Mulhauser, Phys. Lett. A **160**, 71 (1991).

[2] S. S. Gershtein, Zh. Eksp. Teor. Fiz. **43**, 706 (1962). (Sov. Phys. JETP **16**, 501 (1963)).

[3] G. Holzwarth and H. J. Pfeiffer, Z. Phys. A **272**, 311 (1975).

[4] R. Jacot-Guillarmod, F. Bienz, M. Boschung, C. Piller, L. A. Schaller, L. Schellenberg, H. Schneuwly, W. Reichart and G. Torelli, Phys. Rev. A **38**, 6151 (1988).

[5] R. Jacot-Guillarmod, F. Bienz, M. Boschung, C. Piller, L. A. Schaller, L. Schellenberg, H. Schneuwly and D. Siradovic, Phys. Rev. A **37**, 3795 (1988).

[6] F. Mulhauser, H. Schneuwly, R. Jacot-Guillarmod, C. Piller, L. A. Schaller and L. Schellenberg, Muon Catalyzed Fusion **4**, 365 (1989).

[7] F. Mulhauser, R. Jacot-Guillarmod, C. Piller, L. A. Schaller, L. Schellenberg and H. Schneuwly, in *E. M. Cascade and Chemistry of Exotic Atoms*, edited by L. M. Simons, D. Horvat and G. Torelli (Plenum Press, 1990) 217.

[8] H. Schneuwly, R. Jacot-Guillarmod, F. Mulhauser, P. Oberson, C. Piller and L. Schellenberg, Phys. Lett. A **132**, 335 (1988).

[9] H. Schneuwly, Muon Catalyzed Fusion **4**, 87 (1989).

[10] R. Jacot-Guillarmod, F. Mulhauser, C. Piller and H. Schneuwly, Phys. Rev. Lett. **65**, 706 (1990).

Muon Transfer from Excited Muonic Hydrogen to Helium Nuclei

A.V. KRAVTSOV
Petersburg Nuclear Physics Institute
188350 Gatchina Leningrad district
Russia

Abstract. Muon transfer rates from excited muonic hydrogen to helium reach values around 10^{12}/s and can compete with deexcitation processes.

1. Introduction

During the last years muon catalyzed fusion in the most effective d-t mixture is under intensive investigation[1]. Since hydrogen mixtures can contain impurities, it is very important to study muon transfer from muonic hydrogen to other nuclei, because this process should lead to the decrease of the number of cycles of catalysis per muon[2]. For muon catalysis, transfer to helium nuclei is of special interest, because helium is accumulated in the d-t mixture due to the nuclear fusion reactions and tritium decay. Muon transfer from the ground state of muonic hydrogen is strongly suppressed because of the absence of term intersections for $Z \leq 3$ and proceeds mainly via the intermediate quasistationary molecular state $(H\mu He)$[3,4]. One may expect that for the excited muonic hydrogen this suppression is taken off[5] so the muon transfer

$$(H\mu)_n^* + He \longrightarrow (He\mu)_{n'}^* + H \qquad (1)$$

in the course of the muonic hydrogen deexcitation plays an important role in the kinetics of the cascade transitions. This expectation is confirmed by a recent observation[6] of the pronounced pressure dependence of the fraction of muons found in the $H\mu$ and $He\mu$ ground states in the hydrogen-helium mixture.

2. Adiabatic approximation

The muon mass is almost ten times smaller than that of a nucleon. For this reason the velocity of the nucleus is small in comparison with the muon velocity, which allows one to consider the muon motion around the fixed nuclei with the internuclear distance R regarded as a parameter.

The Schroedinger equation for the two-Coulomb-center problem admits a separation of variables in prolate spheroidal coordinates

$$\xi = \frac{r_1 + r_2}{R}, \quad \eta = \frac{r_1 - r_2}{R}, \quad \phi = \text{arctg } (x/y).$$

where r_i is the distance between the muon and the nucleus with the charge Z_i ($Z_2 \geq Z_1$), R is the internuclear distance. Writing the wavefunction as

$$\phi(r) = [(\xi^2 - 1)(1 - \eta^2)]^{-1/2} \, F(\xi) \, G(\eta) \, \exp(im\phi)$$

one obtains for $F(\xi)$ and $G(\eta)$

$$\frac{d^2 F(\xi)}{d\xi^2} + \left[\frac{E R^2}{2} + \frac{a\xi - \lambda}{\xi^2 - 1} + \frac{1 - m^2}{(\xi^2 - 1)^2}\right] F(\xi) = 0 \quad (2a)$$

$$\frac{d^2 G(\eta)}{d\eta^2} + \left[\frac{E R^2}{2} + \frac{b\eta - \lambda}{1 - \eta^2} + \frac{1 - m^2}{(1 - \eta^2)^2}\right] G(\eta) = 0 \quad (2b)$$

where $a = (Z_1 + Z_2) R$, $b = (Z_2 - Z_1) R$ and λ is a separation constant. Solving this eigenvalue problem, one obtains the terms $E_n(R)$, representing the effective potential, in which nuclei move (one should add to the potential also the term corresponding to the Coulomb repulsion of nuclei).

For very small R the system represents the hydrogenlike atom with the nucleus charge $Z_1 + Z_2$. For this reason one may classify terms using spherical quantum numbers of this atom (Nlm). For very large R the term describes the state of the hydrogenlike atom with the muon located either at the nucleus Z_1 (eZ_1-term) or at Z_2 (eZ_2-term) and may be classified with the help of the parabolic quantum numbers of this hydrogenlike atom, $[nn_1n_2m]$ for the eZ_1 and $[n'n'_1n'_2m]$ for the eZ_2 term. Two types of classification (united atom and separated atoms) are related to each other via the correlation diagram[7]. For the eZ_1-term one has

$$l = \begin{cases} m + 2n_2 + n(Z_2 - Z_1)/Z_1 & \text{if } nZ_2/Z_1 = \text{integer} \\ m + 2n_2 + 1 + \text{entier } [n(Z_2 - Z_1)/Z_1] & \text{if } nZ_2/Z_1 \neq \text{integer} \end{cases}$$

$$N = n_1 + l + 1; \qquad n = n_1 + n_2 + m + 1.$$

For the eZ_2-terms

$$l = \begin{cases} m + n_2' & \text{if } n_2' < n'(Z_2 - Z_1)/Z_2 \\ m + n_2' + 1 + \text{entier } [n_2' - n'(Z_2 - Z_1)/Z_2] & \text{if } n_2' \geq n'(Z_2 - Z_1)/Z_2 \end{cases}$$

$$N = n_1' + l + 1; \qquad n' = n_1' + n_2' + m + 1.$$

For large R, the eZ_1-terms can be expressed as

$$E(R) \simeq -\frac{Z_1^2}{2n^2} - \frac{Z_2}{R} + \frac{3}{2}\frac{Z_2 n(n_1 - n_2)}{Z_1 R^2} + O(R^{-3}) \tag{3}$$

A similar expansion for the eZ_2-term can be obtained by the replacement

$$Z_1 \longleftrightarrow Z_2 \text{ and } n, n_1, n_2 \longrightarrow n', n_1', n_2'.$$

3. Quasiclassical approach

The important object in the theory of the adiabatic transitions is the intersection of terms. For sufficiently large R, where the asymptotic expansion (3) works well, the validity condition for the quasiclassical approximation looks like

$$\frac{d\lambda}{dR} \simeq [3Mn(n_1 - n_2)]^{-1/2} \ll 1 \tag{4}$$

and is fulfilled for $n \geq 2$ and $n_1 \neq n_2$ (M is the reduced mass: $M = M_{\text{He}}(M_{\text{H}} + m_\mu)/(M_{\text{He}} + M_{\text{H}} + m_\mu) \simeq 10$ m.a.u.). In this case the transition probability is maximum in the region of the term intersection, where the transition can take place classically.

Let $u_a(R)$ and $u_b(R)$ be two terms under consideration. The exact intersection of terms of the same symmetry is excluded according to the Neumann-Wigner theorem[8], so actually we deal with a pseudocrossing (Fig. 1) and the terms behave like

$$u_{a,b} = \frac{1}{2}\left[u_1 + u_2 \pm \sqrt{(u_1 - u_2)^2 + 4V^2}\right] \tag{5}$$

where V is the matrix element, responsible for the transition. The gap between the terms a and b is

$$u_b - u_a \equiv \Delta u = \sqrt{(u_1 - u_2)^2 + 4V^2} \neq 0 \quad \text{if } V \neq 0. \tag{6}$$

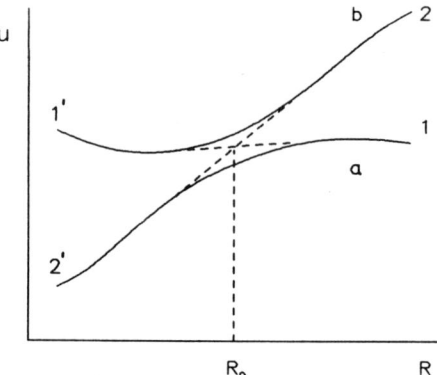

Figure 1:

It is clear that the terms cross each other at the complex point R_c which is the root of the equation

$$(u_1 - u_2)^2 + 4V^2 = 0 \tag{7}$$

Under the natural assumption of linearity of this expression near the zero point, it becomes clear that the terms have a root branch point:

$$u_{a,b} = u_o + \alpha\sqrt{R - R_c} \tag{8}$$

In the two-level approximation of the quasiclassical approach the probability of the transition between the states $u_a(R)$ and $u_b(R)$ associated with the branch point R_c is exponentially small [9]

$$w_1 = \exp(-2\delta) \tag{9}$$

where the Massey parameter δ is expressed through the contour integral of the radial momentum $p(R)$ around the branch point:

$$\delta = |\operatorname{Im} \int_c p(R)dR| = |\operatorname{Im} \int_{R_o}^{R_c} (p_1 - p_2)dR| \simeq$$
$$\simeq \frac{1}{v} |\operatorname{Im} \int_{R_o}^{R_c} \Delta u\, dR|$$

$$p_i(R) = \sqrt{2M(\varepsilon_o - u_i - \varepsilon_o \rho^2/R^2)} \qquad (10)$$

Here $\varepsilon_o = \dfrac{M v_o^2}{2}$ is the energy of the colliding particles at infinite R, ρ is the impact parameter, and $v = (v_1 + v_2)/2$ is the mean velocity in the crossing point. One should take into account only the branch points closest to the real axis, giving the maximum transition probability. Actually the transition region is passed twice, so the total transition probability is

$$w = 2w_1(1 - w_1) = 2 \exp(-2\delta)[1 - \exp(-2\delta)] \qquad (11)$$

The cross section can then be obtained as

$$\sigma = \pi \int_o^{\rho_{max}^2} w d\rho^2 \qquad (12)$$

where ρ_{max}^2 is the maximum impact parameter for which the crossing point (real) R_o lies in the classically accessible region. The reaction rate, reduced to the liquid hydrogen density $N_o = 4.25 \cdot 10^{22}$ cm^{-3}, is then

$$\lambda = \sigma v_o N_o \qquad (13)$$

If the complex branch points are not known, but one knows the term behaviour at real R, one can estimate the Massey parameter, expanding the term difference in (6)

$$u_1 - u_2 \simeq (F_2 - F_1)(R - R_o), \quad F_i \equiv -\frac{du_i}{dR} \qquad (14)$$

(linear Landau-Zener model[9]). In this case the Massey parameter (10) becomes

$$\delta_{LZ} \simeq \frac{\pi(\Delta u_{min})^2}{4v\Delta F} \qquad (15)$$

4. Analytic properties of terms

Solovyov[10] was the first who investigated the analytic properties of the terms of the two-center problem (2) in the complex R plane. All the terms of the given symmetry turned out to be sheets (branches) of a single analytic function, determined everywhere in the complex R plane. The terms are sewn in pairs at the branch points, which form the series. Some of them are not manifested in the term shape at real R.

The closest to the origin are the branch points of the S-series, which connect successively all the terms E_{Nlm} and $E_{N+1,lm}$ for $N \geq l + 1$, i.e. the terms which have $\tilde{n}_2 = n_2$

and $\tilde{n}_1 = n_1 \pm 1$. The S-series is related to the reorganisation of the potential of the radial equation (2a) at $\lambda = (Z_1 + Z_2)R$ and the corresponding reorganisation of the muon wavefunction from the wavefunction of the united atom to the molecular one.

From the point of view of our problem the most interesting are two other branch-point series, determined by the angular equation (2b).

When the term touches the top of the barrier in the effective potential of the angular equation, the T-series arises, which corresponds to the reorganisation of the muon wavefunction from the molecular type to the wavefunction of separated atoms. These branch points connect the terms E_{Nlm} and $E_{N+k,l+k,m}$ ($\tilde{n}_1 = n_1$, $\tilde{n}_2 \neq n_2$), which may belong to either a different muon localisation (eZ_1 and eZ_2) or to the same (for example, both terms may be of the eZ_2 type).

Besides the T-series, at $R > R_T$ the isolated branch points R_I exist, which connect the states with different muon localisation, eZ_1 and eZ_2 ($\tilde{n}_1 = n_1$, $\tilde{n}_2 \neq n_2$). This branch point is related to the coincidence of levels in the two pits of the effective potential of the angular equation (2b). Till 1981 the only crossing points that were considered in the charge exchange problem were those of the R_I type[11].

5. Results and discussion

Table 1 contains a list of terms, connected with the reaction (1), for $n \leq 4$. We consider only terms with $n_1 < n_2$, corresponding to attraction in the initial state. The branch points, corresponding to transitions with $n \leq 3$, were obtained in [12]. For larger n the direct determination of the branch points from eqs. (2) is hardly possible, so we used the quasiclassical equivalent of eqs. (2), obtained in [13], to calculate the terms at real R.

Table 1: Branch points and reduced Massey parameters $\tilde{\delta} = \delta/\sqrt{2M}$ for $n \leq 4$

$H\mu + He$		$He\mu + H$		$H\mu + He \longrightarrow He\mu + H$		
(Nlm)	$[nn_1n_2]$	(Nlm)	$[n'n'_1n'_2]$	$R_c(R_o)$	$\tilde{\delta}$	$\tilde{\delta}_{LZ}$
$5g\sigma$	201	$4f\sigma$	$3'0'2'$	16.6;4.7	0.38	0.30
$8k\sigma$	302	$7i\sigma$	$5'0'4'$	49.7;6.6	0.13	0.11
$7i\pi$	301	$6h\pi$	$5'0'3'$	42.5;5.8	0.12	0.11
$11n\sigma$	403	$10m\sigma$	$7'0'6'$	106.7	–	0.014
$10l\sigma$	412	$9k\sigma$	$7'1'5'$	98.4	–	$9.0\ 10^{-3}$
$10m\pi$	402	$9l\pi$	$7'0'5'$	91.6	–	$4.7\ 10^{-3}$
$9l\delta$	401	$8k\delta$	$7'0'4'$	90.5	–	$8.2\ 10^{-3}$

When calculating the cross section, one should take into account the electron screening. We present the effective potential of the initial state as

$$u(R) = E_{Nlm} + \tfrac{3}{2}n(n_1 - n_2)[\ \mathcal{E}(R) - \tfrac{Z_2}{R^2}],$$

where $\mathcal{E}(R)$ is the electric field of the target atom[14]. The screening correction results in the considerable decrease of ρ_{max} for low energies. The resulting transfer rates are presented in Tables 2 and 3.

As seen from Table 2, the transfer rates, calculated without electron screening, have a pronounced energy dependence (roughly $\lambda \sim \varepsilon_o^{-1/2}$). If one takes into account the electron screening (Table 3), this dependence becomes much weaker due to the decrease of ρ_{max} for low energies[15].

The decrease in λ at the transition from n=3 to n=4 is due to the much smaller Massey parameter for the latter case (see Table 1). This means that for $n \geq 4$ one should take into account the two-step transfer process $eZ_1 \to e\tilde{Z}_2 \to eZ_2$ (for example, $11n\sigma \to 10m\sigma \to 9l\sigma$, Fig. 2). This additional channel would increase the cross section and transfer rate for $n \geq 4$.

Table 2: Muon transfer rates (10^{11}s^{-1}), calculated without electron screening

ε_o (eV)	p^{3}He			p^{4}He		
	n = 2	3	4	2	3	4
0.04	6.8	85	45	6.0	82	45
0.5	2.0	26	14	1.8	25	14
1	1.5	19	10	1.3	18	10
5	0.75	9.9	6.0	0.66	9.5	6.0
10	0.57	7.9	5.4	0.50	7.6	5.3

Table 3: The same as in Table 1 but with electron screening taken into account

ε_o (eV)	p^{3}He			p^{4}He		
	n = 2	3	4	2	3	4
0.04	1.53	12.3	4.6	1.35	11.8	4.6
0.5	1.64	15.2	5.9	1.45	14.6	5.9
1	1.39	14.4	5.9	1.23	13.7	5.9
5	0.74	9.4	4.9	0.65	9.0	4.9
10	0.57	7.6	4.6	0.50	7.3	4.6

Acknowledgements

I am grateful to Drs. D.I. Abramov, V.I. Savichev and A.I. Mikhailov for cooperation.

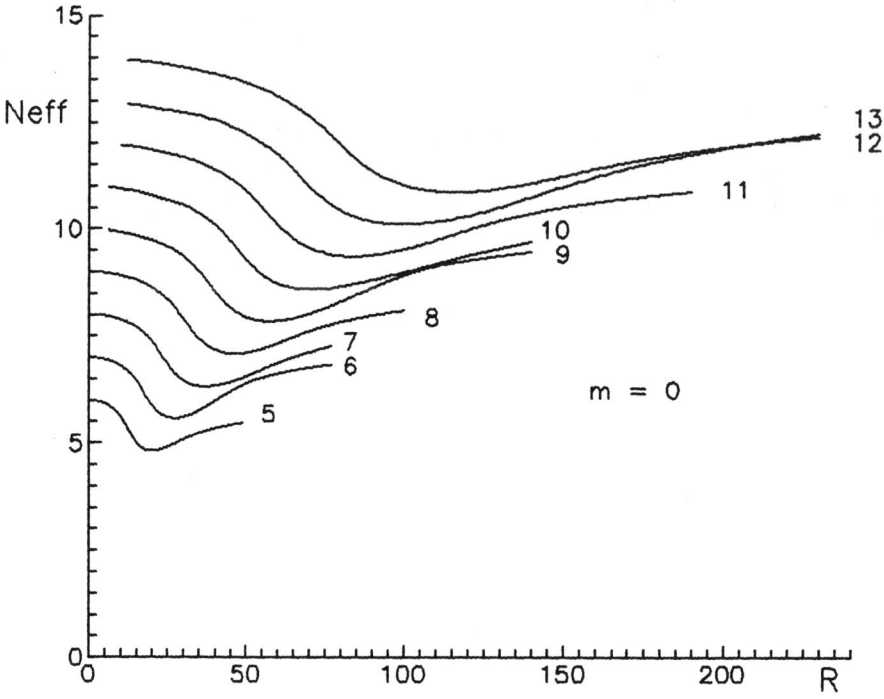

Figure 2: $N_{eff} = \{-9/[2u(R)]\}^{1/2}$ for σ-terms with $n_1 = n_1' = 0$. Numbers at curves denote l-values. Terms with $l = 7, 10, 13$ correspond asymptotically to $(p\mu)_n^*$, with $n = (l+2)/3$. Other terms correspond to $(He\mu)_n^*$, with $n' = l - 1$ for $l = 5, 6$; $n' = l - 2$ for $l = 8, 9$; $n' = l - 3$ for $l = 11, 12$.

References

[1] L. Ponomarev, Atomkernenergie/Kerntechnik 43 (1983) p. 175.

[2] S.S. Gerstein, L.I. Ponomarev, in Muon Physics (eds. V. Hughes, C.S. Wu) Acad. Press. NY: 1975 p. 141.

[3] Yu. Aristov et al., Yad. Fis. 33 (1981) p. 1066 [Sov. J. Nucl. Phys. 33 (1981) p. 564].

[4] V.M. Bystritsky et al., ZhETF 84 (1983) p. 1257 [Sov. Phys. JETP 57 (1983) p. 728], D. Balin et al., Pis'ma ZhETF 42 (1985) p. 236 [Sov. Phys. JETP Lett. 42 (1985) p. 293].

[5] N.P. Popov, Muon Cat. Fusion 2 (1988) p. 207,
M. Bubak, V. Bystritsky, Preprint JINR E 1-86-107, Dubna 1986.

[6] F. Kottmann, in Proc. II Int. Symposium on Muon and Pion Interactions with Matter (Dubna 1987) p. 268.

[7] S.S. Gershtein, V.D. Krivchenkov, ZhETF 40 (1961) p. 1491.

[8] J. Neumann, E. Wigner, Zs. Phys. 30 (1929) p. 467.

[9] L.D. Landau, I.M. Lifshitz, Quantum Mechanics: Nonrelativistic Theory, 3rd ed. (Pergamon, Oxford, 1965).

[10] E.A. Solovyov, ZhETF 81 (1981) p. 1681 [Sov. Phys. JETP 54, (1981) p. 893].

[11] I.V. Komarov, L.I. Ponomarev, S.Yu. Slavyanov, Spheroidal and Coulomb Spheroidal Functions (Nauka, Moscow, 1976).

[12] A. Kravtsov, A. Mikhailov, N. Popov, Preprint LNPI 1488 (1989), ZhETF 96 (1989) p. 437 [Sov. Phys. JETP 69 (1989) p. 246].

[13] D.I. Abramov, S.Yu. Ovchinnikov, E.A. Solovyov, Phys. Rev. A42 (1990) p. 6366.

[14] A. Kravtsov, A. Mikhailov, N. Popov, J. Phys. B 19 (1986) p. 1323.

[15] V.M. Bystritsky, A. Kravtsov, N.P. Popov, ZhETF 97 (1990) p. 73.

Muon Transfer to Elements with Z > 1

Hubert SCHNEUWLY
Institut de Physique de l'Université
Pérolles, CH-1700 Fribourg, Switzerland

Abstract: A series of measurements of muon transfer rates from muonic hydrogen to elements with Z > 1 yielded unexpected discrepancies. In the light of muon transfer results to heteroatomic molecules, these discrepancies have to be taken more seriously than till now. Even if muon transfer to helium proceeds via a molecular state, it belongs probably to the same problematics.

About a quarter of a century ago, E. Zavattini and his italian coworkers performed at CERN the first systematic measurements of muon transfer from muonic protium and deuterium to noble gases [1-3]. The transfer rates were determined by using the method of the muonic X-ray intensities and the method of the time structure of the muon decay electrons. The transfer rates measured under different experimental conditions were normalized either to the molecular (ρ_0 = $2.12 \cdot 10^{22}$ cm^{-3}) or the atomic (ρ_0 = $4.25 \cdot 10^{22}$ cm^{-3}) density of liquid hydrogen. The rates published by the italian group in 1967 and 1969 are given in Table 1 for muon transfer from muonic protium. For a comparison, the results of Basiladze *et al.* (1965) are given in the last column.

Whereas Alberigi Quaranta *et al.* [1] employed the muonic X-ray intensity method, Placci *et al.* [2] used the method of the decay electrons. For Kr and Xe, both methods yield compatible transfer rates. To be more precise, in 1967, the transfer rate to Xe was measured by using the time structure of the muonic µXe(2-1) X-rays, and the transfer rate to Kr was obtained from the relative muonic X-ray intensities in a triple mixture, H$_2$ + Kr + Xe, the transfer rate to Xe being known. Actually, this shows that all three methods yield compatible transfer rates.

The third transfer rate to Xe given in Table 1 is the one of Basiladze *et al.* [4], which is normalized to the atomic density of liquid hydrogen. Already Placci *et al.* noticed explicitely in their paper the discrepancy of Basiladze's transfer rate with the ones obtained by the italian group. However, the discrepancy being of a factor of two, one might, at least for the moment,

TABLE 1: Muon transfer rates from muonic protium to noble gases (status in 1969) in units [$\cdot 10^{11}$ s^{-1}], normalized to the molecular and the atomic densities of liquid hydrogen.

	Alberigi Quaranta et al.		Placci et al.		Basiladze et al.
	mol. N.	atom. N.	mol. N.	atom. N.	
Neon	0.58(14)	1.16(28)	-	-	-
Argon	1.74(30)	3.48(60)	0.73(7)	1.46(14)	1.20(18)
Krypton	3.42(63)	6.84(126)	2.84(28)	5.68(56)	-
Xenon	4.96(46)	9.92(92)	4.41(20)	8.82(40)	4.46(36)

keep the door open for an error in normalization, the transfer rate of the italian group having been confirmed since then in a series of other measurements [5].

If we turn now our attention to the transfer rate to argon, we observe a discrepancy between the two results of the italian group. Placci *et al.* do not comment about this disagreement, but they were certainly very perplexed about it. Some years later, another italian group remeasured the transfer rate to argon by using the time structure method of the muonic X-rays [6]. In four measurements, they determined four rates, in agreement with each other, which agreed also with the rate of Alberigi Quaranta *et al.*, but disagreed with the rates of Placci *et al.* and Basiladze *et al.*, both determined by using the decay electrons. From these disagreements, Iacopini et al. [6] suspected the muonic X-ray method and the decay electron method to give for some unknown reasons different rates. A few years later, a series of measurements performed at PSI showed, that the discrepancies were not due to methodological procedures, and reproduced the rates obtained by Placci *et al.* and Basiladze *et al.* [7, 8]. In 1980, the Munich group published a transfer rate to argon, measured at high pressure, which is almost an order of magnitude higher that the result of Basiladze *et al.* [9]. The present situation is summarized by Table 2.

As can be seen from this Table 2, there are three groups of transfer rates to argon and the differences among them amount to a factor of about 2.5. No correlation could be found between these groups and experimental conditions, except that the highest value was measured at the highest pressure. Very long and very short μp lifetimes yielding the same normalized transfer rates at high and low pressures, the thermalization process cannot be invoked to explain the discrepancies.

Table 2: Experimental muon transfer rates from muonic protium to argon, normalized to the atomic density of liquid hydrogen.

Year	Ref.	pressure [atm]	Λ_{pAr} [$\cdot 10^{11}$ s^{-1}]
1965	[4]	45	1.20(18)
1969	[2]	10	1.46(14)
1988	[8]	100	1.42(16)
1988	[8]	140	1.46(5)
1988	[7]	9.6	1.47(8)
1988	[7]	13.3	1.41(5)
1990	[10]	15	< 1.50
1967	[1]	26	3.48(60)
1982	[6]	2	3.61(220)
1982	[6]	3	3.55(112)
1982	[6]	3	3.77(136)
1982	[6]	4	3.81(167)
1980	[9]	600	9.8(15)

When we published our first results about transfer to argon [7, 8], we were intimately convinced that only transfer rates in agreement with ours were correct, and any respectable scientist, who trusts in his work, would defend the same opinion about his results and those of others. Scientific results have to be reproducible. Thus, in this uncomfortable situation, where one has three groups of muon transfer rates to argon, two conclusions are possible. Either two of the three groups are made of wrong results or there exist three different kinds of argon atoms, which are characterized by three different transfer rates. One might call these three kinds of argon atoms as blue, red and green ones. Actually, in conventional terms, only the first conclusion does make sense.

However, with the transfer rates to neon, we are confronted to a very similar situation. There are also three groups of transfer rates (cf. Table 3). There is, however, an important difference. In the binary H_2 + 0.69%Ne gas mixture, the time distributions of the muonic neon X-rays of all lines of the Lyman series had all the same double-exponential structure as if the muonic hydrogen atoms would have two different lifetimes or, equivalently, as if two different muonic hydrogen atoms would be simultaneously present in the gas mixture [11]. The transfer rate, measured by Alberigi Quaranta et al. [2] using the muonic X-ray intensity method, is reproduced through the shorter and smaller (4%) of the two components [11]. The longer component yields a transfer rate, which is almost twenty times smaller, and which is repro-

Table 3: Experimental (reduced) muon transfer rates from muonic protium to neon.

Year	Ref.	pressure [atm]	mixture	Λ_{pNe} [$\cdot 10^{11}$ s^{-1}]
1967	[1]	26	H_2 + Ne + Xe	1.16(28)
1990	[11]	15	H_2 + 0.69%Ne	1.15(20)
1990	[11]	15	H_2 + 0.69%Ne	0.062(4)
1990	[11]	15	H_2 + Ne + Ar	0.058(48)
1990	[11]	15	H_2 + Ne + Ar	0.062(5)
1991	[12]	15	H_2 + 1.4%Ne	0.079(4)
1991	[13]	15	H_2 + 2.0%Ne	0.081(4)
1991	[13]	15	H_2 + 0.7%Ne	0.084(8)
1992	[14]	15	H_2 + 1.5%Ne	0.080(5)

duced in a triple mixture H_2 + Ne + Ar as well in the time structure of the muonic X-rays as in their relative intensities [11]. In four other binary mixtures neither the double-exponential structure nor the precedingly measured transfer rates could be reproduced [12-14].

The last four transfer rates are only by about 30% higher than the preceding ones. In addition, compared to transfer rates to other elements like nitrogen [10] or oxygen [16], the transfer rate to neon is very small, such that one may suspect impurities in the gas mixtures to simulate higher transfer rates. The advantage of the X-ray time structure and intensity methods is that the elements to which muons are transferred can easily be identified. Muonic oxygen X-rays, appearing delayed with respect to stopping muons, have indeed been observed in the last four measurements. Muonic nitrogen X-rays were absent. The detailed analysis of their time structures and their intensities revealed that the impurities were only very partially responsible for the higher values of the transfer rates [14].

As for argon, we are thus again confronted with three different groups of muon transfer rates. Because of the double-exponential time structure of the muonic neon X-rays in the H_2 + 0.69%Ne mixture, the three groups of transfer rates cannot be due to three different kinds of neon atoms.

These unexpected double-exponential time structures of the muonic neon X-rays have been observed only in the H_2 + 0.69%Ne mixture. Such double-exponential structures are, actually, the general pattern of muonic oxygen X-rays in a whole series of H_2 + SO_2 gas mixtures investigated during the recent years [15,16]. The time distributions of the corresponding sulphur X-rays have single-exponential structures with time constants equal to those of the flatter of the two oxygen components. The transfer rates deduced from these components are reproducible, whereas those deduced from the steeper oxygen component are only reproduced for mixtures with the same SO_2 concentrations. These repeatedly observed time structures in $H_2 + SO_2$ mixtures lead to the hypothesis of two distinct muonic hydrogen atoms in the ground state: the "white" ones, which behave as expected, and the "black" ones, which put our knowledge about muon transfer in question [17]. With the hypothesis, that the "black" ones disappear via an unknown channel with a rate proportional to the hydrogen density, the transfer rates of the "black" μp atoms to oxygen in $H_2 + SO_2$ mixtures become also reproducible. These "black" μp atoms have then been called "ephemeral" ones [18].

Table 4: Experimental (reduced) muon transfer rates from muonic protium and muonic deuterium to helium (^{4}He).

Year	Ref.	pressure [atm]	Λ_{pHe} [$\cdot 10^8$ s^{-1}]
1983	[19]	20	0.36(10)
1988	[7]	15	0.83(14)
1988	[7]	15	0.88(15)
1988	[7]	15	0.92(14)
1989	[20]	5	0.032(13)

Year	Ref.	pressure [atm]	Λ_{dHe} [$\cdot 10^8$ s^{-1}]
1967	[3]	6	<0.2
1984	[21]	92	3.2(3)
1985	[22]	88	3.68(18)
1988	[23]	liq.	13.1(12)
1988	[23]	liq.	6.7(13)
1990	[24]	1350	2.75(22)
1991	[25]	liq.	1.89(55)
1991	[25]	liq.	5.7(26)

Great efforts are presently made to measure with high precision the sticking of muons to helium after the fusion process in dμd and dμt molecules, since the muon sticking is a limiting factor in the recycling process in hydrogen isotope mixtures. The transfer to helium is another limiting factor and has been investigated for the first time already in 1967 by Placci et al. [3]. In this first measurement of the transfer rate from muonic deuterium to helium no transfer was observed.

During the last ten years, in particular after the theoretical prediction of transfer rates to helium of the order of 10^8 s^{-1}, several measured transfer rates have been published. They are listed in Table 4. It is obvious that for the transfer rates from muonic protium as well as from muonic deuterium, one observes the same situations as for the transfer rates to other noble gases. Apart from its very low transfer rate, helium does not seem to be a special case. There are however additional curiosities. In the same experiment, the intensity method and the time structure method did not yield the same transfer rates [23,25]. In a first experiment [23], the transfer rate deduced from the time structure of the muonic helium X-rays was higher by about a factor of two. In a second series of experiments, the results of which are only available through a preprint [25], the intensity method reproduced the rate precedingly obtained by using the same method. The time distributions of the delayed muonic helium X-rays showed double-exponential structures and the transfer rate determined from the slower component was almost an order of magnitude smaller than in the previous experiment.

Except that the muon should be transferred to a molecular state in pµHe, which can be distinguished experimentally from an atomic state in µHe, the problem of muon transfer to helium is not different from the transfer to any other element. Hence there is no hope to determine a normalized transfer rate to helium neither from protium not from deuterium which could be reliably used in gas mixtures like, e.g., those where the fusion cycle is investigated.

It has not yet been shown that the transfer rate to elements different from helium, determined from the ratio of the muonic µZ X-ray intensity, resulting from transfer, to the µp X-ray intensity, is different from the transfer rate obtained from the muonic X-ray time structures. No doubt, the investigation of this difference will be essential for the future understanding of the muon transfer process.

References

[1] A. Alberigi Quaranta, A. Bertin, G. Matone, F. Palmonari, A. Placci, P. Dalpiaz, G. Torelli and E. Zavattini, Nuovo Cimento **B 47**, 92 (1967).

[2] A. Placci, E. Zavattini, A. Bertin and A. Vitale, Nuovo Cimento **A 64**, 1053 (1969).

[3] A. Placci, E. Zavattini, A. Bertin and A. Vitale, Nuovo Cimento **A 52**, 1274 (1967).

[4] S.C. Basiladze, P.F. Ermolov and K.O. Oganesyan, Zh. Eksp. Teor. Fiz. **49**, 1042 (1965) [Sov. Phys. JETP **22**, 725 (1966)].

[5] A. Bertin, M. Bruno, A. Vitale, A. Placci and E. Zavattini, Phys. Rev. **A 7**, 462 (1973).

[6] E. Iacopini, G. Carboni, G. Torelli and V. Tobbiani, Nuovo Cimento **A 67** 201 (1982).

[7] R. Jacot-Guillarmod, F. Bienz, M. Boschung, C. Piller, L.A. Schaller, L. Schellenberg, H. Schneuwly, W. Reichart and G. Torelli, Phys. Rev. **A 38**, 6151 (1988).

[8] F. Bienz, P. Bergem, M. Boschung, R. Jacot-Guillarmod, G. Piller, W. Reichart, L.A. Schaller, L. Schellenberg, H. Schneuwly and G. Torelli, J. Phys. B: At. Mol. Opt. Phys. **21**, 2725 (1988).

[9] H. Daniel, H.-J. Pfeiffer, P. Stoeckel, T. von Egidy and H.P. Povel, Nucl. Phys. **A 345** 409 (1980).

[10] R. Jacot-Guillarmod, F. Mulhauser, C. Piller, L.A. Schaller, L.Schellenberg and H. Schneuwly, *Electromagnetic Cascade and Chemistry of Exotic Atoms* edited by L.M. Simons, D. Horváth and G. Torelli, (Plenum Press, New York, 1990), p. 223.

[11] R. Jacot-Guillarmod, F. Mulhauser, C. Piller and H. Schneuwly, Phys. Rev. Lett. **65**, 706 (1990).

[12] R. Jacot-Guillarmod, F. Mulhauser, C. Piller, L.A. Schaller, L.Schellenberg and H. Schneuwly, Helv. Phys. Acta **64**, 205 (1991).

[13] H. Schneuwly, R. Jacot-Guillarmod, M. Mallinger, F. Mulhauser, P. Oberson, C. Piller, L.A. Schaller and L. Schellenberg, Helv. Phys. Acta **64**, (1991).

[14] P. Baeriswyl, R. Jacot-Guillarmod, M. Mallinger, B. Mischler, F. Mulhauser, C. Piller, W. Reichart, B. Riedo, L.A. Schaller, L. Schellenberg, H. Schneuwly and A. Werthmüller, PSI Nucl. Part. Phys. Newsletter (1991).

[15] F. Mulhauser, H. Schneuwly, R. Jacot-Guillarmod, C. Piller, L.A. Schaller and L. Schellenberg, Muon Catal. Fusion **4**, 365 (1989/91).

[16] F. Mulhauser, R. Jacot-Guillarmod, C. Piller, L.A. Schaller, L. Schellenberg and H. Schneuwly, Helv. Phys. Acta **64**, 937 (1991).

[17] H. Schneuwly, Muon Catal. Fusion **4**, 87 (1989/91).

[18] H. Schneuwly and F. Mulhauser, Phys. Lett. **A 160**, 71 (1991).

[19] V.M. Bystritsky, V.P. Dzhelepov, V.I. Petrukhin, A.I. Rudenko, V.M. Suvorov, V.V. Filchenkov, N.N. Khovanskii and B.M. Khomenko, Zh. Eksp. Teor. Fiz. **84,** 1257 (1983) [Sov. Phys.-JETP **57**, 728 (1983)].

[20] H.P. von Arb, F. Dittus, H. Hofer, F. Kottmann and R. Schaeren 1989 Muon Catal. Fusion **4**, 61 (1988).

[21] D.V. Balin, E.M. Maev, V.I. Medvedev, G.G. Semenchuk, Yu.V. Smirenin, A.A. Vorobyov, An. A. Vorobyov and Yu.K. Zalite, Phys. Lett. **B 141**, 173 (1984).

[22] D.V. Balin, A.A. Vorobyov, An.A. Vorobyov, Yu.K. Zalite, A.A. Markov, V.I. Medvedev, E.M. Maev, G.G. Semenchuk and Yu.V. Smirenin, Pisma Zh. Eksp. Theor. Fiz. **42**, 236 (1985) [Sov. Phys.-JETP Lett. **42** 293 (1985)].

[23] T. Matsuzaki, K. Ishida, K. Nagamine, Y. Hirata and R. Kadono, Muon Catalyzed Fusion **2**, 217 (1988).

[24] V.M. Bystritsky, V.P. Dzhelepov, V.G. Zinov, N. Ilieva-Sokolinova, A.D. Konin, L. Marcis, D.G. Merkulov, A.I. Rudenko, L.N. Somov, V.A. Stolupin and V.V. Filchenkov, JINR Report P1-90-312 Dubna USSR, 1990.

[25] Y. Watanabe, S. Sakamoto, K. Ishida, T. Matsuzaki, P. Strasser, M. Iwasaki and K. Nagamine, preprint 1991.

Hot Muonic Atoms

Kinetic Energies at the Formation and Cascade of μp-Atoms

F. KOTTMANN

Institut für Hochenergiephysik, ETH-Hönggerberg, CH–8093 Zürich, Switzerland

The complex balance of processes occurring at the cascade of exotic H-atoms is usually described by the so-called standard cascade model, but this model neglects variations of the kinetic energy T of the exotic atom during the cascade which are crucial for the analysis of several important experiments. In particular the initial values T_i and the n_i-levels of these atoms just after atomic capture are badly known. Direct measurements of these values are difficult, but the combined analysis of a series of experiments discussed in this contribution tests model calculations of the capture and cascade processes on a sensitive level.

1. Beyond the "standard cascade model"

A more and more refined picture of the complex processes involved in the formation and deexcitation of exotic H-atoms ($\mu^- p$, $\pi^- p$, $K^- p$, $\bar{p}p$, $\mu^- d$, etc.) results from various experimental and theoretical studies performed in the last decade. The present status has recently been summarized by Cohen [1] and Markushin [2]. What is called the "standard cascade model" is a generally accepted description of the deexcitation mechanisms of exotic H-atoms. Based on the early work of Leon and Bethe [3], its general assumptions are as follows:

- The exotic atom is formed at an initial state with principal quantum number $n \approx \sqrt{M/m_e}$ (e.g. $n \approx 14$ for μp) and with a kinetic energy around 1 eV which stays constant during the cascade;

- deexcitation occurs via external Auger effect, Coulomb collision, chemical dissociation, and radiative transitions;

- Stark-mixing of the l-substates and nuclear absorption (for strongly interacting particles) have also to be taken into account.

The success of this model is due to the fact that x ray intensities and most other observables can be well reproduced using rather simple computer codes. Clearly, the cross sections of all cascade processes have at first to be calculated as function of the n and l quantum numbers. In the difficult case of Stark mixing, the cross sections are usually multiplied by a common factor k_{Stark} to get an optimum agreement with the data. This arbitrariness was avoided in the cascade model of Reifenröther et al. [4] who calculated the relevant cross sections *ab initio* by numerical integration, whereas variations of the kinetic energy during the cascade as well as molecular effects were still disregarded.

If the dependence of the deexcitation cross sections on the kinetic energy T is known, T can be treated in the standard cascade model as parameter which has to be optimized to fit the data. As an example, a kinetic energy $T = (1.0 \pm 0.2)$ eV resulted from an analysis of L x ray yields measured in $\bar{p}p$ and $\bar{p}d$ atoms [5]. For similar measurements with πp atoms [6,7], the situation is more intricate, as T increases from about 1 eV at gas pressures of 15 and 40 bar to 4 eV at 3 bar (see sect. 2.3). A number of other experiments is compatible with T-values of one or a few eV: The measurement of diffusion processes of μp_{1S} and μd_{1S} atoms at subatmospheric gas pressures [8], discussed by R. Siegel in this workshop; the lifetime of metastable $2S$ states of μp atoms [2]; or the transfer of pions and muons in mixtures of hydrogen isotopes [9,10].

The standard cascade model — despite of its obvious merits — has severe limits. In particular, the assumption that the kinetic energy is constant throughout the cascade is not at all valid. Elastic collisions decelerate the exotic atom already at excited states. The corresponding transport cross section has been calculated by Menshikov and Ponomarev [11]; the resulting deceleration rate is of the same order as the Stark mixing rate, at least at high levels $n \geq 9$ where it probably exceeds the rates of all deexcitation processes (see fig. 11 of ref. [2]). As discussed in sect. 2.1, it is commonly accepted that the initial kinetic energy at the formation time of the exotic atom lies between 0 and 1 eV. Further deceleration would bring T into a region which is clearly in disagreement with the values deduced from the experiments quoted above. This indicates that the exotic atom is also accelerated during the cascade.

A convincing proof for accelerating mechanisms came from an experiment of a PSI–VIRGINIA collaboration [12] where the Doppler broadening of the neutron time-of-flight spectra from the reaction $(\pi^- p)_{n \approx 3,4} \longrightarrow \pi^0 + n$ was studied in liquid hydrogen. The result, reproduced in qualitative terms, was that half of the πp atoms had kinetic energies below 1 eV and the rest up to 70 eV. A possible explanation of this unexpected distribution was that part of the $n = 5 \rightarrow 4$ and $4 \rightarrow 3$ transitions proceeds via Coulomb deexcitation, where the πp atom attains kinetic energies of about 30 and 70 eV, respectively. The low energy part is attributed to Auger transitions from $n = 5$ to 4 or 3, where nuclear capture occurs; elastic collisions at $n \sim 4$ reduce T to values around 0.5 eV.

Accelerating and decelerating effects have a non-trivial dependence upon target density, for several reasons:

- There is a critical level n_c below which deexcitation proceeds dominantly by radiative transitions. Typical values are $n_c \approx 2$ at $\phi = 1$ (density of liquid hydrogen), $n_c \approx 5$ at $\phi = 10^{-3}$ ($\sim$ 1 bar gas pressure), and $n_c \approx 10$ at $\phi = 10^{-6}$. The kinetic energy can be changed only by non-radiative processes, i.e. only at $n > n_c$.

- The transition rates from (external) Auger effect are extremely high at medium n-levels around 7 (c.f. Fig. 5). The cascade is therefore subdivided into an upper part ($n \geq 9$) where Coulomb and chemical deexcitations possibly increase T up to a few eV, a medium part with several subsequent Auger transitions ($\Delta n = 1$ preferred), and a lower part at $n \leq 5$ (mentioned above when discussing the neutron time-of-flight experiment). Auger transitions enhance T by not more than a fraction of an eV [13]; thus T is only slightly affected during the medium cascade stage. At low densities, T is formed in the upper part of the cascade; at high densities, the lower part has probably a strong influence on T.

- The cross sections of the various cascade processes depend differently on T [2]. For instance, the deceleration process has less relative importance if the kinetic energy increases. A complex interdependence between T and the cascade evolution results therefrom.

There are big uncertainties concerning the first part of the cascade. The distribution of initial n-levels is not known (c.f. sect. 2.1). The cross sections of the various effects are difficult to calculate; their relative importance is thus unclear. Some experimental information comes from measurements of the x ray yields, cascade times, and $2S$-metastability at extremely low densities ($\phi \sim 10^{-6}$), where transitions below $n_c \sim 10$ are purely radiative.

Despite all these difficulties, it will be necessary in the near future to develop a cascade model that takes the evolution of the kinetic energy into account. If such a model is able to describe all existing data in a consistant manner, it will also have some predictive power. Moreover, it will stimulate the analysis of epithermal effects in muon catalyzed fusion (see e.g. ref. [14], and the contribution of V. Markushin to this workshop), and possibly shed some light on the puzzling situation of ground-state muon transfer from H to higher Z elements. As has been pointed out here by S.S. Gershtein, the question of kinetic energies plays an important role in solving the problem of "black and white" muons.

One of the most important tests of any new cascade model will be the results of an experiment performed at PSI using the cyclotron trap [15]. The Doppler broadening of the neutron velocity spectra from πp atoms was measured at much lower densities than in the similar experiment described in ref. [11]. A preliminary qualitative analysis showed surprisingly high T-values of several eV at 18 and 40 bar gas pressure [16].

2. Atomic capture and x ray intensities

2.1 Atomic capture of negative muons in molecular hydrogen gas

There is a big number of theoretical and experimental studies on the atomic capture of exotic particles; an up-to-date survey was given by Hartmann in ref. [17]. The particular problem of $\mu^- + H$ collisions, i.e. slowing-down and capture of μ^- (or π^-) in atomic hydrogen, has been investigated by several authors using different theoretical methods; details and references are summarized in [1]. A consistent treatment of slowing-down and capture processes is crucial, as has been emphasized by Leon [18]. It is a common result of the more recent calculations for atomic hydrogen targets that the muon is captured at kinetic energies below about 20 eV, and that the initial n_i-levels are distributed at $n_i \geq 14$ with probabilities proportional to n^{-3}. Hence the mean n_i-value is considerably larger than what one gets from the well-known rule of thumb $n_i = \sqrt{m_\mu^{\rm red}/m_e} \approx 14$ (corresponding to the optimum overlap of muonic and electronic wave functions).

All models that assume interactions with *atomic* hydrogen are of limited use since *molecular* hydrogen is used in the experiments to be considered here. For this reason, Korenman et al. developed their "microscopic model" where the adiabatic coupled-channel method is used to describe the interaction of a μ^- with a H_2 molecule; the nuclei are assumed to be at fixed positions [19,20]. The cross section of inelastic interaions (i.e. the sum of ionization and capture cross sections) turns out to be about 20% larger for H_2 molecules than for H atoms at interaction energies around 30 eV, and about equal at 10 eV. This sustains the intuitive argument [1] that the H_2 molecule has to be represented by roughly one H atom for low-energy interactions, in contrast to the situation at high energies ($E_\mu \geq 100$ keV, the Bethe-Bloch regime) where the electrons are treated as quasi-free particles, and two H atoms thus represent roughly one molecule.

There are a few cases where the microscopic model can be tested. Using its cross sections for molecular interactions, the slowing-down time of a muon with about 2 keV initial energy was calculated; it agreed reasonably well with the measured values [21,22]. By applying the same methods to the $\mu^- + He$ interaction, the reduced capture ratio

$$A^{He/H_2} = \frac{W_{He}/W_{H_2}}{n_{He}/n_{H_2}} \tag{1}$$

was also calculated, which is the ratio of the probabilities W_i that a muon is captured by a He atom or a H_2 molecule, normalized to the densities n_i of He atoms and H_2 molecules. The resulting value $A^{He/H_2} = 0.80 \pm 0.02$ [20] has a very weak dependence upon concentration and agrees with our direct experimental result $A^{He/H_2} \leq 0.88 \pm 0.10$ (see sect. 3) and the model-dependent value 0.92 ± 0.05 deduced from the DUBNA experiments with pions [23].

In a refined version of the microscopic model, rotational and spin states of the hydrogen molecules were taken into account [24]. It turned out that the inelastic cross sections depend significantly on temperature and spin states. The reduced capture ratio in a H_2-He mixture varies up to 10% when the target temperature increases from 100 to 600 K, as shown in Fig. 1.

The initial distributions of kinetic energy and n- and l-quantum numbers after atomic capture have also been studies in the framework of the microscopic model [20]. As capture occurs only at muon energies $E_\mu \leq 17$ eV, the $(pp\mu e)^*$ complex initially formed has an uniform energy distribution in the interval [0, 0.92 eV], in accordance with momentum conservation. The following scheme shows the possible dissociation channels which lead finally to the formation of a $(\mu p)^*$ atom.

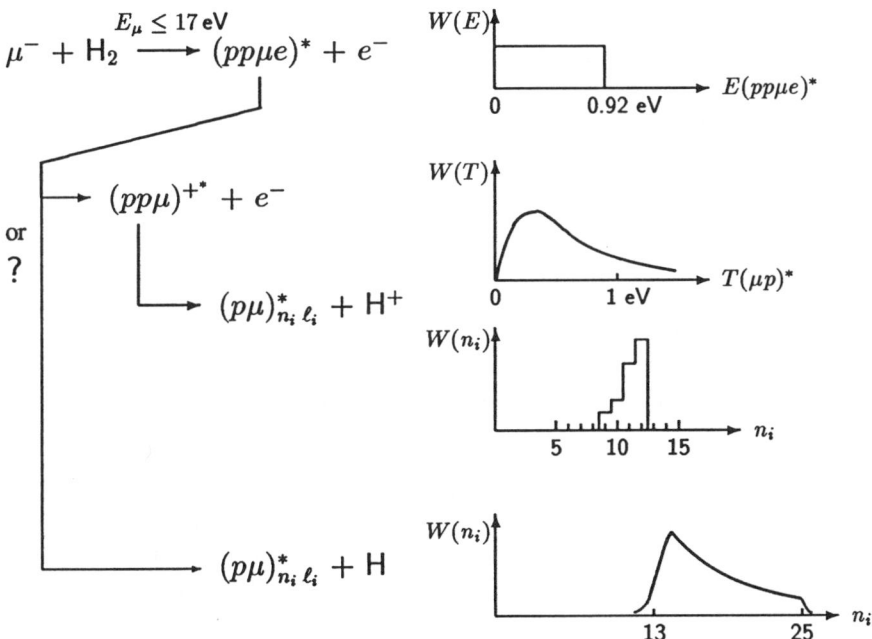

From the two possible decay channels of the $(pp\mu e)^*$ complex, Korenman et al. [20] favour the Auger transition $(pp\mu e)^* \longrightarrow (pp\mu)^* + e^-$, where practically no energy is transferred. The μp atom is then formed after dissociation of the $(pp\mu)^*$ ion. The resulting distribution of the $(\mu p)^*$ kinetic energies (shown in the scheme) has a mean value of about 0.5 eV. The initial levels n_i are distributed around $n_i = 11$ (see scheme), whereas the l_i-states are populated almost according to their statistical weight [20].

No detailed calulation exists for the dissociative decay $(pp\mu e)^* \longrightarrow (p\mu)^* + H$. As the

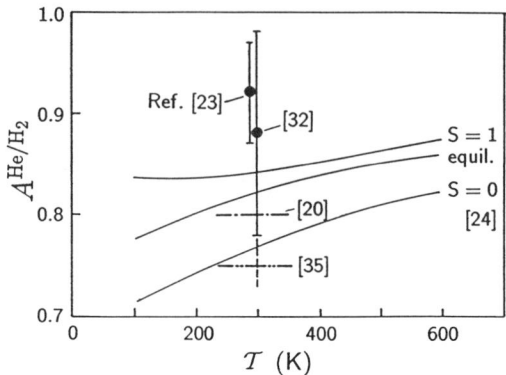

Fig. 1: Reduced capture ratio A^{He/H_2} for muons in a H_2-He gas mixture as a function of temperature T, for ortho- and para-states and their equilibrium concentration, from ref. [24]. The measured values are discussed in the text. The value from ref. [35] was calculated for atomic hydrogen.

dissociation energy is relatively low (≈ 5 eV), the resulting distribution of n_i-levels still extends up to $n \sim 25$, with a mean value around 17. The arguments given in [20] against this dissociation channel are not absolutely convincing. An analysis of measured μp-K_α intensities (presented in sect. 2.2) favours high initial n_i levels. The assumption of dissociative $(pp\mu e)^*$-decay is therefore in better agreement with the data and deserves further studies in order to establish more reliable initial cascade conditions.

2.2 K-line intensities of muonic hydrogen at low gas pressures

There are only a few data on radiative transitions with muonic hydrogen atoms [2]. Precise measurements of the relative K_α, K_β and $K_{>\beta}$ intensities have been performed at very low gas densities, using the "muon bottle" apparatus equipped with large-area xenon gas-scintillation proportional chambers [25]. A typical x ray energy spectrum is shown in Fig. 2. The resulting intensity ratios K_α/K_{tot} (Fig. 3) increase at lower pressures, because the Stark mixing rates decrease and the cascade thus proceeds more through circular states ($l = n - 1$), ending with a K_α transition.

At pressures as low as 10^{-3} bar ($\phi \sim 10^{-6}$), there is a high sensitivity to cascade processes occurring at levels $n \geq 10$, since the transitions below $n_c \sim 10$ are almost purely radiative. To reproduce the data, the cross sections for Stark effect etc. commonly assumed in the standard cascade model had to be suitably extrapolated for high n-levels [25]. The calculated intensity ratios K_α/K_{tot} depend crucially on the Stark-mixing rates and the initial distribution of n_i-levels, whereas they are quite insensitive to the cross sections of Auger and Coulomb effect. For the analysis shown in Fig. 3, two sets of initial conditions were assumed: (i) $n_i = 17$, l_i distributed statistically; (ii) n_i and l_i distributed as predicted in ref. [20]. In each case, the Stark-mixing parameter k_{Stark} was adapted to fit the data as well as possible. For case (ii) the calculation is not in good agreement with the measurements, in particular, if we consider that the experimental errors shown in Fig. 3 include systematic uncertainties common to all data points (e.g. detector efficiencies)

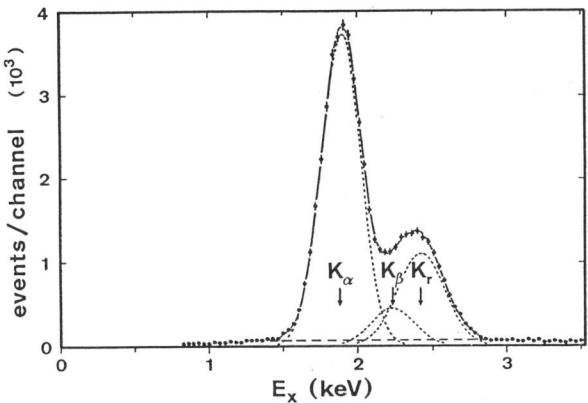

Fig. 2: Energy spectrum of μp K-line x rays at 8 mbar. The three components K_α, K_β, and $K_{>\beta}$ and a flat background are shown. The relative energy resolution is 16% (FWHM) at 1.9 keV.

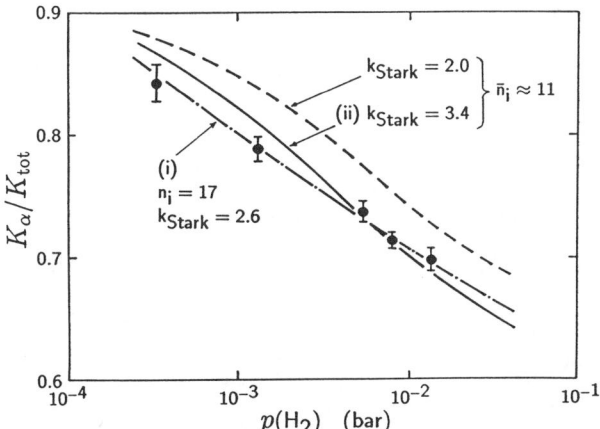

Fig. 3: Measured [25] and calculated intensity ratios K_α/K_{tot} for μp atoms at various gas pressures (1 bar = 10^5 Pascal) at room temperature. The lines (i) and (ii) correspond to the two different distributions of initial n_i-levels described in the text, with k_{Stark} optimized for the data. The dashed line demonstrates the sensitivity on the Stark mixing parameter k_{Stark}.

whereas the relative errors are smaller by about a factor of two. We conclude that the data favour the assumption of high initial n_i-levels with a mean value $\bar{n}_i \geq 16$, and that additional precise measurements of K_α/K_{tot} over a wider range of densities will provide sensitive restrictions on the distribution of initial states.

2.3 Absolute K-line yields of pionic hydrogen at medium densities

A measurement of the absolute yields of πp K-lines has recently been performed by a ETHZ-PSI-NEUCHÂTEL collaboration [26]. A high pion stop density was achieved at medium hydrogen gas densities (corresponding to pressures between 3 and 40 bar at room temperature) using the cyclotron trap [5,27]. The total K-yields are below 20% because most of the cascading pions undergo strong absorption at higher S-states. Table 1 summarizes the measured results together with those calculated with the standard cascade model in the version of Borie and Leon [28]. The optimum Stark parameter was $k_{Stark} = 1.95 \pm 0.20$, in good agreement with the value $k_{Stark} = 1.94 \pm 0.13$ determined from L-yield measurements with $\bar{p}p$ and $\bar{p}d$ atoms in the same apparatus [5]. The values resulting for the mean πp kinetic energy were $T = (4.5 \pm 0.8)$ eV at 3 bar, (1.2 ± 0.2) eV at 15 bar, (1.0 ± 0.2) eV at 40 bar, and (1.0 ± 0.2) eV for the $\bar{p}p$ measurements around 0.03 bar. Probable discrepancies of these values from recent more direct determinations of T by the neutron time-of-flight method [15] (mentioned at the end of sect. 1) indicate that the standard cascade model is not suitable for the analysis of kinetic energies. To know the distribution of T is, however, crucial for the interpretation of the high-precision spectroscopic experiment with πp atoms presently performed at PSI. This again demonstrates the need for better cascade codes.

Table 1: The absolute K-line yields (%) of pionic hydrogen, from ref. [26].

transition	p=3 bar		p=15 bar		p=40 bar	
	exp	theor	exp	theor	exp	theor
2-1	9.23±1.48	8.74	4.17±0.63	4.62	2.49±0.57	2.03
3-1	4.62±0.74	3.82	3.38±0.51	3.41	1.79±0.41	2.15
4-1	3.68±0.59	4.29	1.70±0.26	3.10	0.91±0.21	1.65
rest-1	2.51±0.40	2.46	1.16±0.17	1.00	0.62±0.14	0.46
Y_{tot}	20.04±1.80	19.31	10.41±0.87	12.13	5.81±0.75	6.29

3. Transfer from excited μp^* to μHe^* atoms

It has been demonstrated in the pioneering work of Petrukhin et al. [29,23] that excited-state transfer of pions to heavier nuclei $\pi p^* + Z \longrightarrow \pi Z^* + p$ is a very important cascade process in high-density gas mixtures. The number of πp atoms reaching low n-levels was determined via the reaction $\pi^- + p \longrightarrow \pi^0 + n$, as the π^0 decay is easy to detect. Measurements with H_2–He gas mixtures are particularly suitable to study atomic capture and cascade processes. Corresponding experiments with muons have been performed by adding small amounts of Xe ("triple-gas method") and measuring — at various He concentrations — x ray yields from transfer reactions $\mu p_{1S} + Xe \longrightarrow \mu Xe^* + p$ and, in addition, the muon decay rates [30,31].

It is obviously not easy to disentangle in these measurements the effects of transfer during the cascade and inital capture since neither the capture ratio nor the dependence of transfer upon concentration are known *a priori*. The problem has been overcome in an experiment performed at PSI by an ETHZ group [32] using the "muon bottle" apparatus. The K-line yields of μp and μHe atoms were measured in a H_2–He gas mixture at a total pressure of 5 bar, where excited-state transfer is an important effect, and also at 8 mbar, where transfer is almost negligible. The measuring pressures and concentrations are listed in Table 2. The low-energy x rays were detected by two xenon gas scintillation proportional chambers. A typical energy spectrum is shown in Fig. 4. The data presented here have

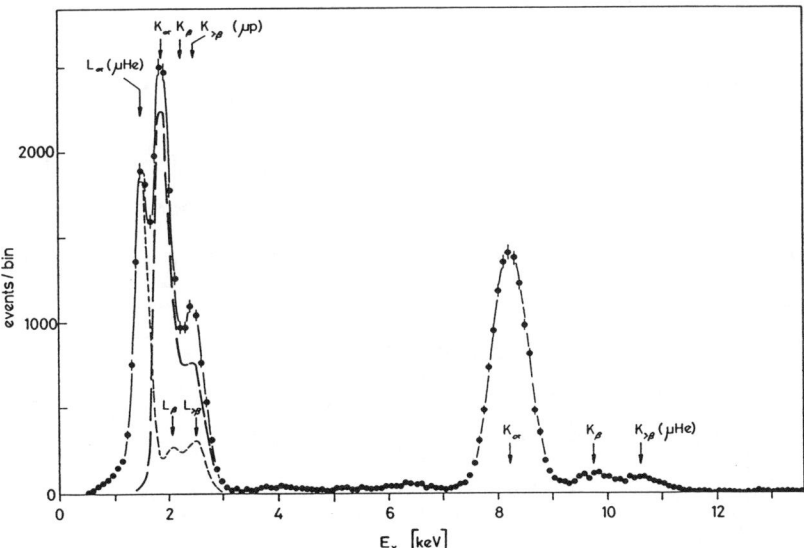

Fig. 4: Typical x ray energy spectrum measured with μ^- stopped in a 4 mbar H_2 + 4 mbar He gas mixture.

been taken in "1X-runs" where only one x ray had to be detected to define an event. At 5 bar, additional "2X-runs" have been performed where in each of the two chambers an x ray was registered. The details are given in ref. [33], where the experimental method is also described in more detail.

The determination of the intensities of μp-K_α (1.90 keV), μp-K_β (2.25 keV), and μp-$K_{>\beta}$ ($\sim$ 2.5 keV) is somewhat difficult because the μHe-L_β (2.05 keV) and $L_{>\beta}$ ($\sim$ 2.5 keV) -lines have to be considered as "background", as is evident from Fig. 4. The problem was mastered at 8 mbar by fixing the (quite small) intensity ratios $K_\beta/K_{tot}(\mu p)$ and $L_\beta/L_{tot}(\mu \text{He})$ at the values measured in pure H$_2$ or He [25,34] and determining $K_{>\beta}/K_{tot}(\mu p)$ and $L_{>\beta}/L_{tot}(\mu \text{He})$ from the simultaneous fit of the energy spectra at both concentrations. (Additional systematic uncertainties have been adopted to include possible deviations at the different concentrations.) At 5 bar, L_β/L_{tot} and $L_{>\beta}/L_{tot}(\mu \text{He})$ was determined from a particular analysis of the 2X-runs: A pure spectrum of μHe L-lines (free of μp-lines) was obtained by requiring one of the x rays to be a μHe-K_α (8.22 keV) transition.

The resulting K-line intensity ratios are summarized in Table 2. The errors include a systematic uncertainty in the detection efficiency of the xenon chambers at 2 and 8 keV.

Table 2: Ratios of muonic K-line intensities measured in H$_2$–He gas mixtures, reduced ratios $R_{1S}^{\text{He}/\text{H}_2}$ of the corresponding ground-state populations, probabilities W that the μ reaches the μp_{1S}-level (or the π undergoes charge exchange), and probabilities q_{1S}^{He} that no excited-state transfer to helium occurs.

particle	μ^-	μ^-	μ^-	μ^-	π^-
Ref.	[32]	[32]	[32]	[31]	[23]
p(H$_2$) [a]	6.4 mbar	4 mbar	2.5 bar	20 bar	40 bar
p(He)	1.6 mbar	4 mbar	2.5 bar	20 bar	40 bar
$n_{\text{He}}/n_{\text{H}_2}$	0.25	1	1	1	1
$\frac{K_{tot}(\mu\text{He})}{K_{tot}(\mu p)}$	0.217 ± 0.024	1.02 ± 0.14	2.61 ± 0.44	–	–
$R_{1S}^{\text{He}/\text{H}_2}$	0.88 ± 0.10	1.04 ± 0.14	2.75 ± 0.47	3.76 ± 0.25 [b]	–
W	0.82 ± 0.02	0.49 ± 0.04	0.27 ± 0.03	0.21 ± 0.01 [c]	0.30 ± 0.02 [d]
q_{1S}^{He}	≈ 1.0	0.92 ± 0.08	0.51 ± 0.07	0.40 ± 0.04	0.57 ± 0.05 [d]

[a] 1 bar = 10^5 Pascal.
[b] calculated from value [c].
[c] corresponding to eq. (9) of ref. [31].
[d] from reaction $\pi^- p \to \pi^0 n$; the πp-1S level is not reached.

To get the reduced ratios

$$R_{1S}^{He/H_2} = \frac{N(\mu He_{1S})/N(\mu p_{1S})}{n_{He}/n_{H_2}} \qquad (2)$$

of μHe and μp atoms reaching the ground state, small corrections of the measured yields $K_{tot}(\mu He)$ were applied in order to consider the effect of non-radiative quenching of μHe-$2S$-states [32]. The probability W that a muon reaches the μp ground state is

$$W = \frac{N(\mu p_{1S})}{N(\mu p_{1S}) + N(\mu He_{1S})} = \frac{1}{1 + R_{1S}^{He/H_2}(n_{He}/n_{H_2})}. \qquad (3)$$

In Table 2, the corresponding values are presented together with typical results from the DUBNA experiments with muons [31] and pions [23]. (These measurements have been performed at various helium concentrations.) W is the product of the probability W_H of atomic capture by hydrogen, and the probability q_{1S}^{He} that the muon does not transfer to helium during the cascade:

$$W = W_H \, q_{1S}^{He}. \qquad (4)$$

W_H is connected with the reduced capture ratio A^{He/H_2} (eq. (1)) by

$$W_H = \frac{1}{1 + A^{He/H_2}(n_{He}/n_{H_2})}. \qquad (5)$$

A^{He/H_2} depends neither on density nor on concentration (possible variations upon the He concentration [19,35] are negligible for this analysis). q_{1S}^{He} depends both on concentration and density, but the corresponding functional dependences are not known since complex balances of cascade processes are involved. The empirical function suggested in ref. 23 is not universal as it neglects the density dependence. The value $A^{He/H_2} = 0.92 \pm 0.05$ deduced from the pion data is therefore questionable. A priori, it follows from $q_{1S}^{He} \leq 1$ that $A^{He/H_2} \leq R_{1S}^{He/H_2}$. From the low-density data we get

$$A^{He/H_2} \leq 0.88 \pm 0.10 \qquad (6)$$

as an upper limit for the initial capture ratio, where the lower 8-mbar value for R_{1S}^{He/H_2} — corresponding to the lower He concentration — was taken, for the following reason: Comparing the 8-mbar data at the two different concentrations, the values given in Table 2 for R_{1S}^{He/H_2} seem to be compatible with each other, but an accurate error analysis shows that their ratio is $0.88/1.04 = 0.85 \pm 0.05$. The significant deviation of this "ratio of ratios" from unity indicates that R_{1S}^{He/H_2} and also q_{1S}^{He} depend upon concentration even at 8 mbar, meaning that there is still some transfer at high n-levels. This is not surprising because collisional cascade processes are important even at the lowest densitites, as is evident from

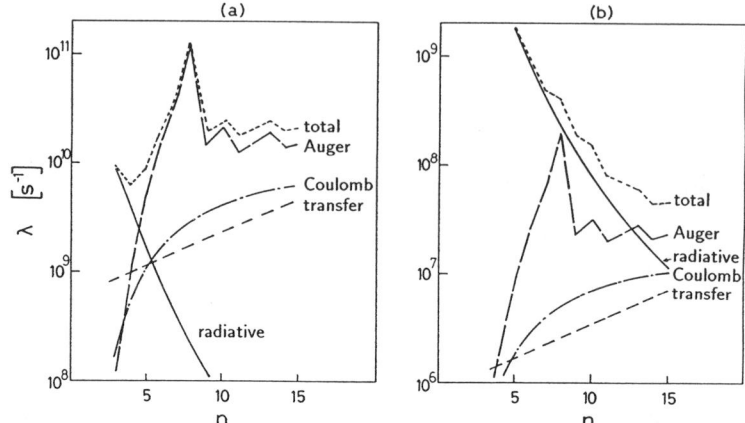

Fig. 5: Averaged deexcitation and transfer rates for μp atoms in a H_2–He gas mixture, in dependence of the n-level. (a) Rates for $p(H_2) = p(He) = 2.5$ bar, (b) rates for $p(H_2) = p(He) = 4$ mbar. The curves originate from [2] and ref. therein; the transfer rates are estimated in order to give qualitative agreement with the data.

Fig. 5. It follows that the capture ratio is probably somewhat lower than the value listed in eq. (6).

A value $A^{He/H_2} = 0.88$ was adopted to get (from eq. (3)–(5)) the probabilities $q_{1S}^{He} = W/W_H$ listed in Table 2, but the errors include a possible value $A^{He/H_2} = 0.75$. Fig. 6 shows q_{1S}^{He} as a function of density for equal partial pressures $n_{He} = n_{H_2}$, together with the theoretical prediction of Bystritsky et al. [36]. The measured q_{1S}^{He}-values can be understood qualitatively from the scheme given in Fig. 5 if one assumes transfer rates an order of magnitude smaller than the total deexcitation rate at $n \sim 5$. As the muon cascades down by Auger transitions favouring $\Delta n = 1$ steps at $n < 9$, there are several subsequent 10%-probabilities for transfer, summing up to the measured values of about 50% for $1 - q_{1S}^{He}$. The pionic value for q_{1S}^{He} is considerably higher than the muonic ones, because the pions annihilate before reaching the ground state ("q_{5S}^{He}", rather than q_{1S}^{He}), and transfer is therefore restricted to $n > 4$ levels.

The calculations given in ref. 36 predict a weak dependence of q_{1S}^{He} on the kinetic energy, but a new study of Solovjov and Ponomarev [37] indicate a significant energy dependence. This probably will allow to determine the kinetic energy of the μp atom at the levels where transfer occurs.

Additional information on the transfer process comes from the μHe K-line intensities measured in the H_2–He mixture at 5 bar. A fraction $1 - W_H \approx 47\%$ of the muons is directly captured by helium, whereas another fraction $W_H(1 - q_{1S}^{He}) \approx 26\%$ forms muonic helium via transfer process. Hence it can be stated that the observed μHe cascade is the sum of a "normal" helium cascade (with a relative probability of $47\%/(47 + 26)\% \approx \frac{2}{3}$) and a

Kinetic Energies at the Formation and Cascade of μp-Atoms

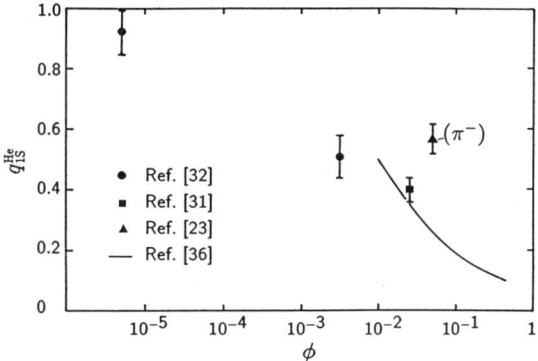

Fig. 6: Measured and calculated values for q_{1S}^{He} as a function of the density ϕ ($\phi = 1$ is the density of liquid hydrogen, i.e. 4.25×10^{22} atoms/cm^3). For simplicity, only the case of equal partial pressures $n_{He} = n_{H_2}$ is shown.

"transfer cascade" ($\approx \frac{1}{3}$). The "normal cascade" has been measured in pure helium gas at 6 bar [34]. We assume that the intensity ratios of the "normal cascade"-part in the gas mixture do not differ significantly from those measured in the pure gas. It is therefore possible to get the intensity ratios of the "transfer cascade" by subtracting the suitably weighted pure-gas data from the mixed-gas data, as is listed in Table 3. It turns out that the "transfer cascade" has a quite high K_β intensity and practically no radiative K-transitions from higher n-levels. This is what one expects qualitatively since the corresponding transfer reaction occurs at medium n-levels, and subsequent internal Auger deexcitations populate even lower μHe states before radiative transitions become dominant.

The next step will be to compare these data with refined cascade calculations. The transfer reaction to helium turns out to be an important tool to study the μp cascade, in particular the kinetic energies at medium n-levels.

Table 3: Muonic K-line intensity ratios measured in pure He gas at 6 bar and in a H_2–He mixture at 5 bar. Suitable subtraction leads to the values for the "transfer cascade", as described in the text.

Condition	K_α/K_{tot}	K_β/K_{tot}	$K_{>\beta}/K_{tot}$
6 bar He	0.57 ± 0.02	0.11 ± 0.01	0.32 ± 0.02
5 bar He+H_2	0.58 ± 0.02	0.21 ± 0.02	0.21 ± 0.02
"transfer cascade"	0.60 ± 0.15	0.42 ± 0.10	0 ± 0.10

References

[1] J.S. Cohen, in *EM Cascade and Chemistry of Exotic Atoms*, edited by L.M. Simons et al. (Plenum, New York, 1990), p. 1.

[2] V. Markushin, in *EM Cascade and Chemistry of Exotic Atoms*, edited by L.M. Simons et al. (Plenum, New York, 1990), p. 73.

[3] M. Leon and H.A. Bethe, Phys. Rev. **127**, 636 (1962).

[4] G. Reifenröther and E. Klempt, Nucl. Phys. A **503**, 885 (1989).

[5] K. Heitlinger et al., Z. Phys. A **342**, 359 (1992).

[6] A.J. Rusi El Hassani et al., in *EM Cascade and Chemistry of Exotic Atoms*, edited by L.M. Simons et al. (Plenum, New York, 1990), p. 105.

[7] A.J. Rusi El Hassani, PhD thesis ETH Zürich No. 9256 (1990).

[8] J.B. Kraiman et al., Phys. Rev. Lett. **63**, 1942 (1989).

[9] A.V. Kravtsov et al., Muon Cat. Fusion **2**, 199 (1988).

[10] A.V. Kravtsov et al., Muon Cat. Fusion **2**, 183 (1988).

[11] L.I. Menshikov and L.I. Ponomarev, Z. Phys. D **2**, 1 (1986).

[12] J.F. Crawford et al., Phys. Rev. D **43**, 46 (1991).

[13] L.I. Menshikov, Muon Cat. Fusion **2**, 173 (1988).

[14] V.E. Markushin, E.I. Afanasieva, and C. Petitjean, Muon Cat. Fusion **7**, 155 (1992).

[15] A. Badertscher et al., "Determination of the Kinetic Energy of Pionic Hydrogen Atoms with a Neutron Time-of-Flight Measurement", Addendum to the PSI-Proposal R 86-05-1 (1991).

[16] E.C. Aschenauer, private communication.

[17] F.J. Hartmann, in *EM Cascade and Chemistry of Exotic Atoms*, edited by L.M. Simons et al. (Plenum, New York, 1990), p. 23.

[18] M. Leon, in *Exotic Atoms '79*, edited by K. Crowe et al. (Plenum, New York, 1980), p. 141.

[19] V.V. Balashov et al., Muon Cat. Fusion **2**, 105 (1988).

[20] G.Ya. Korenman, V.P. Popov, and G.A. Fesenko, Muon Cat. Fusion **7**, 179 (1992).

[21] H. Anderhub et al., Phys. Lett. **101B**, 151 (1981).

[22] P. Hauser et al., "Slowing down of negative muons in gaseous H_2 and determination of the stopping power", in these Proceedings.

[23] V.I. Petrukhin and V.M. Suvorov, Zh. Eksp. Teor. Fiz. **70**, 1145 (1976) [Sov. Phys. JETP **43**, 595 (1976)].

[24] G.Ya. Korenman, S.V. Leonova, and V.P. Popov, Muon Cat. Fusion **5/6**, 49 (1990/91).

[25] H. Anderhub et al., Phys. Lett. **143B**, 65 (1984).

[26] A.J. Rusi El Hassani, PhD thesis ETH Zürich No. 9256 (1990).

[27] L.M. Simons, Phys. Scrip. **T22**, 90 (1988).

[28] E. Borie and M. Leon, Phys. Rev. A **21**, 1460 (1980).

[29] V.I. Petrukhin et al., Zh. Eksp. Teor. Fiz. **55**, 2173 (1968) [Sov. Phys. JETP **28**, 1151 (1969)].

[30] A. Bertin et al., Lett. Nuo. Cim. **18**, 381 (1977).

[31] V.M. Bystritsky et al., Zh. Eksp. Teor. Fiz. **84**, 1257 (1983) [Sov. Phys. JETP **57**, 728 (1983)].

[32] F. Kottmann, in *Muons and pions in matter*, Proc. of the Int. Symposium of Muon and pion interaction with matter, JINR D14-87-799 (Dubna, 1987), p. 268.

[33] H.P. von Arb et al., Muon Cat. Fusion **4**, 61 (1989).

The analysis of the 2X-runs gives evidence for muon molecular x rays at energies around 7 keV due to radiative deexcitation of muonic molecules $(p\mu\,^4{\rm He})^*$, which are formed as intermediate states in the ground-state transfer reaction $\mu p_{1S} + {\rm He} \longrightarrow \mu {\rm He}_{1S} + p$.

[34] H.P. von Arb et al., Phys. Lett. **136B**, 232 (1984); F. Dittus, PhD thesis ETH Zürich No. 7877 (1985).

[35] J.S. Cohen, R.L. Martin, and W.R. Wadt, Phys. Rev. A **27**, 1821 (1983).

[36] V.M. Bystritsky, A.V. Kravtsov, and N.P. Popov, Muon Cat. Fusion **5,6**, 487 (1990/91).

[37] L.I. Ponomarev, private communication.

Slowing Down of Negative Muons in Gaseous H_2 and Determination of the Stopping Power

P. HAUSER, F. KOTTMANN, CH. LÜCHINGER, R. SCHAEREN
Institut für Hochenergiephysik, ETH-Hönggerberg, CH-8093 Zürich, Switzerland

While negative muons are slowing down they excite atoms of the target gas and thus produce scintillation light. By measuring the time distribution of this scintillation light relative to the muon stop we determined the stopping power of μ^- in H_2 in the velocity region $10^{-3} < \beta < 10^{-1}$. Around the stopping power maximum at $\beta \sim 10^{-2}$ we determined dE/dx to an accuracy of ±15%. The maximum stopping power of μ^- attains only 56% of the corresponding proton value.

1. Introduction

Until now there has been only little data about the stopping power dE/dx and the ionisation cross section of heavy negative particles at energies corresponding to the Bohr velocity v=αc.

For antiprotons the stopping power was measured in thin layers of Si [1] and Au [2] at velocities $0.02 \leq \beta \leq 0.08$. In gases, the ionisation cross sections in He and H_2 were measured down to β=0.015 [3] (see also Fig. 4).

For negative muons the stopping power was determined in different solids (Al, Au, MgF_2, C) by the Munich-PSI group [4,5,6]. A few years ago we measured dE/dx in He [7] and more recently in gaseous H_2 which is the subject of this work.

2. Apparatus and measurements

The experiment was performed with the muon bottle apparatus [8] at PSI. The main components of the muon bottle are shown in Fig. 1. Pions of 40 MeV/c momentum are detected by a proportional chamber (DLK) and injected axially into a magnetic bottle filled with the target gas. Some of the muons (originating from pions decaying in flight) have very low momenta in the lab system and are trapped on helical orbits. These muons are then slowed down due to excitation and ionisation of target atoms until they form a muonic atom. Typical slowing down times are a few μs at gas pressures around 1 hPa. The uv light from primary scintillation produced during the slowing down process is detected by photomultipliers coated with a wavelength shifter. The L- and K-line x-rays from the muonic cascade were detected in two large multi-wire proportional chambers (MWPC).

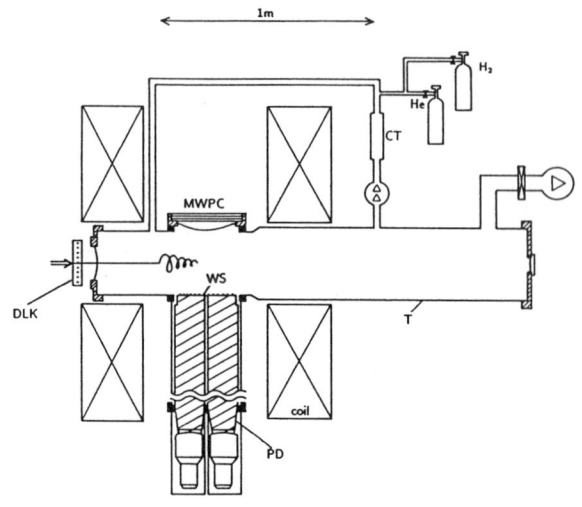

Fig. 1: Experimental set-up: DLK = π^- entrance detector, MWPC = multi-wire proportional chamber, PD = photon detector (wavelength shifter (WS), lucite light guide and RCA 8854 photomultipliers), T = stainless steel tank.

The principle of the experiment was to measure the time distribution of the scintillation light L(t) relative to the muon stop and to deduce from L(t) the energy loss dE/dx. A gas mixture of 80% H_2 + 20% He was used. The muon stop was identified by the detection of a μHe-(K_α, L)-coincidence. Thus only muons captured by He atoms were considered. For our target gas mixture the probability for a muon to be captured by a He atom was ≈ 15% per muon stop, as measured separately [7]. The He admixture had two advantages. The 4 times larger x-ray energies (compared to H_2) from the muonic cascade allowed us to detect also the L-lines (Fig. 2) which was necessary to define a background-free signature for a muon stop in the gas. In addition the cascade time in He is much faster than in H_2 and thus the determination of the

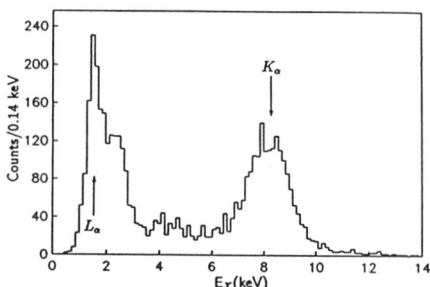

Fig. 2: Energy spectrum of x-rays for events with two coincident x-rays in each detector in 20%He+80%H_2 at P=1.2 hPa.

muon stop time (t=0) is more precise. On the other hand we had to add as little He as possible to limit its influence to the stopping power measurement of H_2. Fortunately He emits less scintillation light than H_2 and thus the admixture of 20% He gave only a minor correction which was calculated from the earlier measurement of dE/dx in pure He [7].

We measured at three different total pressures, 0.4 hPa, 1.6 hPa, and 8 hPa. Fig. 5 shows the measured scintillation light distribution for all 3 pressures. For each event we recorded the times of the uv-photons, typically 1-10 photons per muon stop. By summing up over all events of a given pressure the shown distributions resulted. With decreasing pressure the shoulder of the measured light distributions shifts towards negative times. The peak at t=0 which is clearly visible for the lowest pressure is background which is not correlated to muon stops in the gas.

3. Determination of dE/dx from the scintillation light distribution L(t)

To determine the stopping power dE/dx from the measured scintillation light distribution we have to investigate the slowing down process in more detail. The stopping power can be separated in an excitation- and an ionisation part. Considering the stopping power per molecule $S(\beta)$ we get (see also Ref. [9])

$$S(\beta) = S_{exc}(\beta) + S_{ion}(\beta) \qquad (1)$$
$$= \bar{E}_{exc}\sigma_{exc}(\beta) + (E_{ion} + \bar{E}_e(\beta))\sigma_{ion}(\beta)$$

with the cross sections $\sigma_{exc}(\beta)$ for excitation and $\sigma_{ion}(\beta)$ for ionisation. The mean excitation energy for H_2 is $\bar{E}_{exc} = 12 \pm 1$ eV [10] (corresponding to the excitation levels in H_2) and the ionisation energy is 15.6 eV. The mean kinetic energy of the ionized electron $\bar{E}_e(\beta)$ was measured by Rudd et al. [11] for $p + H_2$ and found in agreement with a classical trajectory Monte Carlo calculation by J. Cohen for $\mu^- + H$ [12]. By multiplying eq.(1) with the muon velocity βc and molecular density N we get

$$\underbrace{\beta(t)c\overbrace{NS(\beta)}^{dE/dx}}_{dE/dt} = Nc[\bar{E}_{exc}\underbrace{\sigma_{exc}(\beta)\beta(t)}_{prop\ \tilde{L}(t)} + (E_{ion} + \bar{E}_e(\beta))\sigma_{ion}(\beta)\beta(t)] \ . \qquad (2)$$

The excitation term in eq. (2) is proportional to the scintillation light $\tilde{L}(t)$ produced per unit of time. Taking into account the efficiency $\varepsilon(\lambda)$ of the wavelength shifter and photomultiplier and the solid angle Ω of the light detector, the light distribution $\tilde{L}(t)$ becomes

$$\tilde{L}(t) = \int \varepsilon(\lambda) \cdot \Omega \cdot \tilde{\sigma}_{exc}(\lambda, \beta(t)) \cdot N \cdot \beta(t) \cdot c \cdot d\lambda \qquad (3)$$

where the cross section $\tilde{\sigma}_{exc}(\lambda, \beta(t))$ includes excitation and deexcitation via emission of a photon with wavelength λ. Since the wavelength spectrum produced by deexcitation of H_2^* is independent of the impact velocity β [10] we can separate the variables of excitation and radiative deexcitation:

$$\tilde{\sigma}_{exc}(\lambda, \beta(t)) \approx A(\lambda)) \cdot \sigma_{exc}(\beta) \qquad (4)$$

The efficiency of our wavelength shifter is nearly constant in the corresponding wavelength region. Thus we have

$$\tilde{L}(t) = \underbrace{\bar{\varepsilon} \cdot \bar{A} \cdot \Omega \cdot c \cdot N}_{const} \cdot \sigma_{exc}(\beta(t)) \cdot \beta(t) \qquad (5)$$

and the energy loss per unit of time becomes

$$dE/dt = Nc\left[\bar{E}_{exc}\frac{\tilde{L}(t)}{const} + (E_{ion} + \bar{E}_e(\beta))\sigma_{ion}(\beta)\beta(t)\right] . \qquad (6)$$

We have now dE/dt as a function of time with the measured contribution of the scintillation light and the contribution from ionisation. For the ionisation cross section we take the data measured for antiprotons [3], and for $\bar{E}_e(\beta)$ the values from Ref. [11,12]. To deduce dE/dx as a function of β we have to iterate the following steps:

- for t=0 we start with a kinetic muon energy corresponding to the capture energy $E_{capt} \approx 15$ eV [12];

- for t<0 we deduce $\beta(t)$ by integrating over the energy loss per unit time as given by eq.(6), $\beta(t) = (\frac{2}{m_\mu} \int_t^0 \frac{dE}{dt}(t')dt')^{1/2}$;

- at $\beta=0.1$ we normalize to the Bethe-Bloch value which at this velocity is valid within a few precent.

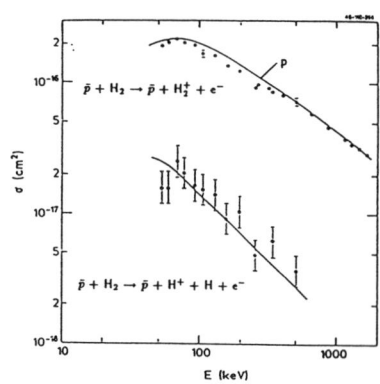

Fig. 3: Ionisation cross sections of $\bar{p} + H_2$ for dissociative and non-dissociative ionisation. The data are taken from Ref. [3].

With that normalization we fix the unknown values of the apparative efficiency $\bar{\varepsilon} \cdot \Omega$ and of the radiative deexcitation term $\bar{A}$. This procedure allows us to determine from the scintillation light distribution $\tilde{L}(t)$ the light distribution as a function of β and thus to fix essentially the energy loss dE/dx(β) and excitation cross section $\sigma_{exc}(\beta)$. However the measured scintillation light distribution $L(t)$ differs slightly from $\tilde{L}(t)$ defined by eq. (5). In $\tilde{L}(t)$ we did not consider variations of the μ^- slowing down times due to different initial energies, variations of the solid angle during the slowing down process, the deexcitation time of an excited H_2^* molecule ($\tau_{deexc} \sim$ 15 ns at 1 hPa, measured separately with an α source), and the time resolution of the x-ray detector. Therefore a more detailed investigation was necessary which considered the mentioned effects. With a Monte Carlo simulation we reproduced the measured data following the μ^- trajectory and using for the slowing down process the iterated values $S_{H_2}(\beta)$ and $\sigma_{exc}(\beta)$ (eq. (6)) and the known He data (S_{He}, σ_{exc}^{He}). The resulting scintillation light distribution was then compared to the measured data (Fig. 4). An optimal fit of L(t) to the data was achieved by suitable adjustments of $\tilde{L}(t)$ and repeating the whole procedure seve-

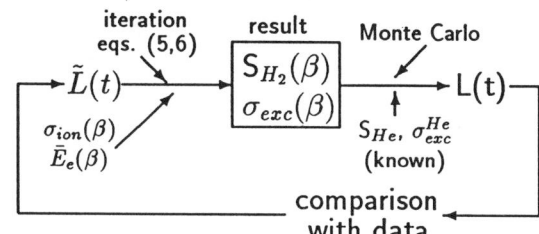

Fig. 4: Data processing to reproduce the measured time distribution L(t) of uv photons.

ral times. Fig. 5 shows the best Monte Carlo results using for all three pressures identical S_{H_2} and σ_{exc}. The agreement of simulated and measured L(t) is good for all pressures.

4. Results and discussion

Before coming to the results of the stopping power $dE/dx(\beta)$ we shall discuss some sources of possible errors. To estimate the accuracy of the measurement we analyzed different types of error sources:

- we investigated deviations of the simulated L(t) from the data up to 3σ ("statistical errors") which caused significant deviations in the maximum of the stopping power (at $\beta = 10^{-2}$).

- we changed the normalization factor at β=0.1 up to $\pm$ 10% relative to the Bethe-Bloch value which gave only small deviations arround the maximum stopping power.

- we admitted 20%-deviations of $\sigma_{ion}(\beta)$ from the antiproton data [3] at $\beta > 10^{-2}$ and free values for $\sigma_{ion}(\beta)$ at $\beta < 10^{-2}$, where no data exist.

In addition we performed a "simple" analysis without using the $\bar{p}$ data for $\sigma_{ion}(\beta)$. In this analysis we assumed $\sigma_{ion}(\beta) \propto \sigma_{exc}(\beta)$ and a constant mean kinetic energy of the ionized electrons ($\bar{E}_e(\beta)$=const). The resulting energy loss $S_{simple}(\beta(t))$ is then proportional to $\tilde{L}(t)$. As before, $S_{simple}(\beta)$ was normalized at $\beta = 0.1$ to the Bethe-

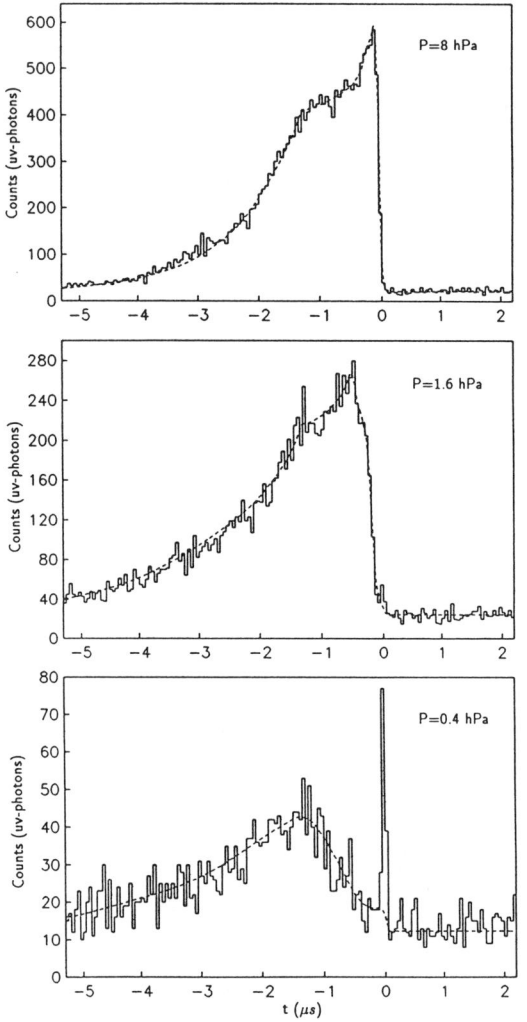

Fig. 5: Measured time distribution of uv photons summed up for all events at a given pressure. The curves correspond to the optimum Monte Carlo fit using identical parameters for all three pressures.

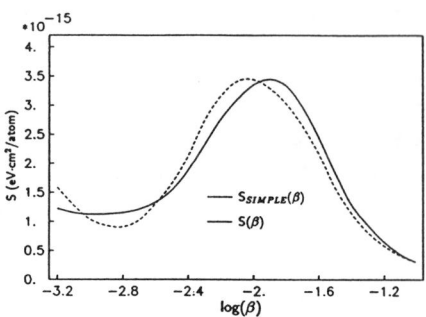

Fig. 6: Comparison between the stopping power $S_{simple}(\beta)$ where no $\bar{p}$ data is used and $S(\beta)$ based on the $\bar{p}$ data of Ref.[3].

Bloch value. Fig. 6 shows a comparison between the stopping power $S_{simple}(\beta)$ and $S(\beta)$ where the $\bar{p}$ data were used. The small deviations between both results indicate that measuring the scintillation light produced during the slowing down process of negative muons leads to a very sensitive determination of the stopping power. In Fig. 7 our final result for H_2 is shown and compared to the proton data from Ziegler et al. [13]. The error bars include the various possible deviations discussed above. The maximum of the stopping power function at $\beta = 10^{-2}$ attained for negative muons is only about 56% of the proton value. The CTMC calculation [12] for negative muons in H-atoms gives for the lowest energies a considerably higher value. This indicates that molecular effects dominate in the lowest energy region.

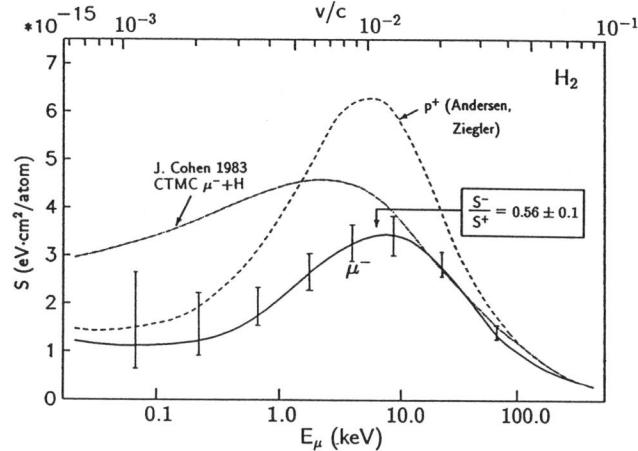

Fig. 7: Stopping power in H_2. The solid line corrresponds to the present result for μ^-, the dashed line to the proton data of Ref. [13], and the dashed-dotted line to the CTMC calculation for μ^- in atomic hydrogen of Ref. [12].

References

[1] L. H. Andersen et al., Phys. Rev. Lett. **62**, 1732 (1989).
[2] K. Elsener et al., PSI Newsletter (1990).
[3] L. H. Andersen et al., J. Phys. B. **23**, L395 (1990).
[4] W. Wilhelm et al., Phys. Lett. **98B**, 33 (1981).
[5] W. Schott et al., to be published in Phys. Rev. Lett.
[6] F. J. Hartmann et al., this Proceedings.
[7] F. Kottmann in *Muons and Pions in Matter*, Dubna 1987, p. 268.
[8] H. Anderhub et al., Phys. Lett. **101B**, 151 (1981).
[9] L. Vegh, Phys. Rev. **A37**, 1942 (1988).
[10] J. M. Ajello et al., Phys. Rev. **A25**, 2485 (1982).
[11] M. E. Rudd, Phys. Rev. **A20**, 787 (1979).
[12] J. Cohen, Phys. Rev. **A27**, 167 (1983).
[13] H. H. Andersen and J. F. Ziegler, in *Hydrogen Stopping Powers and Ranges in all Elements, The Stopping and Ranges of Ions in Matter* (Pergamon, New York, 1977), Vol. 3.

Diffusion of Muonic Deuterium and Hydrogen Atoms

D. J. ABBOTT[1], J.B. KRAIMAN[1], R.T. SIEGEL[1], W.F. VULCAN[1], D.W. VIEL[1],
C. PETITJEAN[2], A. ZEHNDER[2], W.H. BREUNLICH[3], P. KAMMEL,[3]
J. MARTON[3], J. ZMESKAL[3], J.J. REIDY[4], H. WOOLVERTON[4],
F.J. HARTMANN[5]

presented by R.T. Siegel

[1] College of William and Mary in Virginia, Williamsburg, VA 23185
[2] Paul Scherrer Institute, CH-5232 Villigen, Switzerland
[3] Oesterreichische Akademie der Wissenschaften, A-1030 Vienna, Austria
[4] University of Mississippi, Oxford, MS 38667
[5] Technische Universität München, D-8046 Garching, Germany

The analysis of an extensive PSI experiment on the diffusion of muonic hydrogen (μp and μd) atoms in hydrogen and deuterium gas is described. The present state of theoretical calculations of the scattering cross sections for such atoms has progressed such that the analysis of the μd scattering can now present a good statistical fit to the data, although certain physical anomalies remain. The μp analysis still does not yield a good fit, but current development of the scattering theory offers prospects of improved fits to the data.

The "initial" velocity distributions of the muonic hydrogen atoms, defined when the atoms reach the 1s state, are fitted to the diffusion data, and indicate mean energies for the μd and μp of about 1 and 2 eV respectively.

Results of the diffusion experiments are expected to determine the pressure conditions under which the study of muon capture in a statistical mixture of singlet and triplet of μp atoms can be measured, as well as providing information about epithermal effects in muon catalyzed fusion at low pressures.

1. Description of the Experiment

In the present state of beam development, the experimental determination of the properties of muonic hydrogen atoms is possible mainly through the diffusion type of experiment pioneered by Zavattini and collaborators at CERN [1]. These properties include specifically the velocity distribution of the atoms (assumed to have reached the 1s state), and the various cross-sections for scattering of such atoms on the molecules of the (gaseous) hydrogen targets in which the muonic atoms are formed by stopping negative muons. The diffusion experiments carried out by the PSI collaboration in 1988 [2] used an arrangement in which 9 μm thick plastic foils of 10 cm diameter, each coated with 100 Å of Au on both surfaces, were placed in a target chamber with gas at pressures between 47 mbar and 1520 mbar. A negative muon beam of 30-35 MeV/c was stopped in the chamber, with hydrogen at one bar representing about 1% of the total stopping power. Incident muons were signaled by exterior scintillation counters. Muonic hydrogen atoms formed in the interstices between the foils diffused to the foils (or decayed en route), with their arrival at the foil surfaces being followed by transfer to Au atoms. An array of four Ge detectors was used to detect either the Au muonic x-rays following the transfer or nuclear gamma rays from muon absorption in Au. The latter process produces 356 keV gammas from ^{196}Pt with a yield of (35 ± 5)% [3] which were the dominant transfer signals in our detectors. The muon lifetime of ≈ 72 ns in Au results in a corresponding delay in emission of these photons after the transfer.

Time distributions of the ^{196}Pt gamma-rays were obtained for hydrogen and deuterium fillings of the chamber at several pressures, and with two different foil spacings of 0.23 cm ("single spacing") and 0.46 cm ("double spacing"). Fig. 1 shows two such time distributions as directly recorded, i.e., with the effect of muon decay still present. The distribution can be described as an initial peak followed by an extended tail. The time at which the event rate begins to fall is associated with normal traversal of a foil gap by muonic atoms having the mean value of the initial velocity distribution. (The association is perfect if the velocity distribution is a delta function and there is no scattering.) We also note a rise in the D_2 data at early times due to the previously mentioned muon lifetime in Au. This effect is also present in the H_2 data, but at the 94 mbar pressure, background from direct muon stops in the 100 Å Au foil coatings is excessive at times less than 100 ns, so those points do not appear.) From subsequent analysis we have concluded that the two pressures for which data is displayed in Fig. 2 involve moderate amounts of scattering before the muonic atoms strike the foils. Therefore these data are quite sensitive to the initial velocities of the muonic atoms. With that fact in mind, it is striking that the two time distributions are so similar, especially since the data are for single foil spacing in D_2 vs. double spacing in H_2. It thus appears that the average initial velocity (and kinetic energy) of a μp atom is about twice that of a μd at these pressures.

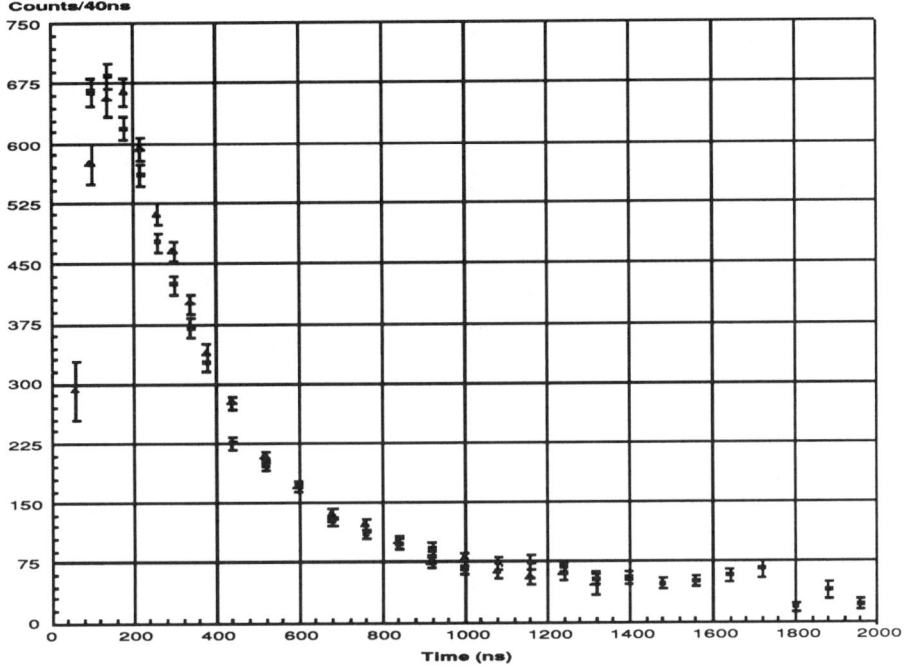

Figure 1. Muonic Hydrogen and Deuterium Time Distributions for Two Conditions
Triangles = μp at 94 mbar, double space; squares = μd at 188 mbar, single space

Data were taken for D_2 with 50 foils at single spacing at 94, 188, 375, 750 and 1520 mbar, and at double spacing with 25 foils at 188, 375, and 750 mbar. Fig. 2 displays these D_2 data along with the fits described below. (In Fig. 2 the effect of muon decay has been removed.) For H_2 the double spacing was alone used at pressures of 47, 94, 188, 375 and 750 mbar. At the higher pressures there were at least 6-8000 events registered over the 2 μs measurement interval, with somewhat fewer recorded at the lowest pressures.

2. Method of Analysis of the Experiment

The purpose of the experiment was to gain information about both the initial velocity distributions for the muonic atoms and also about the scattering processes which affect the motion of the atoms as they progress from their points of formation between the foils to one of the foil surfaces. The scattering of μp (or μd) involves

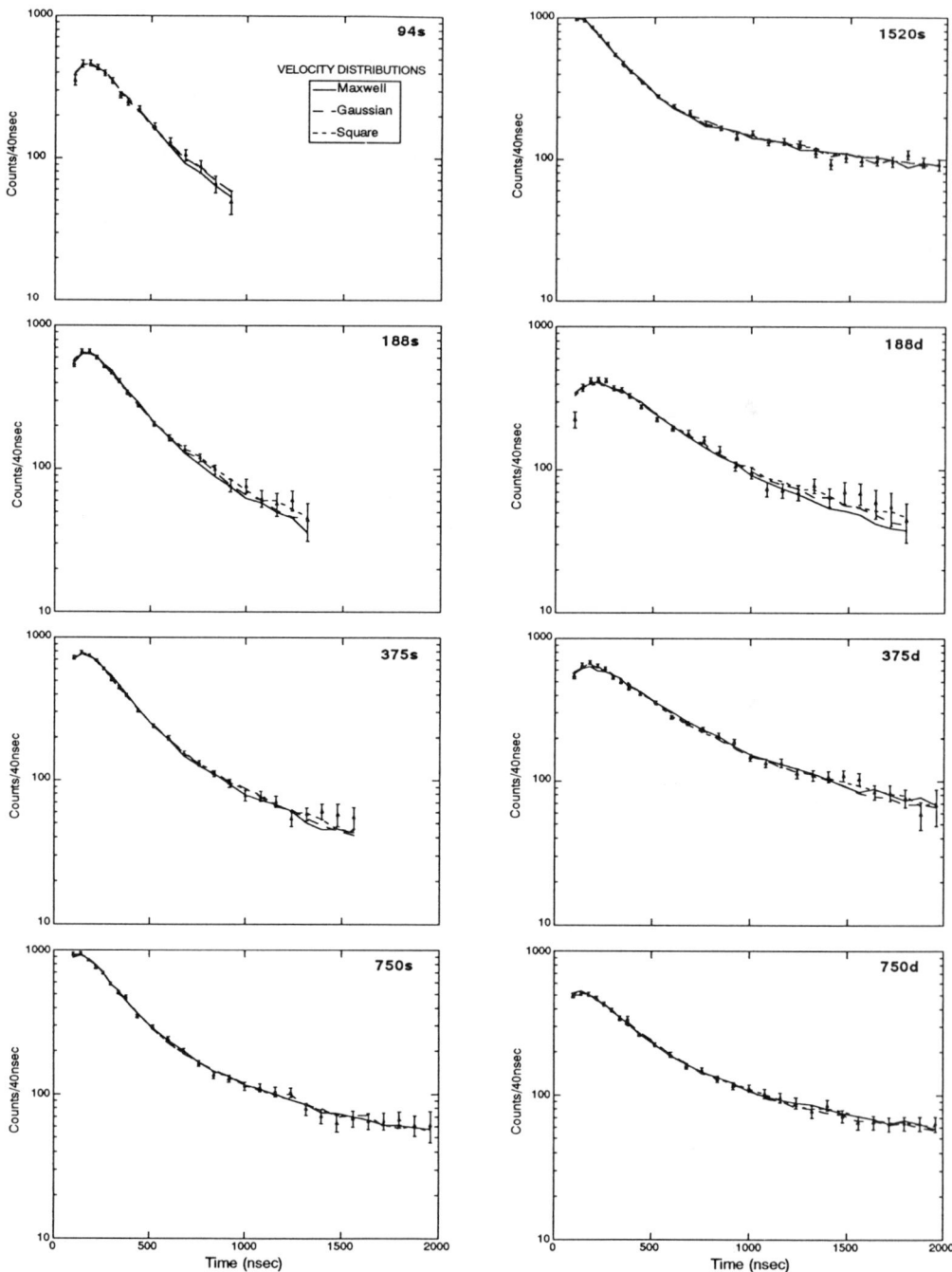

Figure 2. μd + D$_2$ Diffusion Data with Monte Carlo Fits (all conditions)

elastic and spin-flip cross-sections for each of the two hyperfine states, and each of the various cross-sections may have a complicated dependence on both energy and scattering angle. These cross-sections have been the subject of extensive theoretical calculations for many years [4, 5], and the results of these calculations are not readily parameterized. It follows that it is most fruitful to view the experiment as an effort to determine the initial velocity distributions of μp and μd, while simultaneously testing the most recent calculations of the scattering cross-sections. We thus have undertaken to fit the data with various velocity distributions (Maxwell, Gaussian, rectangular, etc.) while using scattering cross-sections as input, subject only to an overall scaling factor. Thus we assume that the scattering is well understood theoretically - in fact this assumption is not wholly correct, for the differential scattering cross-section calculations are still in an incomplete state as of this writing [6].

For each pressure-spacing condition a set of "theoretical" time distributions was generated by Monte Carlo (MC) method for comparison with the experimental data. The MC distributions involved an assumption about scattering cross-section values, with a single free (i.e., varied for each MC) scaling parameter over all energies (.001-50 eV) introduced as a simple test of the correctness of the cross-sections. In addition, a particular velocity distribution was assumed, with its related parameters varied for each Monte Carlo distribution. For each condition a multi-dimensional grid of MC distributions was thus generated, each involving 40,000 muonic atom histories. The experimental data was then compared with the net of MC distributions via the non-linear least-squares fitting program MINUIT, and best fit values obtained for the velocity distribution parameters and the scaling parameter (called the molfac).

3. Present State of the Interpretation

The interpretation of the experiment is strongly dependent on the theoretical values assumed for the scattering cross-sections, as is evident from the fact that at the pressures used the spacing between the foils is of the order of the scattering mean free path. Further, the muonic atoms scatter from D_2 or H_2 molecules, whereas until quite recently only nuclear cross-sections were available. (Thus the scale factor mentioned above is called the molecular factor = molfac, since the nuclear cross-sections were expected to be low by a factor ≈ 2.) Because the theoretical cross-sections for deuterium *nuclei* vary smoothly [4] and have been considered quite reliable (5 -10 %), the μd results have been most thoroughly studied. In Fig. 2 the experimental results for all μd conditions are presented along with the best fits for three types of velocity distribution.

Table 1 shows numerical results, described as follows. The condition 094s denotes 94 mbar D_2 at single spacing, 188d means 188 mbar double spaced, etc. Each type of velocity distribution is normalized to unity and has a mean energy denoted by E. (DOF = degrees of freedom.) Because it is possible that the velocity distribution is affected by collision processes during the atomic cascade of the muon to the 1s state, one type of fit (Free Energy fit) allows all velocity parameters to vary with pressure, i.e. condition. In this case the consistency of single- and double-spaced results is a measure of the correctness of the input assumptions. "Global Fits", for which the velocity parameters are assumed constant over all conditions, were also made, and it is seen that they uniformly gave poorer results than the free energy fits. Of course the free energy fits involve the use of more free parameters (E at the different conditions, rather than a single E), so the quality of fit is expected to be better than for the global fits.

Table 1. Free Energy and Global fits for $\mu d + D_2$
(σ/v = speed width parameter/ mean speed, w = speed fractional half-width)

FREE ENERGY FITS		Maxwell Distribution	Gaussian Distribution		Rectangular Distribution	
Condition	No. Data Pts.	E (eV)	E(eV)	σ/v	E(eV)	w
094s	15	1.23	1.12	0.69	1.25	0.75
188s	20	1.36	1.37	0.61	1.41	0.94
375s	23	1.46	1.49	0.67	1.54	0.83
750s	28	1.35	1.41	0.80	1.48	0.76
1520s	28	1.32	1.30	0.70	1.42	0.92
188d	26	1.63	1.44	0.83	1.35	1.00
375d	28	1.21	1.40	0.58	1.35	0.99
750d	28	1.23	1.20	0.37	1.30	0.93
	Molfac	1.51	1.38		1.46	
	Total DOF	186	179		179	
	Reduced χ^2	**1.41**	**1.04**		**0.99**	

GLOBAL FITS:

Distribution	Energy (eV)	σ/v or w	Molfac	DOF	Reduced χ^2
Maxwell	1.35	----	1.54	194	**1.61**
Gaussian	1.40	0.68	1.34	193	**1.24**
Rectangular	1.46	0.94	1.36	193	**1.25**

The total cross-sections used in the Monte Carlo distributions were all calculated for the molecular case [4]. However, the differential cross-sections used to determine the angular distributions were derived from the theoretical nuclear cross-sections [4]. This may explain the fact that in Table 1 the molfac values always exceed unity, which is the value expected when molecular differential and total cross-sections are used.

It is also evident from Table 1 that the calculated mean energies are only slightly dependent on the assumed type of velocity distribution, and that the best fit is with a rectangular distribution. For this case the free energies lie at about 1.3- 1.5 eV, with little evidence for pressure dependence, and the overall reduced chi-square is satisfactory. However, the molfac remains anomalous. Whether the introduction of newer and presumably more accurate differential scattering cross-sections for ($\mu d + D_2$) will change these conclusions remains to be seen.

The situation with regard to the μp analysis is less satisfactory. The analysis procedure has paralleled that for μd, but the μp nuclear cross-sections are regarded as less reliable than those for μd, and the μp molecular differential cross-sections are also yet to be introduced into our Monte Carlo - MINUIT procedure. It can however be said that the μp mean velocities do appear to be above 2 eV, consistent with the conclusion drawn from Fig. 1 that μp atoms are twice as energetic as μd atoms under the pressure conditions of the experiment.

Acknowledgements

We express thanks to the staff of PSI for their support and their hospitality at this Workshop, and to A. Adamczak, V. S. Melezhik and Prof. L.I. Ponomarev for many enlightening discussions. This work was supported in part by the Austrian Science Foundation, the German Bundesministerium für Forschung und Technologie, the Paul Scherrer Institute, and the U.S. National Science Foundation.

References

[1] A. Alberigi-Quaranta *et al*, Phys. Rev. **177**, 2118 (1969); A. Bertin *et al*, Nuovo Cimento **72A**, 225 (1982)

[2] J.B. Kraiman, *et al.*, Phys. Rev. Lett. **63**, 1942 (1989)

[3] H.J. Evans, Nucl. Phys. **A207**, 379 (1973)

[4] L. Bracci *et al.*, Muon Cat. Fusion **4** (1989) 247, and references therein.

]5] A. Adamczak and V.S. Melezhik, Muon Cat. Fusion **4** (1989) 303

[6] A. Adamczak, Muon Cat. Fusion (to be published)

Hot Muonic Deuterium and Tritium from Cold Targets

G.M. MARSHALL and J.L. BEVERIDGE
TRIUMF, 4004 Wesbrook Mall, Vancouver, B.C. V6T 2A3, Canada

J.M. BAILEY
University of Liverpool, P.O. Box 147, Liverpool L69 3BX, UK

G.A. BEER, P.E. KNOWLES, G.R. MASON and A. OLIN
University of Victoria, Finnerty Road, Victoria, B.C. V8W 2Y2, Canada

J.H. BREWER and B.M. FORSTER
University of British Columbia, 6224 Agricultural Road,
Vancouver, B.C. V6T 2A6, Canada

T.M. HUBER and B. PIPPITT
Gustavus Adolphus College, St. Peter, MN 56082, USA

R. JACOT-GUILLARMOD and L. SCHELLENBERG
Institut de Physique, Université de Fribourg, Pérolles,
CH-1700 Fribourg, Switzerland

P. KAMMEL and J. ZMESKAL
Institute for Medium Energy Physics, Austrian Academy of Sciences,
Boltzmanngasse 3, A-1090 Wien, Austria

A.R. KUNSELMAN
University of Wyoming, Laramie, WY 82071, USA

C.J. MARTOFF
Temple University, Philadelphia, PA 19122, USA

C. PETITJEAN
Paul Scherrer Institute, CH-5232 Villigen, Switzerland

Abstract. Experiments are described which use a solid hydrogen layer to form muonic hydrogen isotopes in vacuum. The method relies on transfer of the muon from protium to either a deuteron or a triton. The resulting muonic deuterium or muonic tritium will not immediately thermalize because of the very low elastic cross sections, and may be emitted from the surface of the layer. Measurements which detect decay electrons, muonic x-rays, and fusion products have been used to study the processes. A target has been constructed which exploits muonic atom emission in order to study the energy dependence of transfer and muon molecular formation.

1. Introduction

When a negative muon is stopped in hydrogen, a richly complex sequence of physical processes is initiated. The probability of a particular outcome depends upon factors such as the density ϕ (usually normalized to the density of liquid hydrogen), the temperature, the concentrations of the heavier hydrogen isotopes c_d and c_t, the impurity concentration, *etc.* Under a certain range of conditions, a particular sequence can lead to the emission of muonic deuterium or tritium from the surface of a solid layer into vacuum.

To be specific, consider the formation of muonic deuterium, μd. If a muon slows in protium containing a small deuterium concentration, it will normally initially form muonic protium, μp. Transfer of the muon from the proton to a deuteron is possible because the increase in reduced mass leads to a slightly larger binding energy. This also leads to an appreciable kinetic energy (about 45 eV) of the μd after transfer. The average time for transfer depends on the deuterium concentration,[1,2] and is of the order of 100 ns for c_d of order 10^{-3}. The μd atom loses energy by elastic scattering with hydrogen, but as the energy moderates to the range of a few eV, the cross section for scattering by a proton is drastically reduced because of the Ramsauer-Townsend (RT) mechanism.[3,4] If the deuterium concentration is not too high, μd can travel for a distance of the order of 1 mm before losing so much energy that the cross section is no longer within the RT region. The muonic atom then thermalizes without further macroscopic displacement.

If the muonic atom reaches the surface of a solid hydrogen layer during its travel, it will be emitted from the layer into the adjacent region. If this region is vacuum, the muonic atom will travel unimpeded until either it reaches another material or the muon decays. Because the RT mechanism exists for muonic tritium (μt) in protium, the same situation arises. Note that these are the only isotopic mixtures where it is expected; for example, there is no analogous RT minimum for muonic tritium in deuterium.

Experiments have been performed at TRIUMF over the past several years with the aim of understanding the factors which control emission of muonic deuterium from a solid layer. Several approaches have been taken to investigate how the muonic atoms interact, using information from the time and position distributions of decay electrons, from muonic x-rays following transfer of the muon, and from the products of muon induced fusion. It was realized that it should be possible to extend the experiments to muonic tritium. The main motivation is to study the energy dependence of muon molecular formation, which is vital in the understanding of muon catalyzed fusion. A cryogenic system has been constructed to allow the use of tritium in small concentrations in solid hydrogen. It also features a second solid hydrogen layer which can be used as a target for muonic tritium emitted from the first layer, enabling certain energy measurements to be based on time of flight.

2. Basic processes and measurements

To understand the processes which precede hot muonic atom emission from solid hydrogen, it was necessary to carry out experiments under conditions of differing target thickness and deuterium concentration. A relatively simple model has been shown to demonstrate qualitatively the dependences observed. More precise comparisons will take place with a monte carlo simulation which is under development.

The emission is easily studied by reconstructing the path of muon decay electrons in order to infer the position of decay. Details of the procedure are given elsewhere,[5] but the important points are as follows. A system of three wire chambers measures spatial coordinates of the electron, and scintillation counters and a sodium iodide crystal determine the time and energy of the electron respectively. The energy measurement is used simply to reject lower energy background and in some cases to improve position resolution by choosing only electrons with energy above a certain threshold; a lower energy decay electron undergoes a larger average scattering angle in material in its path. The time measurement is more important, because muons stopping in heavier elements (such as Au) which make up the cryostat are quickly absorbed. The major source of electrons later than 0.5 μs after the muon arrival time is therefore from muon decay in hydrogen. In addition, the correlation of time and position is a measure of the component of velocity perpendicular to the solid hydrogen layer. The position of decay is estimated by extrapolation of the three electron coordinates to a plane containing the beam axis.

The emission of μd can alter the distribution of decay time for events in which the muon decays in a particular spatial region. The most striking behaviour is shown in Fig. 1, where two such regions have been chosen. Rather than a purely exponential dependence, there exists a structure with a maximum at one to two microseconds, which is due to the emitted μd entering the chosen spatial region at a time considerably later than the arrival of the muon in the hydrogen layer. The sharper but much less intense structure at zero time shows the short lifetime of muons which stop in Au cryostat components rather than in hydrogen. By selecting decays which occur within a particular range in time and space, the backgrounds from muons decaying normally in hydrogen or in the heavier target support materials can be virtually eliminated. For example, a typical choice is the events of Fig. 1 with a decay time between 0.5 and 5.0 μs. The expected rate has been measured as a function of density and target thickness, showing that the emission is a maximum at a concentration of about $c_d = 10^{-3}$. The muonic atoms which are emitted originate mostly from transfer within 1 mm of the surface, in agreement with a simple model.[6] More precise comparisons with theory, which should provide some sensitivity to the energy dependence of some cross sections close to 1 eV, will be forthcoming after development and testing of a detailed monte carlo code.

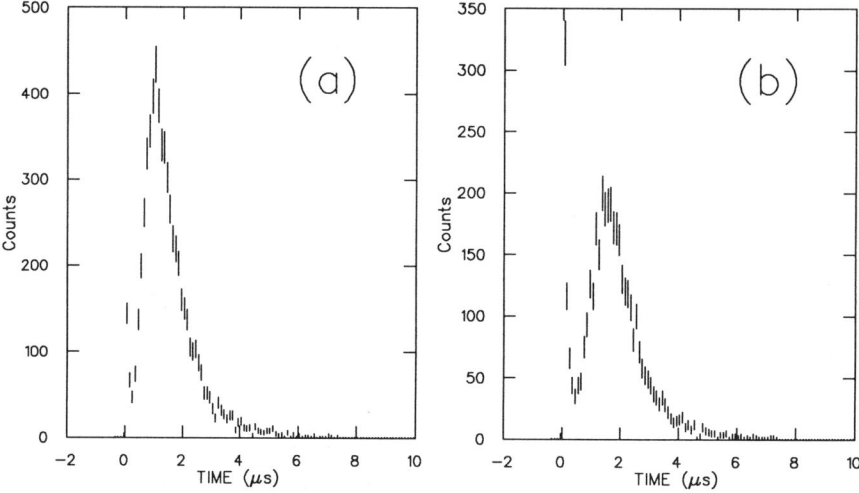

Figure 1: Decay time distribution for muons 15 to 25 mm (a) and 25 to 35 mm (b) from the emitting target

Complementary information has been obtained from muonic x-rays measured by a high resolution germanium detector. The eventual fate of muons is clarified by the observation of muonic radiative transitions either after direct capture or after transfer from muonic hydrogen to heavier atoms. The two processes can be distinguished both by the different time dependences and by the characteristic cascade relative intensities. Muonic x-ray information has been used to monitor possible contamination of the targets and to verify the muon stopping distribution. It has also proved possible to extract reasonably precise estimates of the rates for transfer from protium to deuterium and for muonic molecular formation in protium, as described more completely elsewhere in this volume.[1]

The solid target also provided an opportunity to measure for the first time the characteristics of muon catalyzed dd fusion at a temperature below 10K. A comparatively thick (several mm) layer of solid pure D_2 at 3K was used; of course, no emission was expected, and the muons remained in the deuterium layer. Careful measurements by other groups in the liquid and gas phases at higher temperatures have existed for some time, and have been well understood in terms of hyperfine effects and muon molecular formation from thermalized μd.[7,8] On that basis, no appreciable resonance formation of dμd was expected. However, preliminary results are consistent with an effective molecular formation rate from the $F = 3/2$ hyperfine state of $\tilde{\lambda}_{3/2} = 2.88\pm0.35$ μs^{-1} and an effective hyperfine conversion rate of $\tilde{\lambda}_{3/2 \to 1/2} = 33.3\pm2.4$ μs^{-1}, normalized to liquid hydrogen density. The resonance behaviour is therefore similar to that observed in warmer targets; a comparison is made elsewhere in this volume.[9] The exact interpretation is not complete, and the experiments are being repeated. It is likely that the solid phase of the target modifies the processes substantially, and that the analysis which has been so successful at higher temperatures is no longer adequate.

3. A proposed experiment with tritium

The RT process in a solid layer clearly can be used to create energetic muonic deuterium atoms in vacuum. The typical energy of the atoms is determined by the shape of the Ramsauer-Townsend minimum in the cross section. In the case of muonic deuterium, the distribution of the longitudinal energy (*i.e.*, the energy corresponding to the longitudinal component of velocity) has a most probable value of about 1 eV. Calculations of the RT minimum for muonic tritium in protium show a similar energy dependence, so we expect that muonic tritium will have an energy distribution similar to that for muonic deuterium.

Recent calculations exist[10] for the rate as a function of energy in the interactions

$$\mu t + DX \to [(d\mu t)xee] \qquad (1)$$

where x is p, d, or t and X is the corresponding atomic form. DX is the molecular state of deuterium with X, and e is an atomic electron. The final state is a complex molecule analogous to hydrogen, where one of the "nuclei" is in fact a muonic molecular ion. It is the internal degrees of freedom of the complex molecule which lead to the resonance character of the reaction. It happens that the μt kinetic energies required to satisfy the resonance condition coincide remarkably well with the emission energy spectrum. Furthermore, the calculated rates are large enough to dominate the cross section for elastic scattering of μt by deuterons, the main mechanism for energy loss. This means that the resonant interaction may be observed by passing muonic tritium of the appropriate energy through a thin layer of HD or D_2. The source of μt is the RT emitting hydrogen layer, protium with one part per thousand tritium ($c_t = 10^{-3}$).

The proposed experimental arrangement is shown in Fig. 2. Muons which form muonic protium and are transferred to tritons exit the emitting hydrogen layer. A

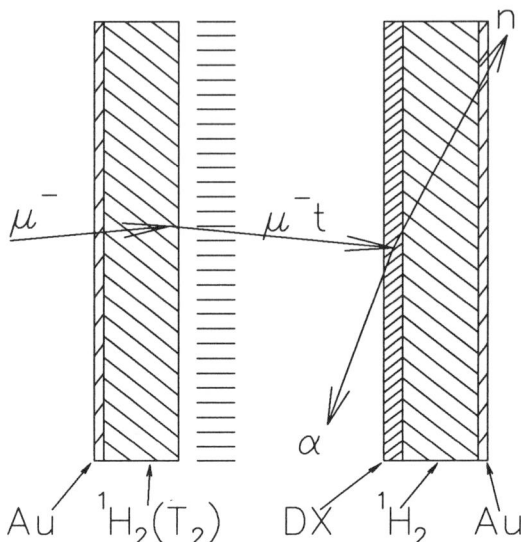

Figure 2: Proposed arrangement of two solid hydrogen layers for measuring the energy dependence of the resonant muon molecular formation rate (not to scale)

second target layer is close to the emitting layer, separated by approximately 20 mm of vacuum. The target layer consists of up to 1 mm of pure protium, covered by a thin overlayer of D_2(or HD, depending on the resonance structure we wish to measure). The thickness of the overlayer is chosen so that the probability of energy loss by the incident μt via elastic processes is small. If the μt has the appropriate resonance energy, it can interact to form the muonic molecule in the overlayer. Fusion will follow immediately to give a measurable fusion product, either a neutron or an alpha particle. Otherwise, the μt atoms may pass through to finally stop in the protium layer, where they give no signal which could be confused with a fusion event.

The energy of the incident μt is determined by its time of flight between the emitting and the target layer. The time of emission is approximately the time of muon arrival, as measured by a scintillation counter in the incident muon beam, because the transfer process is quite fast at the chosen tritium concentration. The time of arrival at the target layer is given by the time of detection of the fusion product, because fusion follows within nanoseconds of muon molecular formation. By far the greatest limitation on the precision of the time of flight energy determination is the unknown angle of emission of the μt. A collimating device is used in the vacuum drift space to limit the angles to a range close to perpendicular with respect to the emitting surface.

4. First results with an improved target system

In order to proceed with the measurement described in the previous section, we have constructed a new cryogenic target system. It incorporates improvements over the initial system defined by the demands of the new experiment plus several years experience with solid hydrogen layer targets.

The features which were considered important for the new system are as follows. Two cold surfaces should be adjacent to each other, with spacing between them variable from less than 10 mm to 40 mm. It must be possible to form solid layers on each surface independently, with no significant cross-contamination. The system should be capable of containing small amounts (up to 10 curies) of tritium safely. Materials in which the muon beam might stop should be restricted to the hydrogen layers and heavy, pure cryostat material (Au or Ag, but not both), in order to eliminate spurious muonic x-rays and longer lived capture neutrons. The thickness of material through which the muon beam must pass before stopping in hydrogen should be minimized, in order that the beam momentum (and therefore the momentum spread) is as small as possible. Control of the temperature should be reliable at about 3K with reasonable liquid helium consumption. Versatility is very important, so that different detector configurations, angular collimators, and target arrangements can be considered.

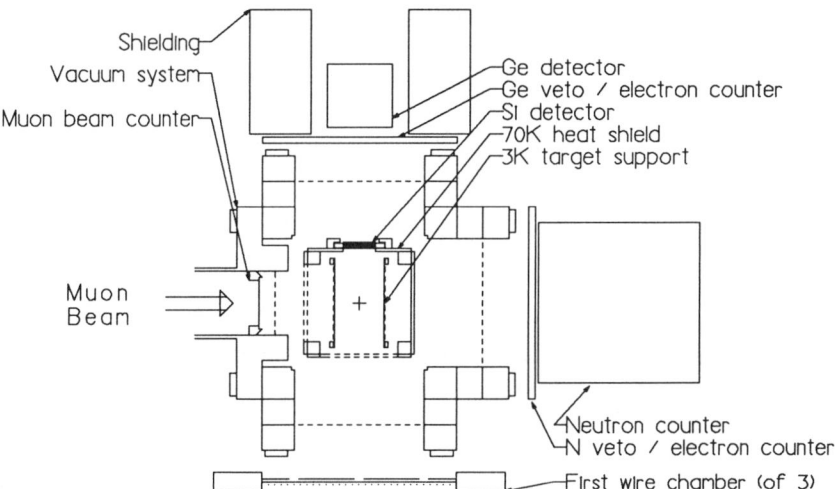

Figure 3: Arrangement of the new cryogenic solid hydrogen target system, showing one of several possible detector arrangements

A top view of the cryostat is shown in Fig. 3. It also shows an arrangement for detectors for decay electrons, muonic x-rays, and fusion products. Solid hydrogen is formed on either or both of two 50 μm metal foils (currently Au) cooled to 3K. The foils are mounted by a clamping system such that their separation may be changed with only minor difficulty. A device known as a diffuser can be inserted from below; it consists of two perforated steel foils attached to a gas feed system. Hydrogen is allowed through the perforations of one foil at a slow rate and the hydrogen gas freezes, building up a layer on the foil. The diffuser can be withdrawn completely, or it can be removed such that an angular collimation device is brought between the two foils. The collimator consists of hexagonal holes close packed inside a circle of size similar to the target. The transparency of the collimator to normally incident particles is greater than 90%. It is machined from silver, thus there is some motivation to make the foils also of silver rather than gold to reduce the types of materials in which muons may stop.

One of the first measurements with the solid target system was to prove that two dissimilar target layers could be formed without significant cross-contamination. The principle of formation of the solid layer is that hydrogen molecules stick almost completely at first contact with a 3K surface. If they do not, it is possible that some molecules may bounce around the diffuser and stick to the other cold surface, which would make the complicated layer structure of the tritium experiment impossible. To test whether this might be a problem, an emitting layer was first deposited on one foil, and the yield of μd in the vacuum space between the two foils was measured. Then a substantial layer of pure D_2 was deposited on the opposite foil, and the measurement was repeated. If a significant amount of D_2 contaminated the surface of the first foil, μd arriving at the surface would lose energy by scattering on deuterium and the yield would be drastically reduced. No significant reduction was observed, establishing that less than 1% of the deuterium was deposited on the wrong surface.

The intensity reduction of μd during passage through a thin overlayer of D_2 is due essentially to scattering. At 1 eV the μd scattering cross section by a deuteron is typically of order 10^{-19} cm^2, which is several orders of magnitude larger than that for scattering by protons in the RT minimum. The effect of a deuterium overlayer on both the intensity and the longitudinal energy E_l of the emitted atoms has been studied with the new target system. The longitudinal energy is derived from the time of flight before decay, assuming the mass of μd.

The intrinsic motivation for studying E_l is twofold. First, we can hope to extract from measurements the cross section as a function of energy for a range of μd kinetic energy around 1 eV. To determine reliable values, or even to test for consistency with published calculations,[11] a comparison with detailed monte carlo calculations is clearly essential. The computer program which is being developed includes the effects of transfer, molecular formation, scattering, and emission, all of which are necessary in this case. Second, it has become clear that the properties of the emitted μd can be controlled to some extent due to energy loss in the overlayer. Fig. 4 shows longitudinal energy distributions based on muons decaying between 10 and 30 mm

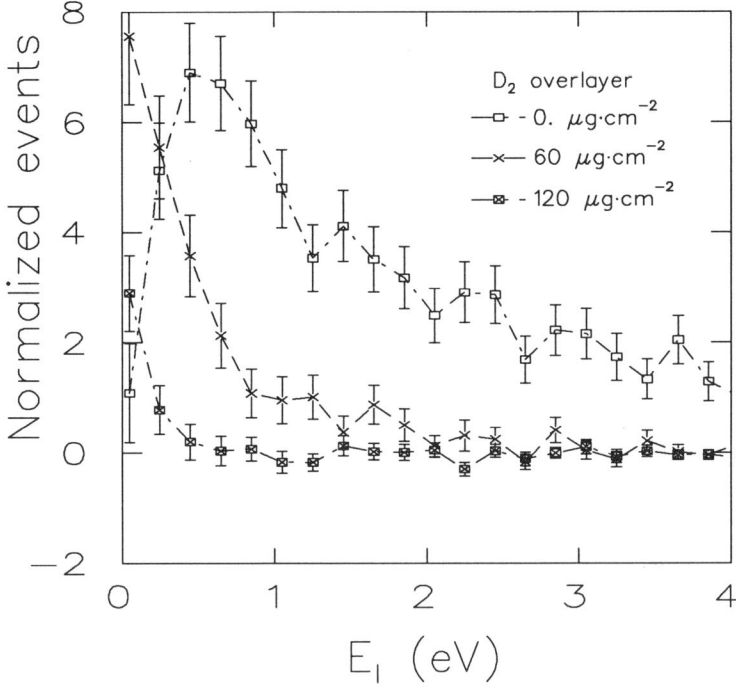

Figure 4: Longitudinal energy distributions for different D_2 overlayer thicknesses, for muonic deuterium which decays between 10 and 30 mm from the emitting surface

from an emitting layer, for two overlayer thicknesses as well as for no overlayer. In all cases, the distributions have been normalized and a similar one from a pure protium layer has been subtracted. The reduction in intensity is obvious, but it is interesting to note the shift in the energy spectrum of the emitted μd. Keeping in mind that the strongest resonances in the reactions of (1) are in the range 0.2–0.6 eV, it may be advantageous to use an overlayer to adjust the spectrum of emitted energies. The first experiments with muonic tritium emission will show whether such control is practical.

5. Conclusions

The emission of muonic deuterium from solid hydrogen is well established and understood semi-quantitatively, based on accepted cross sections and transfer rates. However, detailed comparisons with monte carlo calculations will make more precise estimates possible. The properties of muonic tritium emitted from a solid layer may be very similar, but it is essential to make measurements in order to examine any differences. Several applications exist for hot muonic hydrogen atoms in vacuum. A few of the possibilities are being pursued, but it is probable that others exist which have not yet been considered.

Acknowledgements

The authors wish to acknowledge the assistance of Dr. W.N. Hardy and Dr. C. Winter for assistance with the cryostat design. Major parts of the cryogenic system were manufactured by Quantum Technology Corporation, Surrey, B.C. Able technical support was provided by C.A. Ballard and K.W. Hoyle. This research was supported by the Natural Sciences and Engineering Research Council (NSERC) of Canada.

References

[1] R. Jacot-Guillarmod et al., these proceedings.
[2] A. Bertin, M. Bruno, A. Vitale, A. Placci, and E. Zavattini, Phys. Rev. A **7**, 462 (1973).
[3] C. Chiccoli, V.I. Korobov, V.S. Melezhik, P. Pasini, L.I. Ponomarev, and J. Wozniak, INFN (Bologna) preprint INFN/BE-91/09 (1991).
[4] James S. Cohen and Michael C. Struensee, Phys. Rev. A **43**, 3460 (1991).
[5] B.M. Forster et al., in *Proceedings of the 5th International Conference on Muon Spin Rotation, Relaxation, and Resonance (μSR 90), Oxford, 1990*, edited by S.F.J. Cox, G.H. Eaton, D. Herlach, and V.P. Koptev, Hyperfine Interactions **65**, 1007 (1990); G.M. Marshall et al., in *Proceedings of an International Symposium on Muon Catalyzed Fusion μCF-89, Oxford, 1989*, edited by J.D. Davies, (Rutherford Appleton Laboratory Report RAL-90-022).
[6] G.M. Marshall et al., in *Proceedings of the Workshop on the Future of Muon Physics, Heidelberg, 1991*, (to be published in Zeitschrift für Physik C – Particles and Fields).
[7] J. Zmeskal et al., Phys. Rev. A **42**, 1165 (1990).
[8] L.I. Menshikov et al., Zh. Eksp. Teor. Fiz. **92**, 1173 (1987) [Sov. Phys. JETP **65**, 656 (1987)].
[9] P. Kammel, these proceedings.
[10] M.P. Faifman and L.I. Ponomarev, Phys. Lett. B **265**, 201 (1991).
[11] V.S. Melezhik and J. Wozniak, Phys. Lett. A **116**, 370 (1986).

Detection of Hot Muonic Hydrogen Atoms Emitted in Vacuum Using X-Rays

R. JACOT-GUILLARMOD
Université de Fribourg, Chemin du Musée 3, CH-1700 Fribourg, Switzerland

J.M. BAILEY
University of Liverpool, P.O Box 147, Liverpool L69 3BX, UK

G.A. BEER, P.E. KNOWLES, G.R. MASON and A. OLIN
University of Victoria, Finnerty Road, Victoria, B.C. V8W 2Y2, Canada

J.L. BEVERIDGE and G.M. MARSHALL
TRIUMF, 4004 Wesbrook Mall, Vancouver, B.C. V6T 2A3, Canada

J.H. BREWER and B.M. FORSTER
University of British Columbia, 6224 Agricultural Road, Vancouver, B.C. V6T 2A6, Canada

T.M. HUBER
Gustavus Adolphus College, St. Peter, MN 56082, USA

P. KAMMEL and J. ZMESKAL
IMEP, Austrian Academy of Sciences, Boltzmanngasse 3, A-1090 Wien, Austria

A.R. KUNSELMAN
University of Wyoming, Laramie, WY 82071, USA

C. PETITJEAN
Paul Scherrer Institute, CH-5232 Villigen, Switzerland

Abstract. Negative muons are stopped in solid layers of hydrogen and neon. Muonic hydrogen atoms can drift to the neon layer where the muon is immediately transferred. We found that the time structure of the muonic neon X-rays follows the exponential law where the rate is the same as the disappearance rate of μ^-p atoms. The ppμ-formation rate and the muon transfer rate to deuterium are deduced.

1. Introduction

By slowing down via Coulomb interaction with atomic and molecular electrons, muons are stopped in a layer of hydrogen at 2.5 K containing a small amount of deuterium. The μ^-p can disappear by muon decay, ppμ-mesomolecular formation or muon transfer to deuterium.

After the charge exchange reaction μ^-p+ d $\to$ μ^-d+ p, the muonic deuterium atom has a kinetic energy of about 45 eV. It is then slowed down to a few eV (Ramsauer-Townsend effect) by elastic scattering on protons.

The emission of these "hot" μ^-d atoms in vacuum has been extensively studied in terms of deuterium concentration and layer thickness dependences [1]. Because of their high velocity ($\sim$ 1 cm/μs), the μ^-d can drift a few centimeters away from the hydrogen layer before muon decay occurs. Therefore the decay e$^-$ imaging system (DEIS) is a very convenient device for studying this process. The velocity of residual μ^-p atoms should be very low ($\sim$ 0.1 mm/μs) compared to the μ^-d's, but this type of emission cannot be completely excluded. This can however not be studied by using the DEIS because of its unsufficient spatial resolution.

To study the specific emission of μ^-p and μ^-d atoms we need another technique. The method is to freeze a neon layer on the surface of hydrogen. Muonic hydrogen atoms emitted in this direction will instantaneously release their muon and induce the formation of an excited muonic neon atom. Characteristic X-rays will be emitted and observed with a germanium detector.

The yield of muonic hydrogen emission (μ^-p and μ^-d) outside the layer (neglecting the drift time) is :

$$\frac{dN}{dt}(t) \propto N_{\mu p}(t) \propto e^{-\lambda t}$$

with

$$\lambda = \lambda_0 + (1 - c_d)\lambda_{pp\mu} + c_d \lambda_{pd} \qquad (1)$$

where $N_{\mu p}(t)$ is the number of remaining μ^-p atoms at time t, and λ their total disappearance rate.

2. Measurement

The experimental apparatus is shown in Fig. 1. The target preparation was made by first freezing the hydrogen on the gold foil and measuring the μ^-d emission with the DEIS. Simultaneously, the X-rays were observed to identify the presence of impurities, if any. Then, neon was added on the hydrogen layer surface.

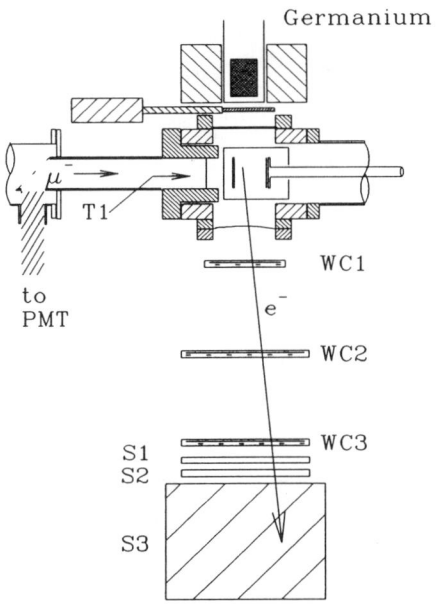

Figure 1: Experimental apparatus

This technique was applied for three targets presented in Table 1.

Table 1: Composition of the three targets. The $\oplus$ sign indicates that the neon is frozen on the hydrogen surface and not mixed. 1000 torr $\cdot$ l H_2 is 4.2 mg $\cdot$ cm^{-2}.

Target #	Composition		
I.	1506 torr $\cdot$ l H_2	$\oplus$	194 torr $\cdot$ l Ne
II.	1454 torr $\cdot$ l (H_2 + 145 ppm D_2)	$\oplus$	193 torr $\cdot$ l Ne
III.	725 torr $\cdot$ l (H_2 + 1150 ppm D_2)	$\oplus$	175 torr $\cdot$ l Ne

3. Data analysis

The total energy spectra obtained with a natural isotopic concentration of D_2 (target II), before (2a) and after (2b) the neon layer deposition is shown in Fig. 2. The absence of low Z element X-rays in Fig. 2a, and especially those from neon, indicates clearly that the impurity contamination in hydrogen is negligible. Most of the neon line intensity of Fig. 2b is due to μ^- stopping directly in the neon layer.

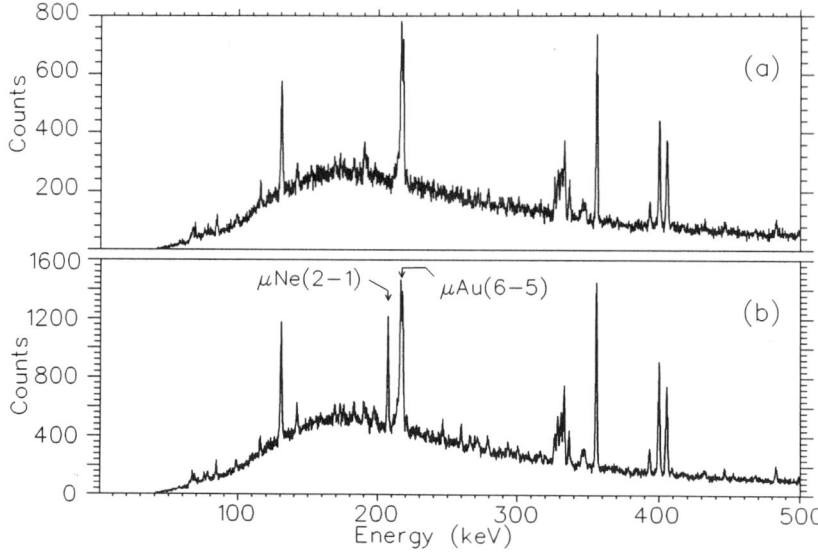

Figure 2: Part of the X-ray energy spectra without any time cut measured with target II (cf. Table 1). Upper (a) and lower (b) spectra have been obtained before and after neon deposition, respectively.

The specific contribution of the transfer process from muonic hydrogen can be observed by requiring a delayed time cut, as shown in Fig. 3 for the same data. It is essential to observe that the huge μAu(6-5) peak of Fig. 2 is then completely absent.

Table 2: Intensity of the μNe(2-1) line measured for different time cuts. By assuming an exponential dependence of the delayed intensities, the lifetime $\tau = 1/\lambda$ of the μ^-p is deduced.

Target	c_d (ppm)	μNe(2-1) intensity for each time cut					Lifetime τ (ns)
		Prompt	Del I	Del II	Del III	Del IV	
I	0	9116	170	260	246	33	303
		(125)	(42)	(37)	(33)	(33)	(40)
II	145	5032	228	311	171	27	210
		(91)	(34)	(32)	(26)	(24)	(21)
III	1150	7057	365	110	20		56
		(100)	(36)	(28)	(23)		(9)

The complete absence of the delayed muonic gold X-rays is *a priori* very intriguing because one would expect that delayed muonic gold X-rays would be apparent after μ^-d emission in direction of the gold foil. The intensity of the muonic transition μAu(6-5) by direct μ^- stop is about twenty times more than via μ^- transfer, because the circular transitions are much more favoured for the former.

One can then state that the delayed muonic neon X-rays in Fig. 3 follow the emission of muonic hydrogen into the adjacent neon layer where muon transfer eventually occurs.

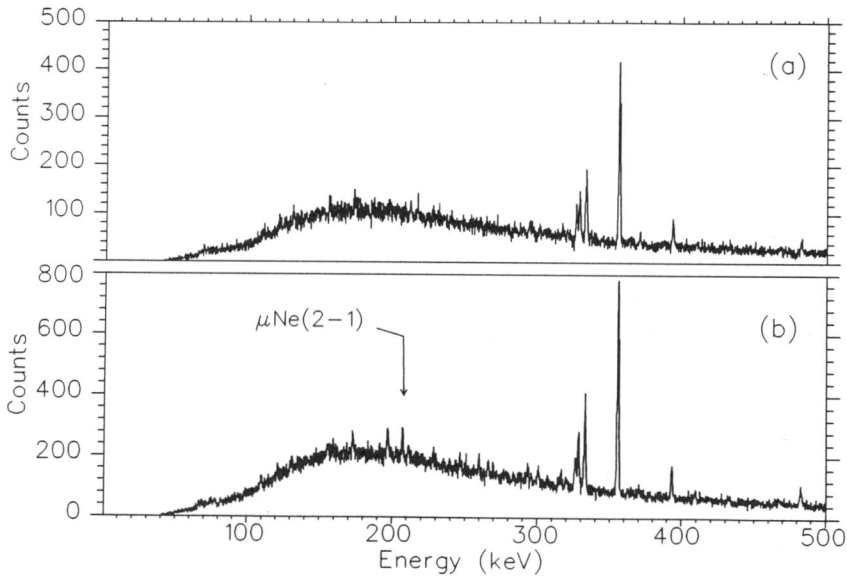

Figure 3: Part of the delayed X-ray energy spectra measured with target II (cf. Table 1). Upper (a) and lower (b) spectra have been obtained before and after neon deposition, respectively.

The counts of the μNe(2-1) line have been grouped in four time bins: from 40 to 120 ns after μ^- signal (this spectrum will be called Del I), from 121 to 339 ns (Del II), from 340 to 1539 ns (Del III) and from 1540 to 7931 ns (Del IV). The fitted intensities of μNe(2-1) in each time bin and the deduced lifetimes are reported in Table 2.

Using Eq. 1, one obtains a set of three linear equations, each one corresponding to one value of a deuterium concentration c_d, with two free parameters $\lambda_{pp\mu}$ and λ_{pd}. One finally gets :

$$\lambda_{pp\mu} = (2.18 \pm 0.32) \cdot 10^6 s^{-1}$$

$$\lambda_{pd} = (0.98 \pm 0.19) \cdot 10^{10} s^{-1}$$

These values are in fair agreement with those measured at comparable experimental conditions i.e. with liquid hydrogen during the sixties [2-4]. The rate λ_{pd} however differs from the value which is generally accepted for room temperature [5], $(1.68 \pm 0.26)\cdot 10^{10}\text{s}^{-1}$, although recent calculations [6] do not predict such energy dependence.

4. Acknowledgements

The support of the Natural Sciences and Engineering Research Council (NSERC) of Canada is gratefully acknowledged. One of us (R. J.-G.) wishes to thank the Swiss National Science Foundation for financial support.

References

[1] See G.M. Marshall et al. contribution in this volume.
[2] V.P. Dzhelepov, P.F. Ermolov, E.A. Kushnirenko, V.I. Moskalev, and S.S. Gershtein, Sov. Phys. JETP **15**, 306 (1962).
[3] E.J. Bleser, E.W. Anderson, L.M. Lederman, S.L. Meyer, J.L. Rosen, J.E. Rothberg, and I-T. Wang, Phys. Rev. **132**, 2679 (1963).
[4] G. Conforto, C. Rubbia, E. Zavattini, and S. Focardi, Il Nuovo Cim. **33**, 4281 (1964).
[5] A. Bertin, M. Bruno, A. Vitale, A. Placci, and E. Zavattini, Phys. Rev. **A7**, 462 (1973).
[6] C. Chiccoli, V.I. Korobov, V.S. Melezhik, P. Pasini, L.I. Ponomarev, and J. Wosniak, Muon Cat. Fusion **7**, 87 (1992).

Kinetics of Muon Catalyzed Fusion in H/D/T Mixtures

V.E. MARKUSHIN
Kurchatov Atomic Energy Institute
Pl. Kurchatova, Moscow 123182, Russia

Abstract. The results of kinetics calculations of muon catalyzed fusion in H/D/T mixtures at low deuterium and tritium concentrations are presented and characteristic features of the dt branch of the μCF cycle are discussed. The importance of the epithermal effects in the $dt\mu$ cycle is demonstrated and the theoretical framework for future experimental data analysis is outlined.

1. Introduction

Using triple H/D/T mixtures in muon catalyzed fusion provides a wide range of opportunities for measuring the kinetic rates and reaction yields for different branches of the fusion cycle (see [1 − 3] and references therein). By dissolving the D/T mixture with hydrogen one can, for instance, slow down (at the same density) the 'fast' $dt\mu$ formation and make it possible to observe the time evolution of a single cycle instead of multiple recycling. The detailed investigation of the fusion cycle is very important for testing the recent theoretical results for muonic atom collisions and muonic molecule formation (see [4, 5] and references therein). Since the analysis of the μCF data necessarily involves the use of a kinetics model, a comprehensive description of kinetics processes at different conditions (density ϕ, temperature T, H, D, and T fractions C_p, C_d, C_t) proves to be essential for current and future experiments.

2. Epithermal effects in μCF kinetics.

Our discussion will be focused on the $dt\mu$ cycle in triple mixtures with $C_t \ll C_d \ll C_p$, the conditions being typical for the μCF collaboration experiment at PSI [2, 3, 6]. At small C_d and C_t the atomic capture stage is dominated by the μp formation, and the μt atoms are mainly produced in the transfer reactions $\mu p \to \mu t$ and $\mu p \to \mu d \to \mu t$ in a very hot

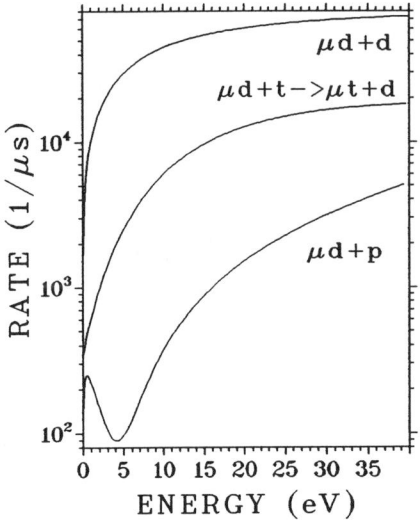

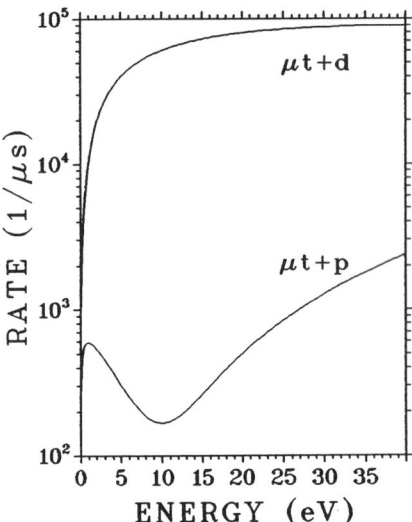

Figure 1: The energy dependence of elastic collisions and muon transfer rates for μd atoms.

Figure 2: The energy dependence of elastic collisions rates for μt atoms.

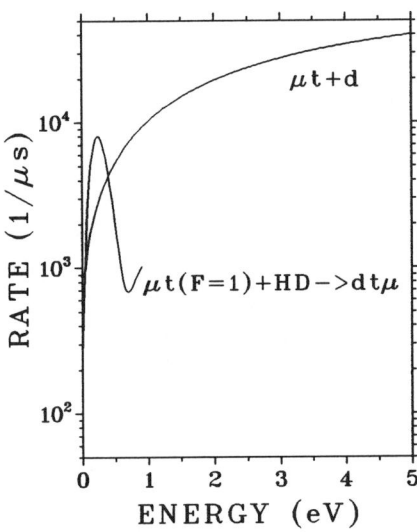

Figure 3: The energy dependence of the $dt\mu$ formation rate at $T = 300$ K.

state with kinetic energy of some tens of eV. Since the rates of the elastic collisions $\mu t + p$ and $\mu d + p$ are significantly reduced by the Ramsauer-Townsend effect (see Figs. 1,2), the time scale of muonic atom deceleration is mainly determined by $\mu t + d$ and $\mu d + d$ collisions even at rather small deuterium concentrations. The recent calculations [4] show that the resonant $dt\mu$ formation in $\mu t + HD$ collisions can successfully compete with the deceleration for hot μt atoms in the energy region $E = 0.1 - 0.5$ eV (see Fig. 3).

Another epithermal effect can be expected from the strong energy dependence of the μd - μt transfer rate (Fig. 2) [7]. Being rather small for thermalized atoms this rate rapidly grows with increasing μd energy. As a result, the fast $\mu d \to \mu t$ transfer (before the μd is thermalized) can be comparable with the direct $\mu p \to \mu t$ transfer.

High probability of the epithermal $dt\mu$ formation in the triple mixture with a bulk of hydrogen meets the goal of achieving a high fusion yield at moderate temperature and reduced tritium fraction. It is one of the important problems in μCF kinetics to understand the relationship between the epithermal and thermal molecule formation and the competition between the $dt\mu$ branch and the dead end channels of $pt\mu$ and $pd\mu$ formation.

3. Monte Carlo simulation of kinetics

For the purpose of a detailed theoretical description of the μCF kinetics a new version of Monte Carlo computer code was developed [8] which can be used for any kinetics scheme with the energy dependence of the transition rates and the hyperfine structure of muonic atoms and molecules taken into account.

The collision rates and the final energy distributions for elastic scattering, spin-flip and muon transfer reactions were calculated using the T-matrixes from refs. [7, 9, 5]. The $dt\mu$ resonant formation rates as functions of μt kinetic energy for a given target temperature were taken from [4]. The calculations have been performed for a wide range of experimental conditions ($\phi = 0.05 - 1$, $C_d = 0.001 - 0.9$, $C_t \ll C_p$, T=30–300 K). Taking into account the present day uncertainties in the molecule formation calculations we have investigated the effects of variation of the epithermal $dt\mu$ formation rates.

Here we present the results for the dt fusion cycle in the gaseous H/D/T mixture corresponding to the μCF collaboration experiment at PSI with the LNPI ionization chamber (run 1989) [2, 3]: $C_d = 9.5\%$, $C_t = 0.045\%$, $\phi = 0.17$, $T = 300$K. The details of kinetics calculations will be published elsewhere.

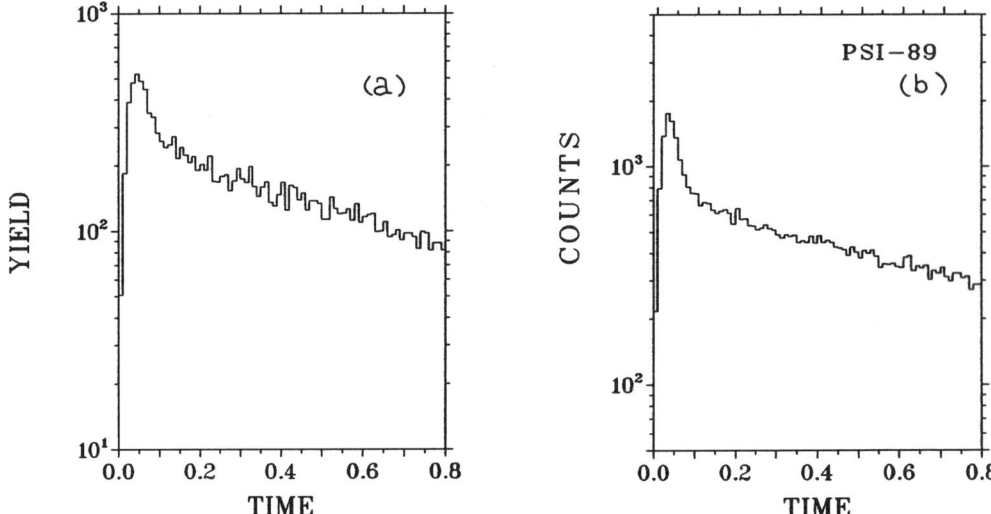

Figure 4: The dt fusion time spectrum: (a) theory, (b) PSI-89 experimental data.

4. Results of calculations

The theoretical dt fusion time spectra are shown in Fig. 4 in comparison with the experimental data from the PSI-89 run, the calculations having been performed with the epithermal $dt\mu$ formation rates from [4] reduced by a factor $R = 0.25$. This tuning allows one to get a good agreement between kinetics calculations and the experiment, but even without any tuning of the kinetic parameters the theoretical results turn out to be in qualitative agreement with the data. The main features of the fusion time spectrum, namely a sharp peak at small time followed by a quasistationary-like regime, have been found for a wide range of experimental conditions considered. This multicomponent time structure results from the contributions of different paths to the dt fusion cycle:

$$\mu \to \mu p \to \mu d \to \mu t \to dt\mu \qquad (1)$$

$$\mu \to \mu p \to \mu t \to dt\mu \qquad (2)$$

$$\mu \to \mu d \to \mu t \to dt\mu \qquad (3)$$

$$\mu \to \mu t \to dt\mu \qquad (4)$$

The calculated contributions of the 'fast' and 'slow' μt atoms to the dt fusion time spectrum are shown in Fig. 5.

The epithermal formation manifests itself as a peak (spike) at small time in the dt fusion time spectrum. Similar time structures were previously observed in binary DT mixtures at low densities (see [10, 11] and references therein).

Two kinetics paths contribute to the peak: the direct pt transfer (2) and the two step transfer (1) with the hot intermediate μd atom. The time structure of the peak is mainly determined

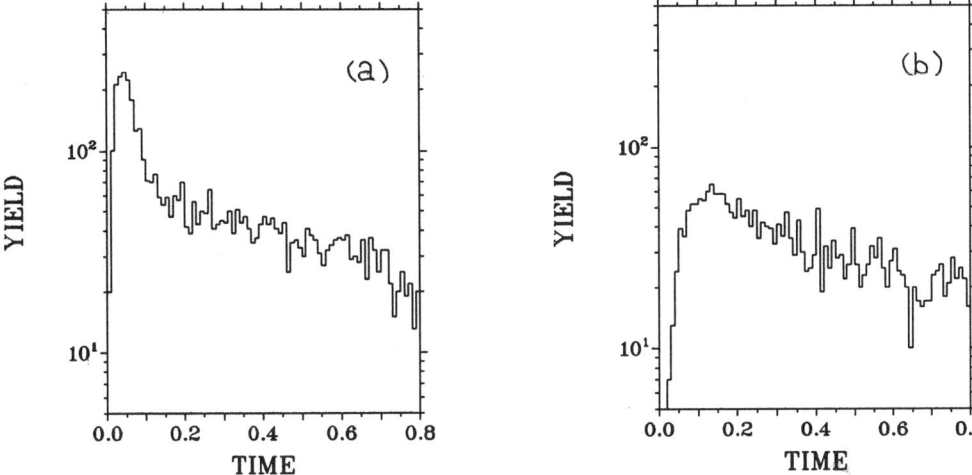

Figure 5: The contribution of the 'fast' and 'slow' μt atoms to the dt fusion time spectrum: (a) $E_{\mu t} > 0.05$ eV, (b) $E_{\mu t} < 0.05$ eV.

by the μt deceleration and, therefore, is sensitive to the deuterium concentration. The fusion yield depends on the competition between the muon transfer to tritium and deuterium, and with decreasing C_d–to–C_t ratio the spike contribution to the total yield increases.

Under the conditions of the PSI-89 experiment the two step transfer path (1) dominates the total dt fusion yield. Since the μd deceleration rate is much larger than the dt transfer rate, most of the dt transfer takes place after the μd atoms have slowed down. The quasistationary-like regime corresponds to the slow transfer from thermal μd to μt which is followed by fast μt deceleration and resonant $dt\mu$ formation (both epithermal and thermal). The time slope of the dt fusion in the quasistationary-like regime is mainly determined by the μd disappearance rate.

The absolute dt fusion yield from the quasistationary region Y_{QS} depends on the competition between the slow dt transfer, the $pd\mu$ formation and the muon decay. Using the experimental value Y_{QS} and the balance equation one can determine the dt transfer rate at low energy.

The $dt\mu$ formation probability as a function of μt kinetic energy is shown in Fig. 6. The relationship between the epithermal and thermal formation as well as the fusion time structure depends on the competition between hot $dt\mu$ formation and deceleration. This dependence is illustrated in Figs. 6,7 where the fusion Green functions (the fusion time spectrum for the μt atom with energy $E = 40$ eV at instant $t = 0$) are shown for different formation rates together with the μt energy distributions at the moment of $dt\mu$ formation. A more detailed discussion of the theoretical results in comparison with the experimental

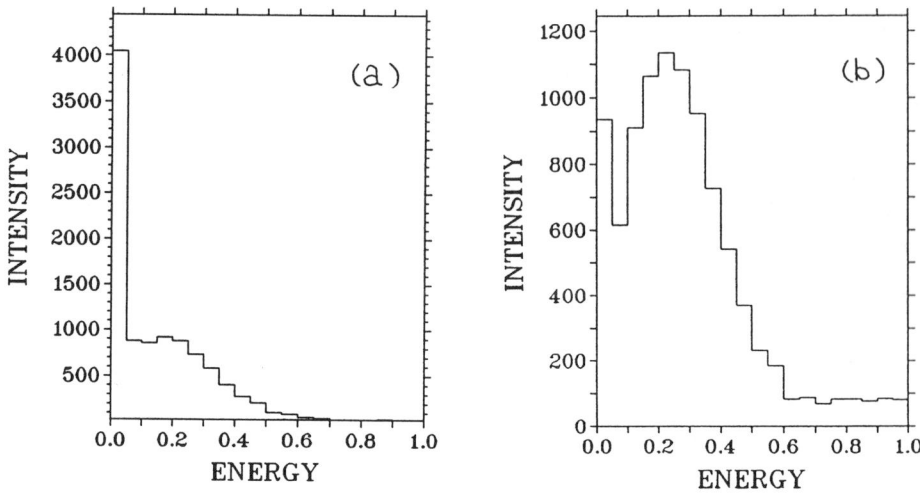

Figure 6: The energy distribution of μt atoms at the instant of $dt\mu$ formation: (a) reduced epithermal formation ($R = 0.25$), (b) formation rates from [4] ($R = 1$).

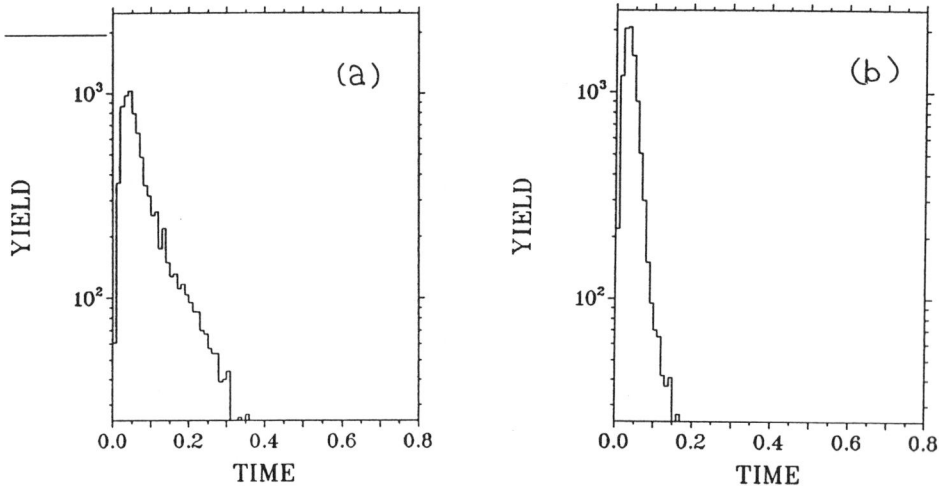

Figure 7: The dt fusion Green functions: (a) reduced epithermal formation ($R = 0.25$), (b) formation rates from [4] ($R = 1$).

data [3, 6] will be published elsewhere.

5. Conclusion

With the proper choice of experimental conditions the time-space characteristics of the μCF reactions are shown to be sensitive to the energy dependence of the rates of the molecular formation, the muonic atom deceleration and the muon transfer. The calculations performed demonstrate that the present theoretical results for the kinetics rates of the molecular formation [4], the muon transfer, and the muonic atom scattering [5, 7, 9] are in qualitative agreement with the μCF data. However, some tuning of the molecular formation rates is required in order to get a fair agreement between the kinetics calculations and the observed time spectra. The more refined calculations of the muonic molecule formation rates would be valuable for further kinetics studies.

The analysis of a multicomponent structure in the dt fusion time spectra measured in the PSI experiment on muon sticking confirms the theoretical prediction that the $dt\mu$ formation in $\mu t + HD$ collisions plays an important role in triple mixtures [4], both epithermal and thermal processes contributing to the molecule formation.

Detailed calculations aimed at a possible application of triple mixtures for high fusion yields are in progress.

Acknowledgments

The author would like to express his gratitude to Prof. L.I. Ponomarev and Dr. C. Petitjean for valuable discussions and E.I. Afanasieva for fruitful cooperation. This work is supported in part by the Paul Scherrer Institute.

References

[1] V.E. Markushin, E.I. Afanasieva, C. Petitjean, Muon Catalyzed Fusion **7** (1992) 155.
[2] C. Petitjean et al., Muon Catalyzed Fusion **5/6** (1990/91) 261.
[3] T. Case et al., Muon Catalyzed Fusion **5/6** (1990/91) 327.
[4] M.P. Faifman and L.I. Ponomarev, Physics Letters **B265** (1991) 201.
[5] C. Chiccoli et al. Muon Catalyzed Fusion **7** (1992) 87.
[6] K. Lou, et al., this volume.
[7] M. Bubak and M.P. Faifman, Preprint JINR E4-87-464 (1987).
[8] E.I. Afanasieva and V.E. Markushin, to be published.
[9] J.S. Cohen and M.C. Struensee, Phys. Rev. **A43** (1991) 3460.
[10] P. Kammel et al., Muon Catalyzed Fusion **3** (1988) 483.
[11] M. Jeitler et al., Muon Catalyzed Fusion **5/6** (1990/91) 217.

Accelerator Plans and New Experimental Methods

Future Plans at TRIUMF

M.D. HASINOFF
Physics Dept., University of British Columbia
6224 Agricultural Road, Vancouver, B.C. Canada V6T 1Z1

Abstract. The TRIUMF KAON project is briefly described along with an overview of the basic physics program with special emphasis on the possibilities for high flux muon beams at KAON. An update of the current funding situation is also presented.

1. Introduction

The TRIUMF KAON (Kaon-Antiproton-Other hadrons-Neutrino) Factory has been described in detail in the original proposal [1] and subsequent reports [2]. The 100 μA – 30 GeV accelerator will consist of two fast-cycling synchrotrons and three storage rings with the TRIUMF cyclotron injecting the initial 100 μA beam at 450 MeV. This will provide an increase in intensity of roughly 100 over what is currently available in this energy region. Such a large increase in intensity for the secondary beams will provide opportunities for many new experiments at the "precision" or "intensity" frontier. These experiments can place limits on the masses of heavy exchange particles significantly higher than those which can be reached by direct production at the "energy" frontier. One example of such a limit from a "precision" experiment is the $\approx$ 400 GeV limit placed on the mass of a right-handed W-boson in the muon decay polarization experiment by a Berkeley/TRIUMF/Northwestern group at TRIUMF more than 10 years ago. The proposed physics program at KAON is extensive. It includes the following possibilities:

- rare decays of kaons, hyperons, pions and muons
- CP violation in the K and Λ systems

- hadron spectroscopy and interactions
- neutrino scattering and oscillations
- hypernuclear studies
- $K^{\pm}$ and $\bar{p}$ scattering
- spin physics and symmetries
- low energy muon science

2. Accelerator layout

The accelerator complex for KAON consists of two rapid-cycling synchrotrons (B,D) interleaved with three storage rings (A,C,E). The five rings are arranged in two separate tunnels as shown in Fig. 1. A 450 MeV – 100 μA H$^-$ beam will be extracted from the present TRIUMF cyclotron and accumulated over 20 msec periods in the Accumulator ring before being accelerated to 3 GeV in the 50 Hz Booster synchrotron situated in the same 216m circumference tunnel. A Collector ring, in the main 1 km circumference "race-track" tunnel will then store 5 Booster pulses before injecting them into the 30 GeV Driver synchrotron which operates at 10 Hz to produce bunches of 3.5 μsec width every 100 msec. The Driver synchrotron can produce either a fast extracted

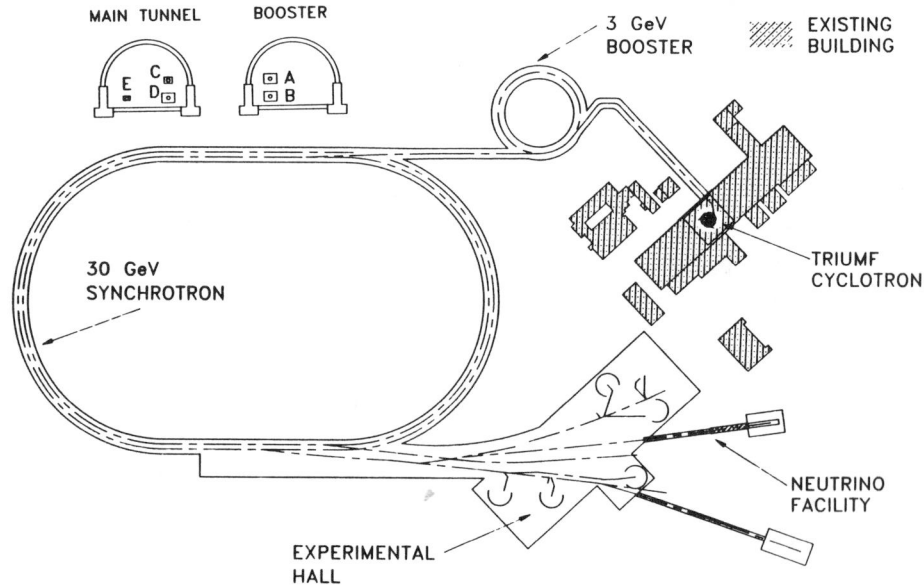

Fig. 1. Accelerator complex for KAON.

Future Plans at TRIUMF

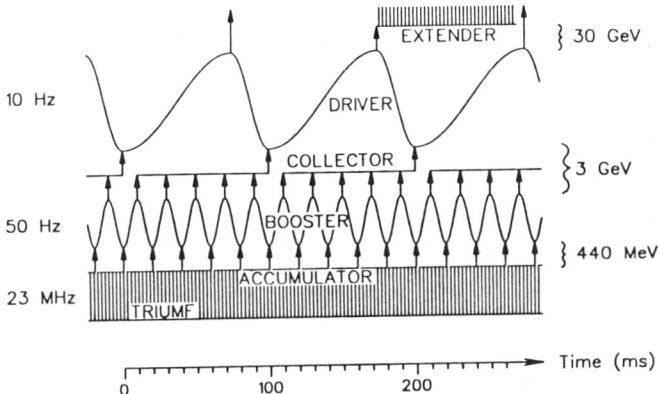

Fig. 2. Energy-time sequence for the five rings. This system of accelerators and storage rings allows them to run continuous acceleration cycles without flat tops or bottoms.

beam for neutrino studies or a slow extracted beam by using an Extender ring which is also located in the main tunnel. The energy-time plot in Fig. 2 shows how the 5 rings transform the DC beam from the TRIUMF cyclotron into the appropriate structure for acceleration in the synchrotrons and then back to a slow extracted, nearly DC, beam after the Extender ring. The preservation of a sharp RF μstructure (1 ns pulses @ 62.9 MHz) in the slow extracted beam will allow fast beam timing which should be a significant advantage for many experiments. The long straight sections in the main tunnel provide space for the beam transfer between the machines and also for the slow extraction process. Detailed calculations indicate beam losses during extraction to be below 0.2%, which is nearly a factor of 5 better than in current machines. The entire complex of accelerators should also be able to accelerate polarized protons with only minimal losses in polarization ($\leq 15\%$ overall) as a result of the very carefully tuned synchrotron lattices and the inclusion of several Siberian snakes in the Driver and Extender rings [3].

3. Beam line layout

A proposed layout for the primary and secondary beam lines is shown in Fig. 3. In this design three production targets will be viewed by 6 secondary channels to provide a full momentum range of charged particles ($K^{\pm}, \pi^{\pm}, \mu^{\pm}, \bar{p}$) up to 20 GeV/c. The most important design goal for the secondary beams is quality (or purity) which will be achieved using large acceptance RF separators for high momentum beams and conventional 2-stage DC separators for low momentum beams. High intensity, low energy π/μ beams will be extracted from each production target in the backward direction. A fourth production target will provide a K^0 beam. The neutrino facility

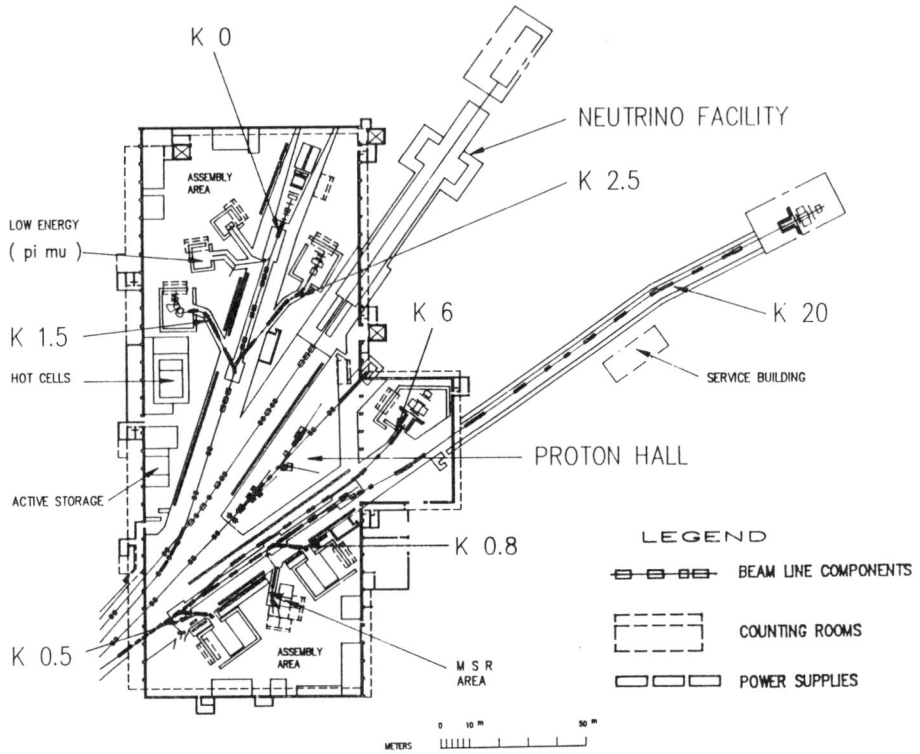

Fig. 3. Proposed layout of the experimental areas.

will be situated outside the main hall and will receive the fast extracted beam via its own beam line. The detailed properties of the various secondary channels are listed in Table 1 while the expected $K/\pi/\overline{p}$ beam fluxes for a 100 μA - 30 GeV beam are given in Table 2.

4. Low energy muon beams

The low energy π/μ channels are perhaps of particular interest to the participants of this workshop. The $K/\pi/\overline{p}$ cross sections are strongly forward-peaked; hence the take-off angles for these secondary channels are close to 0°. This leaves the floor area in the backward direction at each production target available for slow π/μ beams. At least two areas have been tentatively reserved for slow muons in the proposed layout of the experimental hall – a superconducting solenoidal μ^- channel and an elaborate μ^+ channel with several ports situated behind a *kicker* magnet [4]. The expected π^-/μ^- yield at 90° from a 100 μA, 30 GeV beam striking a 7.5 cm tungsten target has been estimated to be equivalent to $\simeq$ 18 mA @ 585 MeV on a 6.0 cm carbon target [5]. This would represent nearly an order of magnitude increase in intensity over that expected at PSI at the proposed operating current of 1.5 mA. This increased flux could be utilized for both rare decay or other particle physics experiments where the present intensities are inadequate, as well as to produce higher-quality, higher-luminosity μ^- beams for μ^-SR.[1] The estimated μ^- fluxes and luminosities at KAON are compared with other existing or proposed facilities in Table 3.

Table 1. Anticipated $K, \pi, \bar{p}$ intensities for a 100 μA 30 GeV beam on a 6 cm Pt target at KAON.

Channel	Momentum GeV/c	K^- 10^6/s	K^+ 10^6/s	π^- 10^9/s	π^+ 10^9/s	$\bar{p}$ 10^6/s
K20	21	0.75	29	0.16	0.95	0.05
	18	2.4	43	0.35	1.05	0.35
	15	5.9	62	0.60	1.50	1.7
	12	9.2	52	0.90	1.90	5.0
	9	7.9	23	0.70	1.30	10.5
	6	2.3	4.2	0.78	1.20	11.5
K6	6	15	34	1.9	3.6	23
	3	2.5	4.5	3.2	5.0	43
K2.5	2.5	66	119	16	24	110
	2.0	39	76	21	30	91
	1.5	14	27	25	36	52
	1.25	5.4	9.7	27	37	26
K1.5	1.5	193	366	49	69	81
	1.2	52	93	36	49	25
	1.0	18	31	27	36	8.3
	0.8	3.7	6.3	18	23	1.9
K0.8	0.8	99	203	87	113	7.1
	0.65	32	59	63	80	2.6
	0.55	10	19	44	55	1.0
K0.55	0.55	41	80	80	101	1.5
	0.50	21	44	67	82	0.93
	0.45	9.2	21	50	61	0.53
	0.40	3.8	9.4	33	44	0.30

---

[1] In the latter case the increased intensity by itself probably cannot be readily utilized because of the long muon lifetimes in most targets.

Table 2. Beam line specifications at KAON.

Channel	Momentum GeV/c	Solid Angle msr	Momentum Acceptance $\Delta p/p$ in %	Length m	Type of Separation
K20	20–6	0.1	1	160	RF, 3 cavities, 2.8 GHz
K6	6–2.5	0.08–0.30	3	110	RF, 3 cavities, 1.3 GHz
K2.5	2.5–1.25	0.5–2.0	4	54	DC, 2 stages
K1.5	1.5–0.75	2.0	4	30	DC, 2 stages
K0.80	0.80–0.55	6.0	5	18	DC, 2 stages
K0.55	0.55–0.40	8.0	6	14	DC, 1 stage, extra optics

Table 3. Comparison of flux and luminosities for high intensity, slow μ^- facilities.

Beam line	Momentum (MeV/c)	μ^- Flux (s^{-1})	μ^- Luminosity $(s^{-1} cm^{-2})$
PSI(1992)[+]			
μE1	125	6×10^7	10^7
μE4	50	10^6	4×10^4
KAON (1999)[++]			
Backward	100	$\sim 10^9$	$\sim 10^8$
Backward	40	$\sim 10^8$	$\sim 10^7$
LAMPF-PSR(1996)[*]			
SCMC	30	10^7	2×10^5
LAMPF(1992)[**]	40	3×10^6	10^5

[+] 1.5 mA @ 585 MeV — CW
[++] 100 μA @ 30 GeV — 75% CW or 3.5 μsec @ 10 Hz
[*] 200 μA — 270 nsec @ 40 Hz
[**] 1 mA @ 800 MeV — 800 μsec @ 120 Hz

More elaborate muon facilities including a surface muon storage ring with cooling and bunching have also been proposed although the poor optics of a μ^- beam from a decay channel might limit its usefulness solely to μ^+ beams. However, the increased μ^+ intensity would certainly be difficult to utilize without a fast *kicker* magnet and several additional low energy μ channels to simultaneously share the available μ^+ beam.

For π^+/μ^+ beams at $p \leq 200$ MeV/c the increase in flux is expected to be only a factor of 3 over that at the upgraded (1.5 mA) PSI since the increase in the π^+ cross section between 580 MeV and 30 GeV is not quite as large.

5. Muon physics at KAON

Many workshops have concentrated on the physics potential for KAON in the fields of rare π/K decays, CP violation, neutrino physics, hadron spectroscopy, antiproton physics, and even low energy muon physics. Since most of the participants of this workshop are currently working in muon physics I will focus on a few of the more fundamental muon experiments which could benefit from the increased flux of μ's available at KAON.

Since the number of seconds/year is fixed, even the most ingenious of experimenters must eventually resort to an increased beam flux when attempting to improve upon existing rare decay limits. Of course, in many cases, such a large increase in primary μ^- flux cannot be utilized directly since many backgrounds also increase (at least linearly) with the beam rate. However, such an increased flux permits more elaborate *OPTICS* or collimation so that the final μ^- beam is much "cleaner" or has better luminosity, or both. It can also be used in conjunction with thinner and smaller stopping targets as well as higher resolution spectrometers which help to reduce the intrinsic background. New technologies, such as Phase Space Compression, currently under investigation by Taqqu at PSI [6] offer additional promise for cleaner μ^- beams which can stop in ultra-thin targets suitable for high precision experiments. In this way even those "precision" experiments which are not limited by currently available fluxes can also benefit significantly from the increased μ^- flux promised at KAON.

- $\mu^- \to e^-$ conversion:
  This is one of the classic muon rare decay/conversion experiments which sets stringent limits on Lepton Number Conservation. The TRIUMF TPC collaboration established an upper limit of 4.6×10^{-12} for this process limiting the mass of any mediating Higgs scalar boson to 22 GeV. For a leptoquark the bounds are even higher–22 TeV for a pseudoscalar interaction and 118 TeV for a vector interaction. The SINDRUM II collaboration [7] at PSI intends to lower this Branching Ratio limit by at least one order of magnitude and there is a proposal at the Moscow Meson Factory for a 10^{-16} experiment. Clearly, given the inventiveness shown by the experimenters over the past 20 years, the increased μ^- flux available at KAON might lead to even lower experimental bounds on this rare process.

- $\mu^+ e^- \to \mu^- e^+$ conversion:
  The PSI experiment [8] with the muonium-antimuonium conversion spectrometer (MACS) will attempt to reach an upper limit $G_{M\overline{M}} \leq 10^{-3} G_F$ where G_F is the normal Fermi coupling constant of the weak interaction. This would raise the lower limit on the mass of any doubly-charged Higgs boson well above the

14 GeV limit presently established by e^+e^- collider experiments. Even more sensitivity should be available with KAON's μ fluxes.

- μ decay MICHEL parameters at the "Precision" Frontier:
  It is now 10 years since a Berkeley/TRIUMF/Northwestern collaboration measured Pξ in a muon decay asymmetry experiment at TRIUMF to set a lower limit of 400 GeV on the mass of any right-handed W-boson. No more stringent limit has yet been obtained although there is now a new proposal at TRIUMF by a Russian/Canadian collaboration [9] to remeasure the Michel parameters and push this limit well into the TeV region. Presumably this apparatus could also be designed to utilize the higher luminosity of ultra slow μ^+ beams at KAON.

- T-violation in μ^- capture:
  To date Time Reversal violation has not been observed in any system other than the neutral KAON system. Searches for T-violation in the nuclear β decay or in $K^+ \to \pi^0 \mu^+ \nu_\mu$ have thus far been negative [10,11]. Certain extensions to the standard model predict T-violation terms proportional to m_ℓ so that μ-capture should be more than 200 times as sensitive as β-decay. Odd-Parity T-violation correlations such as $(\vec{\sigma}_\mu \cdot \vec{J}_{\text{final}} \times \vec{k}_{\text{recoil}})$, in the reaction $^{12}\text{C}(\mu^-, \nu_\mu)^{12}B_{gs}(1^+)$, or even-Parity T-violation correlations such as $(\vec{\sigma}_\mu \cdot \vec{k}_\gamma \times \vec{q}_{\text{recoil}})(\vec{k}_\gamma \cdot \vec{q}_{\text{recoil}})$, in the reaction $^{16}\text{O}(\mu^-, \nu_\mu)^{16}\text{N}(1^-)$, respectively, could benefit greatly from the higher fluxes and detailed time structure of the μ^- beams at KAON [12].

- $\mu^- p \to \nu_\mu n \gamma$ and the Induced Pseudoscalar Coupling:
  The radiative muon capture (RMC) reaction is very sensitive to the induced pseudoscalar coupling of the semi-leptonic weak interaction, especially in the μp singlet state. Because of the extremely low RMC rate (5×10^{-3} s^{-1}) from the μp singlet state, this experiment has not yet been attempted. However, such an experiment to measure both the high energy γ and the low energy n in coincidence using a low density gaseous target has been proposed at PSI [13]. About 700 RMC events could be obtained in a 70 day running period at an incident μ^- rate of 3×10^7 μ^-/s. This would lead to a determination of g_P to better than 2 % excluding the unknown systematic errors. An even more accurate result might be possible using the μ^- beam at KAON.

- Muonium (or μ^- ^{4}He) Lamb shift:
  The 2S - 2P energy difference in muonium can, in principle, provide a pure QED test free of complications from the quark substructure of the proton. Unfortunately the existing TRIUMF and LAMPF experiments finished far from the 10–20 ppm level at which the H atom results disagree with theoretical calculations. In μ ^{4}He the Lamb shift between the $2S_{1/2}$ and $2P_{1/2}$ states is ≈ 1.38 eV which can be reached by LASER spectroscopy techniques as shown

by Zavattini [14] in the pioneering experiments at CERN several years ago. This could provide a test of QED calculations of both the vacuum polarization and the ^{4}He^{++} charge radius if sufficient precision can be obtained using the high intensity enhanced luminosity, low background μ^- beams with improved time structure from KAON.

6. Status of KAON funding @ April '92

The original KAON proposal was presented to the Canadian federal government in 1985/86. After much discussion an $11M Project Definition Study (PDS) was funded in July '88 to review the overall design and prototype various accelerator components. In addition, there was a review of the scientific justification and the construction schedule and cost estimates for the entire project as well as the capabilities of Canadian industry. International contributions for 1/3 of the capital costs ($708M in 1989 funds) were also solicited from the United States, Japan and Europe during this time. A summary of the important developments since the presentation of the PDS report in May '90 is given below:

- May 24/90 - PDS report made public with another strong endorsement for the proposed accelerator design and physics at KAON.
- Aug. '90 - The PDS report was received by the Prime Minister's National Advisory Board on Science and Technology (NABST). Although the final NABST report was never released, it is thought that NABST did not disagree with the high scientific priority of KAON but was concerned with the large amount of federal funds required for KAON.
- Sept 24/90 - B.C. Government offers to pay a full 1/3 of the capital costs of KAON and indicates some willingness to negotiate a small contribution towards the annual operating expenses (estimated @ $90M/year in total).
- Sept 19/91 - Canadian federal governement offers to pay $236M towards the construction costs of KAON.
- Oct 17/91 - New government elected in B.C. This caused a delay of 3 months while the new government examined the state of the provincial finances and reviewed the case for KAON.
- Jan 25/92 - B.C. government names Eric Denhoff as the chief B.C. negotiator for KAON.

Once the basic funding is in place (t=0) it will require 6 years for construction of the accelerator complex and experimental facilities. Full beam current of 100 μA will be available 2 years later. Many conferences and workshops on KAON's physics have been held over the past few years, in Vancouver and around the world. We anticipate having a series of workshops to generate the Letters of

Intent from the various working groups in the first year following the agreement to proceed with construction. Detailed proposals will be reviewed by a KAON Experimental Evaluation Panel, KEEP, about 2 years after t=0.

Although KAON would require a large increase in the expenditures for scientific R&D in Canada it will still not raise the funding level up to that found in the U.S., Japan or Europe. The Canadian government currently spends only ≈$2/capita on Research and Development in particle & nuclear physics versus ≈$6/capita in Europe and the U.S. Even if the proposed increase in operating funds for TRIUMF/KAON were to be paid entirely by the federal government the total Canadian expenditure on particle & nuclear physics/per capita would still be less than $5/citizen.

Thus we are confident that, recognizing the ultimate benefits to the Canadian economy as a result of investment in basic science projects, both levels of government in Canada will soon agree on the formula for apportioning the annual operating costs and begin construction. We look forward to beginning the very exciting KAON experimental program before the year 2000.

Acknowledgements

I would like to thank J. Brewer, J.-M. Poutissou and E. Vogt for many useful ideas and discussions used in the preparation of this paper.

References

[1] KAON Factory Proposal, TRIUMF (1985).
[2] KAON Factory Studies(1,2,3), TRIUMF (1990).
[3] U. Wienands, Proc. of a Workshop on Science at a KAON Factory, Vancouver, 1990, edited by D.G. Gill (TRIUMF, 1990) Vol. 2 Sess. 6.
[4] J.H. Brewer, ibid, Sess. 9.
[5] E.W. Blackmore, TRIUMF internal report TRI–DN91–K171 (April 1991).
[6] H.K. Walter, Proc. of a Workshop on Science at a KAON Factory, Vancouver, 1990, edited by D.G. Gill (TRIUMF, 1990) Vol. 2 Sess. 9.
[7] PSI proposal R-87-03.
[8] PSI proposal R-89-06.
[9] TRIUMF proposal E614.
[10] P. Herczeg, Hyperfine Int. **43** 77 (1988).
[11] N.C. Mukhopadhyay, Proc. of Int. Conf. on Weak and Electromagnetic Interactions in Nuclei, Montreal, 1989, edited by P. Depommier (Éditions Frontière, 1989) p.51.

[12] J. Deutsch, Proc. of a Workshop on Science at a KAON Factory, Vancouver, 1990, edited by D.G. Gill (TRIUMF, 1990) Vol. 2 Sess. 9.
[13] PSI proposal R-86-06.
[14] G. Carboni *et al.*, Phys. Lett. **73B** (1978) 229.

Future Plans at PSI

H.K. WALTER, W. JOHO, U. SCHRYBER
Paul Scherrer Institute
CH-5232 Villigen PSI, Switzerland

Abstract. The upgrade program of the 600 MeV cyclotron of PSI from 250 μA to 1.5 mA proton current is described. The experimental program in Nuclear and Particle Physics, the progress of the spallation neutron source and the plans for a synchrotron light source are discussed.

1. Introduction

The PSI accelerator facility [1] consists of 3 isochronous cyclotrons, all quite unique in their beam performances (see table 1). The main accelerator is a 600 MeV ring cyclotron, in operation since 1974, which is normally fed with 72 MeV protons by Injector 2. Both stages were built by PSI.

Table 1: Performances of the PSI cyclotrons (end 1991)

Cyclotron	Energy	Intensity	beam power
Ring	600 MeV p	500 μA	300 kW
		pol. 10 μA	
Injector 2	72 MeV p	1500 μA	110 kW
Injector 1	72 MeV p	200 μA	15 kW
		pol. 10 μA	
	ions:$(q/A)^2$		
	130 MeV/N	≤ 1 μA	

A specialty of these two cyclotrons is the relatively clean extraction process, with losses below 0.1%. This is obtained by having completely separated turns even at high beam currents (see figure 1).

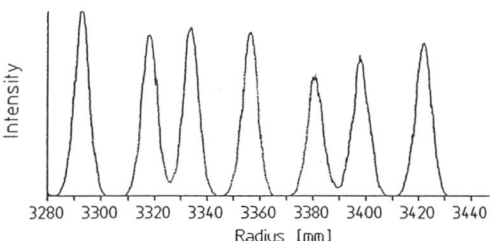

Figure 1: The last 7 turns in the Injector 2 for a 1.5 mA beam. The measurement was done with a radial beamprobe holding a 50 μm thick carbon fiber. The average radial gain is 21mm/turn and the average beamwidth is 12mm, dominated by the strong longitudinal space charge forces at this high current level. Nevertheless the turns are still clearly separated and allow to keep the extraction losses below 0.1%.

Injector 1, a K=130 MeV cyclotron built by Philips, is mainly used now for nuclear research with light ions and polarized protons and deuterons [2]. Every 4th week it is dedicated for eye tumour treatment with 65 MeV protons (OPTIS). For a few weeks per year this cyclotron is still used as an injector for the acceleration of 10 μA of polarized protons to 600 MeV to further produce polarized neutrons in the Nucleon Area. Figure 2 shows a general layout of the accelerator facility at PSI.

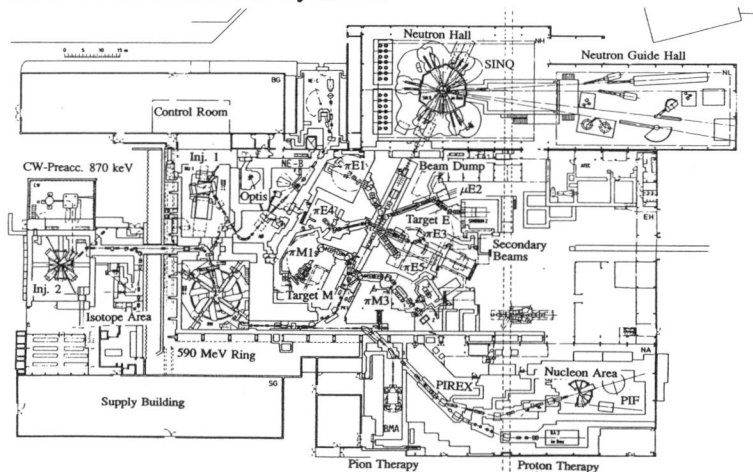

Figure 2: Layout of the PSI cyclotron facility. The 600 MeV-ring and injector 2 will deliver 1.5 mA of protons in the future. The target stations M and E were renewed and adapted to the intense beam in 1985 and 1991 respectively. Two new facilities are under construction: the spallation neutron source SINQ and the cancer therapy beam with 250 MeV protons.

Future Plans at PSI

While the experiments in nuclear and particle physics were dominating in the past, the emphasis has shifted recently towards the applications of nuclear methods at our facility, a trend that will certainly continue in the future. The main applications are:

- Solid state physics using the μSR-facility in πM3 and other beams.
- Curative treatment of tumors with pions (BMA) and protons (at 65 MeV, and soon at 250 MeV as well).
- Defect physics and material tests using proton beams at 72 MeV and 600 MeV (PIREX, PIF).
- Production of radioisotopes, used in PET experiments and for radiopharmaceutical products.

The 600 MeV protons from the ring are guided onto two consecutive target stations "M" and "E". Behind target "E" the protons are stopped in a high power beam dump. In the future the beam will be refocused and guided on the target of the spallation neutron source (SINQ), which is now under construction [3]. For a full exploitation of this new facility, proton currents ≥ 1 mA are required. In the proton channel following the main ring, an electrostatic beam splitter allows to peel off 20 μA from the main beam for the proton areas.

2. The upgrade program

Until the end of 1989 the main ring was routinely operated at beam currents up to 250 μA. The maximum current was 370 μA, limited by the available RF-power in the ring. This is only a fraction of the potential intensity limit of the main ring, which is estimated to be at 1.5 mA.

Starting in 1990 PSI experienced an 18 month long shutdown, during which time the main activities were:

- rebuilding the 600 MeV target E for a max. beam current of 2 mA.
- creating new secondary beamlines for pions and muons.
- preparing the proton beamline between target E and the future spallation neutron source SINQ.
- start of the RF upgrade program in the ring cyclotron [4], enabling a new intensity record of 0.5 mA at the end of 1991.
- replacement of all injection- and extraction elements of the main ring, including their local radiation shields.
- extensive beam developments on injector 2, bringing the intensity record to 1.5 mA, with losses of only 1.3 μA!

2.1 Accelerators. In injector 2 the beam width at extraction (averaged over 5 revolutions) is used as a simple criterion for the horizontal beam quality and energy spread. A large beam width results in high beam losses and jeopardizes the extraction of high beam currents. Figure 3 shows the improvement of the beam width versus the beam current from 1988 to 1991. The average beam width increases with beam current. This is attributed to the effect of longitudinal space charge forces. Also shown is the original acceptance limit of the 600 MeV-ring. This limit will be raised in steps until 1994 by increasing the cavity voltage, as part of the ring upgrading program.

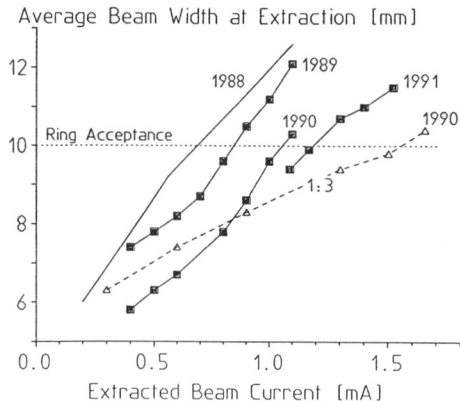

Figure 3: Beam width at extraction from injector 2 as a function of beam current. The dotted line shows the acceptance of the 600 MeV-ring cyclotron prior to the upgrading program. The dashed line shows the results from pulsed beam experiments, where 2 out of 3 micropulses were suppressed.

The main components of the ring are 8 separated magnets and four acceleration cavities at 50 MHz, leading to an average energy gain per revolution of 1.7 MeV. A flattop cavity operating at 150 MHz results in separated turns at extraction even for a wide phase width, giving an extraction rate close to 100%. Additional RF-power is needed to increase the cavity voltage by 50%. This reduces the number of revolutions and increases the turn separation at extraction. The current limit given by longitudinal space charge forces [5], probably the dominating effect for this type of cyclotrons, is thus drastically reduced. At the highest current envisaged, the 50 MHz RF-power needed per cavity amounts to 520 kW! In practice this is achieved with one power amplifier using the Siemens tetrode RS2074. Some problems arise at high beam currents, because the flattop cavity absorbs power from the beam which might jeopardize phase and amplitude control of the cavity when the absorbed beam power exceeds the wall losses in the cavity. During the shut down 1990 one of the four 50 MHz-cavities was equipped with a new amplifier chain, enabling an increase in energy gain per turn by 10%. This allowed us to deliver in 1991 beam currents of 0.5 mA through the ring and on target "E" with extraction losses in the ring in the order of 0.5 μA. These losses are below the limit which we consider to be tolerable for a safe operation and maintenance of the cyclotron.

2.2 Beams. During the years 1985-1991 an extensive upgrading program has been carried out on the 600 MeV-proton beamline with the following aims:

- Adaptation of targets and beam dump to the megawatt beam in the future (thermal load, shielding, handling of components)
- Installation of the first section of the proton beam transport system to the spallation neutron source SINQ.
- Improvement of the secondary beam lines and installation of the new high acceptance, low energy beam line πE5.

In order to achieve these goals, a new mechanical design of the target station and of the beam dump has been applied. The main features are:

- All the elements and their local shieldings are mounted on support stands which are precisely positioned on ground plates, allowing a self-centering installation of the elements. Except for the support stands and some positioning pins there are no additional fastenings. All elements can thus be installed and removed exclusively in the vertical direction with the crane. There is no need for local mechanical work on the highly activated components.
- The connection between beam pipes and target vacuum chambers are made by means of inflatable all metal elements which do not require any clamping.
- All the power-, cooling- and signal connections are brought through the local shielding to a working platform 2.5 m above the proton beam axis, where a low dose rate is expected after the beam is turned off.
- The rotating carbon targets, beam monitors, beam collimators and elements of the beam dump are designed as vertical insertion devices. They can be removed into remotely controlled shielded containers.

Whereas the layout for the secondary beams has not been changed at the target station "M", several changes were made at the target "E": the target length was reduced from 10 to 6 cm of carbon which allows to recover about 60% of the protons for the operation of the SINQ; the two secondary beam lines, collecting pions produced in the forward direction, were converted into two independently operating systems by replacing the old extraction system by two sets of half-quadrupoles, arranged at 8 degrees to the left and right of the proton beam. Starting about 30 cm downstream of target "E", a set of tapered collimators removes the scattered protons. We expect to lose 10% of the protons by absorption in target "E", and 30% by scattering into the collimators. The beam dump, designed for a beam power of 1.2 MW, is used when the SINQ is not in operation.

Figure 4 shows a cut-away view of the spallation neutron source SINQ, which is under construction, and which expects its first proton beam during 1995.

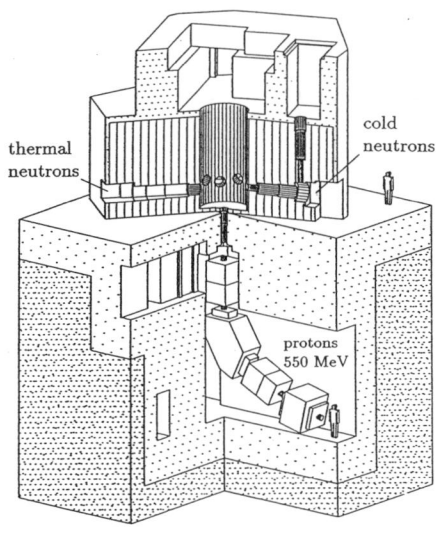

Figure 4: Schematic view of the Spallation Neutron Source SINQ. The 550 MeV proton beam is guided vertically into the target vessel (not shown), where neutrons are produced and moderated. Channels through the 7m shielding surrounding the target guide these neutrons to dedicated experimental stations. Cold neutrons are produced with the help of a special liquid deuterium moderator.

The SINQ- and Neutron guide halls have been finished and the source itself is being assembled. Also the work on five basic research instruments is under way.

The main features of the secondary beam lines at target "E" are given in table 2. The basic optical design of the two muon channels and of the πE1 beam has not been changed. However we have removed the entrance windows of the πE1 and πE3 beam lines in order to be able to accept low energy muons ($p_\mu \geq 5$ MeV/c). In addition, the vertical beam line πE3 has been completely re-designed and can now be used for experiments requiring high intensity, low energy pions and muons as well as for the low energy pion spectrometer which requires a high intensity, high resolution dispersed beam. A novel beam line, the πE5 beam, has been installed. It has a very large acceptance of 150 msr and extracts pions and muons in the backward direction. Since its first bending magnet deflects the proton beam, two compensating magnets had to be added upstream in the proton beam line.

In figure 5 the expected flux of pions and muons from this new beamline is plotted as a function of particle momentum.

Table 2: Summary of positive pion and muon beams. Negative beams have about 4 times lower intensity due to the smaller production cross sections, and cloud μ^- beams are ~ 50 times less than surface μ^+ (p<30 MeV/c).

π^+	momentum MeV/c	max. flux at 1 mA	$\delta p/p$ FWHM	spot size cm^2
πE1	120-600	2×10^9 at 300 MeV/c	0.3 %	1.5 × 2
πE3	40-180	8×10^7 at 180 MeV/c	0.1 %	10 × 4 dispersive
πE5	30-120	1.5×10^{10} at 120 MeV/c	2 %	6 × 4
πM1	120-500	2×10^8 at 300 MeV/c	0.1 %	2 × 1.5
πM3	120-500	8×10^8 at 300 MeV/c	0.2 %	1.7 × 2

μ^+	mom. MeV/c	max. flux at 1 mA	$\delta p/p$ FWHM	spot size cm^2	polar.
μE1	40-125	2×10^8 at 125 MeV/c		3 × 2	75 %
μE4	30-100	4×10^6 at 50 MeV/c	3 %	6 × 4	75 %
πE3	5-30	3×10^7 at 28 MeV/c	1 %	2 × 3	> 95 %
πE5	5-30	2×10^8 at 28 MeV/c	2 %	6 × 4	> 95 %
πM3	5-30	4×10^6 at 28 MeV/c	0.4 %	2 × 2	> 95 %

The dismantling of the target station and beam dump after 20 years of operation has been a very challenging task. We had to remove 500 t of activated material with dose rates up to 400 Sv/h. In localized regions the contamination exceeded the allowed value by a factor of 10^4. About 10^{15} Bq were handled during these operations. 300 t of this material could be re-used, while the remaining 200 t were enclosed in concrete boxes and re-installed as shielding. 12'000 t of slightly activated concrete and iron shielding were rearranged. The sum of the absorbed doses of the 200 persons involved in the work was kept at the low value of 470 mSv and no incorporation of contamination was monitored. The highest dose received by a single person was 30 mSv. Other activities besides moving lots of shielding blocks were: Installation of about 3'000 new cables with a total length of 77km; Installation of 5km cooling pipes with 200 valves, creating a cooling capacity of 5MW around the target station and beamlines.

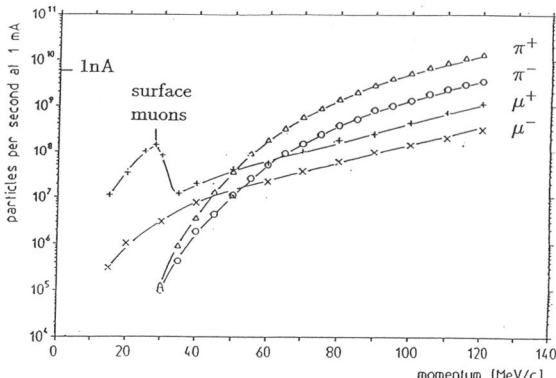

Figure 5: Expected particle fluxes at the final focus of the $\pi E5$ channel for a 1 mA proton beam. The length of the channel is 10.4m, the solid angle 150 msr, the momentum acceptance 10% FWHM and the final spot size 6.4cm^2.

3. Experimental program of the Physics Division

In table 3 and 4 we list the main experimental programs which are performed by about 17 experimentalists and 10 doctorands of PSI in collaborations with about 400 scientists from around the world at the PSI accelerators and at external facilities respectively.

3.1 Experiments at PSI. Compared to other Meson Factories (LAMPF, TRIUMF, KEK, Rutherford) the specialty of PSI are very high intensity, high quality, DC muon beams. They are used either to search with high sensitivity for processes which are forbidden in the so-called Standard Model of Particle Physics or for precision measurements of its parameters. Figure 6 shows the detector SINDRUM II which is being used to search for anomalous muon-electron conversion in nuclei. A preliminary new upper limit for the branching ratio.

$$B_{\mu e} < 4.4 \cdot 10^{-12} \quad (90\% \text{ C.L.})$$

has been obtained [6]. A new superconducting solenoid, the pion-muon converter (PMC) [7] is being designed and will be installed in the new $\pi E5$-area to further increase the sensitivity by two orders of magnitude.

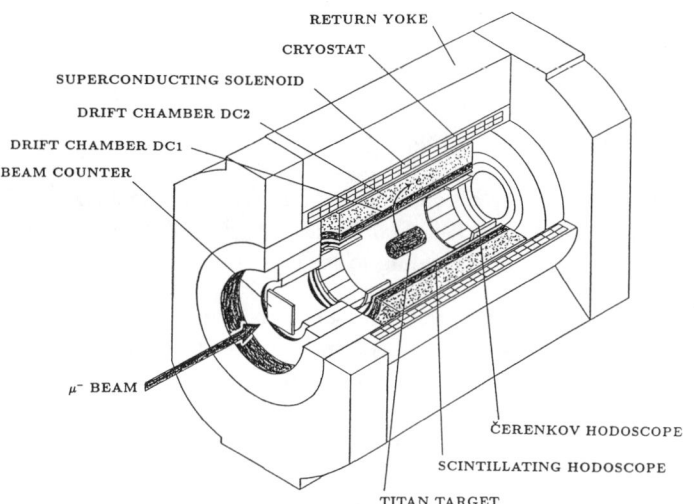

Figure 6: Cut-away view of the SINDRUM II detector (not to scale).

Table 3: Experimental program in the particle physics at the PSI-ring accelerator

Experimental Program	Facilities	Collaborations
Rare decays $\mu - e$-conv., $M - \bar{M}$-conv.	SINDRUM I & II	PSI, ETHZ, UNIZ, Aachen, Heidelberg, Tbilisi, Dubna, Swierk, Yale
Muon capture in light systems	Cyclotron trap and special	PSI, Virginia, ETHZ, Munich, Vienna, Neuchatel, Frascati, Trieste, Louvain, KIAE Moscow
Neutrino masses	πE1-channel and special	PSI, ETHZ, UNIZ, Virginia
Muon catalyzed fusion	μCF apparatus	PSI, Vienna, LANL, Munich, Neuchatel, Berkeley, KIAE Moscow, Gatchina
Slow muon production	Cyclotron trap and special	PSI, Munich, ETHZ, Heidelberg
Pion scatt. on polarized targets	SUSI, polarized targets	PSI, TRIUMF, Karlsruhe, Regina Saskatchewan, Maryland, Hannover
Low energy pion scatt. and DCEX	LEPS	Karlsruhe, Tübingen, LANL
Pion absorption	LADS	PSI, Basel, Karlsruhe, LANL, Maryland, MIT, Albuquerque, Zagreb
Polarized neutron scatt.	NA2-Area	PSI, Geneva, Freiburg/D, Saclay

The spectrometer SINDRUM I is shown as part of figure 7. After it had been used to find a new upper limit for the decay $\mu \to 3e$ [8] of

$$B_{\mu 3e} < 1.0 \cdot 10^{-12} \quad (90\% \text{ C.L.})$$

it is now put together to search for the conversion of muonium into antimuonium with a sensitivity factor of 1000 higher than before [9]. Sophisticated spectroscopy of 10 eV positrons is done with the low field solenoids extending to the right in figure 7.

Most of the muon experiments are reported during this conference, including PSI's attempts to develop new beams of very low energy and very small phase space. But also the pion beams at PSI are widely used. Old lady SUSI and new polarized targets are doing a good job of measuring sensitive polarization variables, the low energy pion spectrometer LEPS and high-tec (cyclotron trap, crystal spectrometer, CCD-detectors) pionic X-ray spectroscopy are testing QCD in the confinement regime and the 4π-spectrometer LADS is disentangling the complicated multibody final states after pion absorption. Not listed is a new pion beta decay experiment, which will use pions stopped inside a 4π crystal calorimeter of CsI (pure). Last but not least the program of polarized neutron scattering using the new beamline to the NA2 area has to be mentioned. The neutron facility has been commissioned at the end of last year and the intensity and polarization was found to correspond to the expectations.

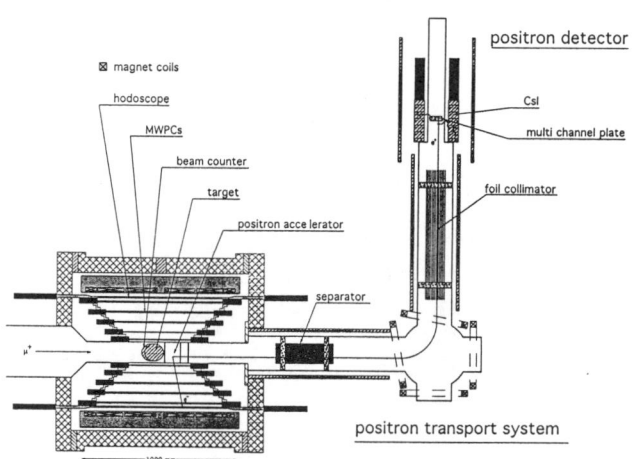

Figure 7: Set up of the new muonium-antimuonium experiment under way at PSI. The refurbished SINDRUM I detector is used to identify the fast Michel particles. Atomic positrons are accelerated to approximately 10 keV and guided in a magnetic field of 0.1 T to a position sensitive multichannel plate detector. 511 keV γ's from positron annihilation are detected in a set of twelve CsI crystals surrounding the MCP.

3.2. Experiments at external accelerators. It is widely recognized that particle physics gets boosted by experiments at various energies. Proton decay at rest in underground laboratories is testing Grand Unification scales, rare decay experiments allow to severely constrain beyond standard model physics, as do experiments at the highest energies available. PSI has a tradition to support Swiss universities in experiments at external accelerators. Table 4 lists five major activities which have survived out of a very diversified list of smaller involvements. The former collaboration of the EIR with L3 for the hadron calorimeter has been extended into a collaboration for the Forward/Backward extension of this detector to cope with LEP 200 conditions. The H1 proposal of building a new microvertex detector is being evaluated by DESY. PSI is also ready to share work together with Swiss universities for detectors to be used in the LHC-project. Presently we are evaluating R&D activities, which are adapted to the expertise of our scientists and the capabilities of our infrastructure.

Table 4: Experiments at external facilities

Experiments	Facility	PSI-part
CP-LEAR	LEAR/CERN	MWPC's
Rare K-decays	E865 at BNL	Muon chambers, beam chambers
L3 F/B-system	L3/CERN	Muon chambers, alignment system
H1	HERA/DESY	Microvertex Detector
Detectors	LHC/CERN	Participation in R&D

4. New projects at PSI

As mentioned in the introduction the accelerators are being used by a more and more diversified community for research in nuclear, particle and solid state physics, material science and life sciences. With the decision, not to support the project of a B-factory at PSI the priorities of PSI have been set to strengthen solid state physics (SINQ), solid state technology (PSIZ) and general energy research. In the life science division strong support is given to a new proton therapy station, which will on medium term supersede the pion therapy program at the piotron. Also the astrophysics program (X-ray astronomy on satellites) is continued to be supported, in line with the recommendations reached by a workshop in ZUOZ on behalf of the Schweizer Wissenschaftsrat [10].

Two new independent but connectable initiatives are being discussed, one in nanotechnology, the other being the project of a Swiss Light Source (SLS) [11]. Whereas the former lies within a reorientation of PSIZ, the SLS project (figure 8) could be the future light-house of PSI. Envisaged is an intense VUV to X-ray machine with very flexible parameters allowing several modes of operation. Among them: very small emittance (~ 1 nm) at lower energies; high flux, high photon energy operation at the maximum possible energy of around 2 GeV; a bypass option for production of very short pulses with possible operation of a single pass high gain FEL. Under study is the incorporation of superconducting bending magnets into the lattice to increase the critical photon energy into the 10 keV range. The lattice accommodates two long straight sections (~ 20 m each) for flexible use of insertion devices including the production of switchable circularly polarized light.

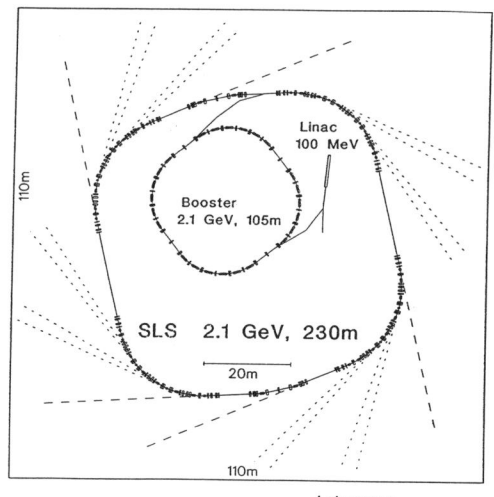

Figure 8: Preliminary layout of the SLS facility with linac, booster and storage ring. Shown are the photon beamline from insertion devices in two 20 m long and three 7 m long straight sections (- - -) and the twin beamlines from eight superconducting bending magnets (...).

Conclusion

With the upgrading of the PSI accelerators the worlds most powerful cyclotron will allow very challenging nuclear and particle physics experiments for at least another decade. A sound mixture between in-house and external experiments will guarantee PSI to remain a center of competence, estimated by Swiss universities and by Swiss industry as a pool of educated trainees.

References

[1] U. Schryber, Proc. 12th Int. Conf. on Cyclotrons and their applications, 1989, Berlin, World Scientific, 93 (1991).
[2] P.A. Schmelzbach, Proc. 1. European Part. Accelerator Conf. (EPAC), Rome 1988, World Scientific, 1370 (1989).
[3] G.S. Bauer, SINQ-Status report, Proc. ICAN-11-Meeting, KEK, Tsukuba, Oct. 1990.
[4] P. Sigg. 12th Int. Conf. on Cyclotrons, Berlin, 212 (1989).
[5] S. Adam, IEEE NS-32, 2507 (1985).
[6] A. v.d. Schaaf, Nucl. Phys. **A546,** 421c, (1992)
[7] W. Dzhordzadze et al., Design study for a high intensity stopped muon channel, SINDRUM II note, in preparation.
[8] W. Bertl et al., Nucl. Phys. **B260,** 1 (1985).
[9] B.E. Matthias et al., Phys. Rev. Lett. **66,** 2716 (1991).
[10] M. Bourquin et al., On the Interface of Astrophysics with Nuclear and Particle Physics, Report on behalf of the Schweizerischer Wissenschaftsrat, Aug. 1992.
[11] A concept for a Swiss Light Source (SLS), PSI-PR-92-24, Sept. 1992.

Laser Polarized Muonic Helium

P. A. SOUDER
Syracuse University
Syracuse, NY 13244

Abstract. We have developed laser techniques to polarize targets of muonic helium and plan to use these targets to measure the spin dependence of nuclear muon capture.

1. Introduction

Measuring the spin dependence of the capture of negative muons by ^{3}He is an important method for determining the weak interaction between the muon and the nucleon. The spin dependence is sensitive to the induced pseudoscalar coupling g_P of the muon to the proton as well as to possible second class currents[1,2]. However, a prerequisite for a successful experiment is the production of a sample of highly polarized muonic helium atoms.

The subject of polarizing muonic helium is complex. Indeed, even the term *polarized muonic atom* may be interpreted from two totally different viewpoints. One way views the muon as a heavy electron, and the muonic atom is the system comprised of a nucleus and a muon. The polarization of this atom is that of the muon. The nucleus, if it has nonzero spin, will also be polarized due to the nucleus-muon hyperfine interaction. The other way views the muon as part of the nucleus, and it is the atomic electrons that are polarized. Due to the muon-electron hyperfine interaction, the muon will also be polarized. An important feature of muonic helium is that it may involve both meanings of *polarized muonic atom* simultaneously.

In general, there are three methods for producing polarized muonic helium. The most obvious is to stop a beam of polarized negative muons in a helium target. The result is a polarized muonic *ion*, which is a muonic atom in the first sense. The problem with this technique is that low polarizations result.[3,4]

A second method, which yields higher polarizations at the cost of experimental complications, is to stop the muons in a *polarized nuclear target*. Again, the first meaning of muonic atom applies. This method has been shown to yield high polarizations for ^{209}Bi.[5] We have recently used this method for ^{3}He, and will describe our results below.[6]

The third method, which we have recently demonstrated,[7] exploits the fact that muonic helium may be neutralized to form a muonic atom in both viewpoints cited above. We stop unpolarized muons in a target of helium with a small admixture of polarized Rb. The muonic ions formed by the stopping muons pick up a polarized electron from the Rb, forming polarized muonic atoms. The spin of the electron is transferred to the muon, and that spin in turn is transferred to the ^{3}He nucleus. Spin-exchange collisions with the Rb further enhance the polarization of the system. The method also works for muonium and ^{4}He. We refer to the method as a *polarized atomic target*.

2. Stopping muons in polarized nuclear targets

The more complex polarized atomic target method is a variation of the polarized nuclear target method which we describe first. For our polarized nuclear targets, we use the method of spin exchange with polarized Rb.[8] The targets are 2.5 cm diameter glass cells filled with about 8 atm of He and a small amount of Rb metal. When the cell is heated to about 200°C, Rb vapor with a density of about 10^{15} atoms/cc results. A polarized Ti-Sa laser beam polarizes the Rb to nearly 100%. About 70 torr of N_2 must be added to the cell to de-excite the Rb by collisions; otherwise the Rb would emit light that would depolarize other Rb atoms. The spin of the Rb is slowly transferred to the ^{3}He during collisions, a process which takes more than 10 hours. With such a long time constant, it is essential to avoid any depolarizing collisions with the He. This requirement restricts the materials used and also demands exquisite cleanliness.

To simplify our experiment, we polarized our targets first and then carried them into the beam. This was possible due to the $\sim$ 30 hr lifetime of the polarizations of the targets. Initial polarizations were in the range of 50-65%, and the polarizations averaged 46% during the run. The spins of the He were held with a 2 G field which could be reversed to flip the target polarization. This method also meant that the target cells were cold and thus almost devoid of Rb vapor, preventing unwanted reactions.

The polarization of the muons in the target was measured by a typical μSR apparatus. A special feature was the small size of the target; it was only 8 atm of He and had little stopping power. We used a low-momentum muon beam[9] at the Stopped Muon Channel at the Los Alamos Meson Physics Facility (LAMPF). Care was taken to minimize the material in the beam. The target cells were 100μ thick. Incoming muons were detected by a coincident pair of 40μ scintillators. Decay electrons were detected by a pair of scintillation counter telescopes parallel $N \uparrow\uparrow$ and antiparallel $N \uparrow\downarrow$ to the magnetic field.

The experimental asymmetry

$$A_{exp} = \frac{N \uparrow\uparrow - N \uparrow\downarrow}{N \uparrow\uparrow + N \uparrow\downarrow} \tag{1}$$

was thus measured. The muon polarization normalized to the target polarization, P_μ^N, was obtained from A_{exp} by correcting for the fraction of stops in the gas, target polarization, etc. We obtain a result of $P_\mu^N = 5.6 \pm 0.6\%$, which is more than a factor of 2 smaller than predicted by calculations that neglect the effects of collisions during the cascade.[10,11] We note that when polarized muons are stopped in ^{4}He, the asymmetry is a factor of 3 or so smaller than predicted[3,4,12]

3. Polarized atomic targets

More recently, we have developed a new method for polarizing muonic helium which we refer to as a polarized atomic target. Since the muon only lives 2.2 μsec, the polarization must be achieved on that short time scale instead of the tens of hours required for Rb to polarize He. Thus processes with large cross sections, on the order of the size of the area of an atom, must be used. The basic reaction we use is spin exchange:

$$(\mu^{-3}\text{He e}^-) + \text{Rb} \uparrow \rightarrow (\mu^{-3}\text{He e}^-) \uparrow + \text{Rb}. \qquad (2)$$

This reaction only occurs for *neutral* muonic helium (μ^3He e$^-$). However, stopping negative muons in helium gas yields muonic helium *ions* (μ^{-3}He)$^+$. Thus an electron donor must be added. Xe is the usual choice,[3,4] but has the disadvantage that it depolarizes the Rb when used in the required quantity. Thus we tried CH$_4$:

$$(\text{CH}_4) + (\mu^{-3}\text{He})^+ \rightarrow (\text{CH}_4)^+ + (\mu^{-3}\text{He e}^-). \qquad (3)$$

Several percent of CH$_4$ is expected to be required because the reaction only occurs for epithermal muonic helium. Once muonic helium is thermalized, it forms molecular ions which are inert in the presence of Xe and also presumably in the presence of the CH$_4$. On the other hand, since Rb has a much lower ionization potential, it should react with the thermal muonic ions. Thus we have

$$\text{Rb} \uparrow + (\mu^{-3}\text{He})^+(\text{He}) \rightarrow \text{Rb}^+ + (\mu^{-3}\text{He e}^-) \uparrow + \text{He}. \qquad (4)$$

This reaction has the additional advantage that the neutral muonic helium is formed already partially polarized.

The only modification of the apparatus required to study this new method was that the cell had to be hot enough to have the ample Rb vapor density and the laser had to strike the target simultaneously with the muon beam. We measured the asymmetry as before, but studied it as a function of the time after the muon was formed.

Preliminary results are shown in Figure 1 for both ^{3}He and ^{4}He with and without CH$_4$. The results are dramatic. For long times in the ^{4}He target, the raw asymmetry (corresponding to stops in the He gas) is almost 20%. Assuming an analyzing power of the muon decay to be 33%, we obtain a muon polarization of over 50%. This is consistent with a polarization of 100% for the region of the target with polarized Rb. Using methane as a donor also works, but the polarizing time is longer, suggesting that the spin exchange cross section is smaller than the charge exchange cross section. Similar results are seen for the ^{3}He, but the asymmetries are smaller for two reasons. First, the muonic ^{3}He has spin 1 instead of spin 1/2 for the muonic ^{4}He. Thus more collisions are required to provide additional angular momentum. Second, 25% of the muonic ^{3}He is in the singlet state which cannot be polarized.

4. Spin dependence of muon capture by helium

Measurements of the spin dependence of negative muons captured by ^{3}He into the ground state of ^{3}H, $\mu^- + N_i \to \nu_\mu + N_f$ can provide unique information about the charged weak current,[1,2] given by

$$J_\mu = \bar{u}_f \Big(g_V \gamma_\mu + \frac{g_M}{2M_P}\sigma_{\mu\nu}q^\nu + g_A \gamma_5 \gamma_\mu - g_P q_\mu \gamma_5 \frac{2M}{m_\pi^2}\Big) u_i. \qquad (5)$$

Here g_V, g_M, g_A and g_P are isovector form factors, which depend upon the four momentum transfer Q^2, and $M = \frac{1}{2}(M_f + M_i)$. The form factor g_P is important for muon capture because the mass of the initial state lepton, the muon, is significantly different from that of the final state lepton. Since there is no pseudoscalar coupling in the Standard Model for free quarks, g_P arises from the fact that the quarks are confined in the proton by QCD forces.

The reaction is highly spin dependent. This fact may be exploited experimentally by determining the initial hyperfine state described by the quantum numbers F and M_F, and the direction of the recoil ^{3}H, which is exactly opposite $\hat{k}$. The initial populations are defined by $P(F, M_F)$. It is convenient to define a vector polarization $P_V = P(1,1) - P(1,-1)$, a tensor polarization $P_T = P(1,1) + P(1,-1) - 2P(1,0)$, and a departure from the statistical singlet-triplet mixture $\Delta = P(1,1) + P(1,0) + P(1,-1) - 3P(0,0)$. Then we may define a decay rate

$$\frac{d^2\omega}{d\hat{k}} = \frac{R}{4\pi}(A + B \cdot P_V(\hat{k}\cdot\hat{z}) + C \cdot P_T[(\hat{k}\cdot\hat{z})^2 - \frac{1}{3}] + D \cdot \Delta). \qquad (6)$$

Here the average rate $\Gamma = RA$ for an unpolarized initial state defines the constant R, and the constants A, B, C, and D are functions of the from factors.

The constant A, which is relatively insensitive to g_P, has been measured[13,14] by using unpolarized ^{3}He. Little other experimental information exists on this system. With the large polarizations we have achieved, it may be possible to obtain large P_T as well as P_V and thus obtain both B and C. This provides redundancy or checks some of the underlying assumptions such as the absence of second class currents.

We are presently developing a drift chamber which will be part of the polarized atomic target in order to detect the direction of the resulting tritons. Preliminary results using a thermal neutron source to produce the reaction

Laser Polarized Muonic Helium

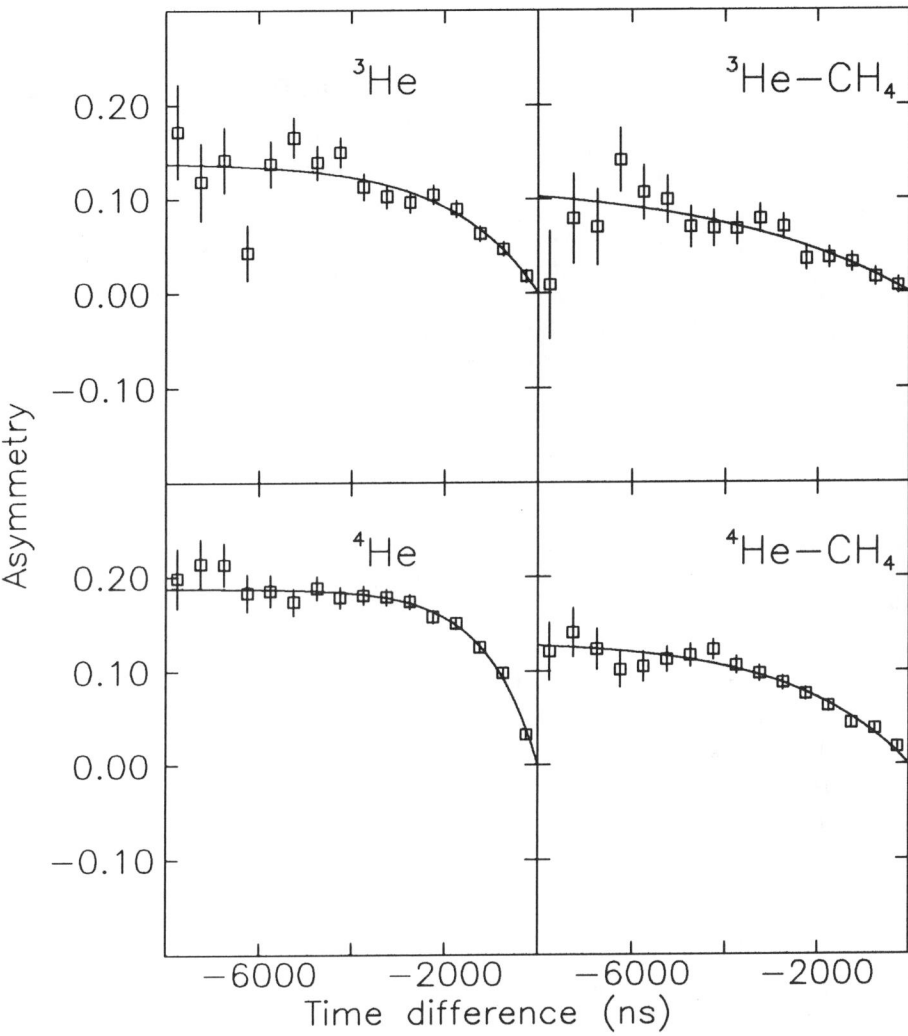

Figure 1: Asymmetries as a function of the time between the detection of a decay electron and a muon stop for the atomic polarized targets. Note that times thus defined are negative. The top two curves are for ^{3}He, and the bottom are for ^{4}He. The curves on the left used Rb as the electron donor whereas the curves on the right are obtained using CH_4. The data are corrected to include only stops in the helium. These results are preliminary.

$$n + {}^3\text{He} \to p + {}^3\text{H} \tag{7}$$

are promising. We are planning to test this detector at LAMPF this summer, and measure the spin dependence of muon capture on ^{3}He in the near future.

Acknowledgements

I would like to thank my colleagues on this work, including L. Han, R. Holmes, J McCracken, and J. Xu (Syracuse University), A. S. Barton, P. Bogorad, G. D. Cates, M. Gatzke, and B. Saam (Princeton University), and D. Tupa (LAMPF). This work is supported in part by the U.S. Department of Energy under Grant No. DE-FG02-84ER40146.

References

[1] B. R. Holstein Phys. Rev. C **4**, 764 (1971)
[2] W-Y. P. Hwang, Phys. Rev. C **17**, 1799 (1978).
[3] P. A. Souder et al., Phys. Rev. Lett. **34**,1417 (1975).
[4] P. A. Souder et al., Phys. Rev. A **22**, 33 (1980).
[5] R. Kadono et al., Phys. Rev. Lett. **57**, 1847 (1986).
[6] N. R. Newbury, Phys. Rev. Lett. **67**, 3219, (1991).
[7] A. S. Barton et al., to be published.
[8] W. Happer et al., Phys. Rev. A **29**, 3092 (1984).
[9] R. Holmes et al., Nucl. Instrum. and Meth. in Phys. Res. A **303**, 226 (1991).
[10] Y. Kuno, K. Nagamini, T. Yamazaki, Nucl. Phys. A **475** 615 (1987).
[11] A. P. Bukhvostov, and N. P. Popov, JETP **19**, 1240 (1964).
[12] See N. P. Mukhopadhyay, Phys. Rep. **30C**, 1 (1977) and references therein.
[13] L. B. Auerbach et al., Phys. Rev. **138**, B127 (1965).
[14] D. R. Clay et al., Phys. Rev. **140**, B586 (1965).

The Cyclotron Trap and Low Energy Muon Beams

L.M. SIMONS
Paul Scherrer Institute
CH 5232 Villigen
and
F. KOTTMANN
Institut für Hochenergiephysik, ETH-Hönggerberg
CH 8093 Zürich

Abstract. Future specialized experiments with muonic atoms require high intensities of low energy muons. A method is investigated in which the Cyclotron Trap is used as an efficient device to decelerate muons in thin foils. By applying an electric field to the stop region, some are extracted before stopping. Thus muon beams in the energy range of 5-40 keV will be produced with intensities of 10^5/s in the near future at PSI.

1. The need for low energy muon beams

Muonic atoms have been studied now at accelerators for almost 40 years and used as tools in different fields of physics. For the sake of illustration a few examples are mentioned:

The determination of the muon mass or of Q.E.D. corrections by precise measurements of the binding energies of muonic atoms and muon capture studies from the ground state are examples for experiments belonging to particle physics [1][2]. In a typical nuclear physics application, the measurement of muonic X-rays is routinely used to determine nuclear shape parameters [3]. Another example is an application from materials analysis: the measurement of energies of muonic atoms can provide an elemental analysis of bulk matter. Via measuring the intensities, even the chemical composition can be deduced from muonic X-ray measurements [4].

For most past experiments the amount of target material needed to form a sufficient amount of muonic atoms was typically of the order of grams. This is prohibitive for many interesting experiments like the studies of muonic X–rays and muonic capture in low pressure hydrogen isotopes, in which collisional processes would not affect the atomic cascade. In nuclear physics it would be interesting to extend the measurements to rare and hence expensive isotopes where only a minute amount of target material would be available. In material analysis, a muon beam of variable energy from the keV range upwards would be ideal to separate surface from bulk effects.

In order to produce beams which are useful for this type of experiments, the gap has to be bridged between conventional beam energies in the MeV range and the beam energies needed, which are typically below 40 keV.

2. Approaches at PSI

All approaches to produce low energy muons are restricted by the fact that the deceleration process from MeV to keV should take less time than the life time of the muon. The simplest way to achieve this is decelerating muons in matter by the normal energy loss caused by ionization. It is conceivable that after moderation of a muon beam, an electrostatic filter could select only the low energy muons. The efficiency of this method, however, is rather poor. Even from the beam at PSI with the highest expected flux of low energy muons, not more than a few thousand negative muons can be expected below 20 keV in a beam spot of 4x4 cm^2 with a divergence of almost 300 mrad [5].

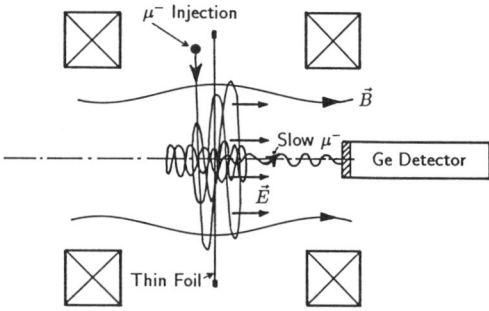

Figure 1: Scheme used at PSI

Therefore at PSI, an extension of the degrader method was considered. Suppose muons of several MeV energy are brought into a magnetic mirror field where they can swing forth and back between the magnetic mirrors. With a suitable foil arrangement in the median plane the muons will loose energy on each traversal through the foil structure. With decreasing energy the probability of scattering muons out of the confinement regime into the loss cone will increase. Thus, some of the low energy muons leave the boundary region before they are stopped in the foil structure or have decayed.

This process can be considerably enhanced by applying an electric field parallel to the axis of the magnet. Muons having left a foil with low energy will gain enough momentum into the axial direction to leave through the loss cone. It can be expected, and it was later proved by simulation studies, that the number of extracted muons is rather insensitive to the foil thickness in a broad range between several $\mu g/cm^2$ up to about 200 $\mu g/cm^2$, since with thickness the stopping ability increases but the extraction efficiency decreases.

The availability at PSI of a superconducting magnet, the so-called cyclotron trap [6], with mirror field characteristics shortened the time between planning and first experiments considerably. Moreover, the special shape of the magnetic field of the cyclotron trap permitted different approaches for the injection of muons to the mirror field which in turn lead to different scenarios with alternative merits.

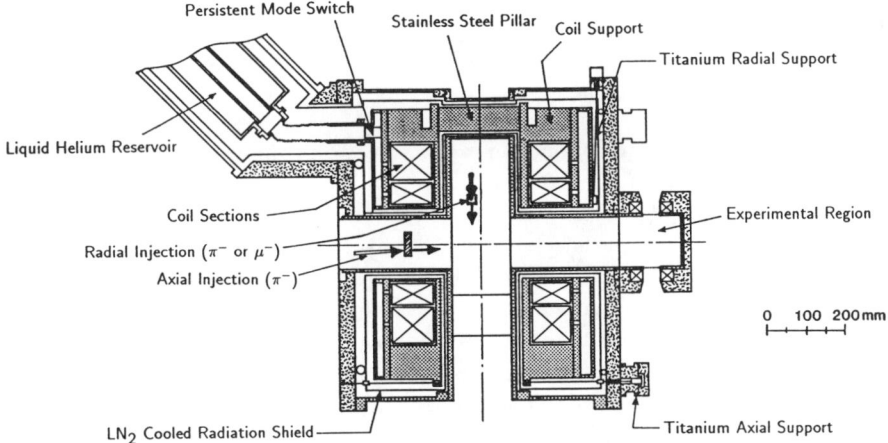

Figure 2: The cyclotron trap

2.1 Axial injection. There are several possibilities to fill the magnetic mirror with muons. Well known is the magnetic bottle technique which uses the decay of pions having the magic momentum of about 40 MeV/c. Pions of this momentum are injected along the axis of the magnetic mirror by degrading a higher momentum pion beam at the entrance of the mirror field. Traversing a decay length of about 2 m, they decay into muons with energies between zero and 17 MeV. The low energy muons are emitted almost perpendicularly to the axis and therefore will be captured in the magnetic field. This technique has the advantage of a relatively simple injection scheme and a sufficient number of extracted muons as shown by the simulation studies summarized in the next chapter. With an axial extension of the mirror field by a magnetic channel these muons can be transported to an experimental region. A drawback arises from the fact that on the axis of the magnetic field will also be high energy muons, pions and electrons. Hence, a special filter arrangement has to be added to the system, which limits the momenta of the particles to low momenta only and additionally reduces the electron contamination. All this prevents an intermediate use of the extracted muons.

2.2 Radial injection of muons. The injection scheme routinely used to transport particles into the cyclotron trap magnetic field is to inject them radially in the median plane. The geometry of the coils is designed such that a weak focussing cyclotron field is established over a radius of 143 mm. For the present application muons with momenta of 30 MeV/c are injected radially into the outer region of the cyclotron field at about a radius of 120 mm. Then the beam is decelerated to momenta of about 20 MeV/c by a suitably shaped moderator/scintillator arrangement. The magnetic field strength is arranged such that particles of these momenta are captured by the cyclotron field. Then the particles are decelerated further by the energy loss due to the gas inside the target chamber. For the extraction studies a different technique has to be applied, because the required high voltage would immediately lead to a discharge inside the gas. Hence, the gas has to be replaced by a foil in the median plane, or a foil structure parallel to the median plane thick enough to decelerate the muons during their life time.

Compared to the axial injection, the radial injection is more complicated and will, as it turns out from computer simulations mentioned later, yield a factor of ten lower intensity of extracted muons. On the other hand optimization studies of injection and extraction can be performed without any major change of the apparatus. In particular the optimization of the high voltage and the foil structure can be done with X-ray detectors mounted axially near the stop region. The results of these studies are independent of the injection modus and hence representative of all kinds of injection schemes. It should be mentioned also that the foil can be replaced by low pressure gases at any time. The arrangement with detectors near the stop region allows then for immediate experiments [7],[8].

2.3 Radial injection of pions. A combination of the merits of the two aforementioned methods is provided by the radial injection of pions. Pions with momenta of about 100 MeV/c can be injected into the cyclotron field and decelerated to momenta below 60 MeV/c with a moderator at a mean radius of 100 mm. The pions not stopping in the walls or the injection moderator will decay into muons. Only a few of these muons with suitable momenta are accepted by the potential-well formed by the cyclotron field and transported to the center.

This method yields higher muon fluxes than the axial injection scheme. On the other hand, it will certainly entail higher background than produced during the radial injection of muons. The detailed investigation of the background will be a matter of experimental studies during planned test runs, which should provide the necessary parameters for the beam guiding and purification system, which may be required.

3. Results of the computer simulations

Simulation studies have been performed with the parameters of existing or planned beams, for the magnetic field of the existing cyclotron trap and for different electrical field and foil configurations in the center.

For the study of radial muon and axial pion injection, particles were followed until extracted to a plane outside the magnetic mirror. These studies showed that the stopping and the extraction efficiencies can be factorized. Therefore, for the simulation of radial pion injection, the decay muons were followed only until they stopped. The calculations were performed using the calculated beam parameters of the πE5

Table 1: Computer simulations for 1 mA proton current

Injection scheme	Momentum	Injected particles/s	B-field in center	extracted muons/s	beam spot
Axial pions	100 MeV/c	10^9	4.3 T	6×10^4	$1.5 \times 1.5 cm^2$
Radial muons	40 MeV/c	8×10^6	0.9 T	2.4×10^4	$5 \times 5 cm^2$
Radial pions	120 MeV/c	2.5×10^9	3.0 T	4×10^5	$5 \times 5 cm^2$

channel due to be commissioned in summer 1992. The highest number of extracted muons can be expected for the radial injection of pions, but these muons are distributed over a much bigger beam spot than those produced by the axial injection of pions. The bigger beam spot associated with the radial injection of muons can be reduced by guiding the muons into a field region of higher strength. As mentioned before, working with radially injected muons allows for immediate tests of extraction schemes and even for physics experiments without the need of building an extraction channel.

The radial injection of pions attracts much interest since it does not suffer from the unavoidable high background typical of the axial injection. There will be background from higher energy muons from pion decay and also from electrons. In a set-up presently under consideration, the axial field is extended by matching a split coil axially to the cyclotron magnet. The extracted low energy muons can then be stopped in a target volume inside the split coil magnet. Detectors will be mounted radially thus minimizing the effect of high energy background on the axis.

4. First experimental studies

In a first test at the πE1-channel, a 2μm thick mylar foil was installed in the median plane of the cyclotron trap. A Ge detector was mounted axially in order to monitor the X-rays emitted from the mylar foil. A maximum of 15% of the incoming muons could be stopped compared to the prediction of about 20%. In a subsequent test, which took place after this workshop, extraction was established. From the otherwise stopped muons about 2% could be extracted by charging the foil, which was metallized with a $80\mu g/cm^2$ Ti layer.

The extracted muons were detected by stopping them in the Be-window of the Ge detector and measuring the muonic Be X-rays. The measured number of extracted muons was in agreement with the result of computer simulations.

5. Outlook

Since the numbers measured agree with the computer simulations, we are confident that the special approach at PSI is physically sound and worth development at beam lines with higher intensities. Moreover, replacing the present cyclotron trap with a magnetic system optimized for muon extraction could possibly yield fluxes up to an order of magnitude higher than presented above. Work in this direction is underway.

7. Acknowledgement

It is a pleasure to thank all the members of the low energy muon community (μ^+ and μ^-) for all their effort in making the first studies a success.

8. References

[1] E. Zavattini, Contribution to this workshop

[2] J. Deutsch, Contribution to this workshop

[3] C. J. Batty, E. Friedman, H. J. Gils, and H. Rebel, Adv. Nucl. Phys. **19**, 1 (1989).

[4] H. Daniel, F.J. Hartmann, E. Köhler, U. Beitat, J. Riederer, Archaeometry **29**, 110 (1987).

[5] E. Morenzoni, private communication.

[6] L. M. Simons, Phys. Scripta **T22**, 90(1988).

[7] D. Abott, B. Bach, J. Missimer, R. Siegel, L. M. Simons, D. Viel, PSI–proposal R91–10.1

[8] P. DeCecco et al. PSI–proposal R92–06.1

Laser Spectroscopy of Muonic Hydrogen with a Phase Space Compressed Muon Beam

D. TAQQU
Paul Scherrer Institute
CH-5232 Villigen-PSI, Switzerland

Abstract. Laser experiments involving muonic atoms are optimally done with muon beams of low energy and small dimensions. The recently developed phase space compression (PSC) method can be used to provide muon beams of a few mm width and about 10 keV energy. Application to high resolution spectroscopy of the muonic hydrogen atom is presented. A 10^{-5} precision for the 2s-2p splitting and a 10^{-7} precision for the ground state hyperfine splitting can be obtained with available laser systems.

1. Introduction

The laser spectroscopy of muonic atoms involves a handful of experiments whose realization will lead to a significant advance in precise tests of QED. A real breakthrough in this research field has not yet been obtained because of the small stopping volumes required (at relatively low densities) in order to achieve sufficient transition probability with readily achievable laser power. This unfavorable situation was one of the main motivations for proposing and developing phase space compression (PSC) methods. Their implementation leads to high quality beams of the lowest energies and allows the laser to be triggered by the muon detection in the last PSC stage. High transition probabilities in small stopping targets can be achieved with conventional low repetition rate lasers [1].

Following the initial PSC proposal the development of various specific components (trajectory detectors and pulsed high voltages) allowed the successful test of primary PSC operations [2]: fast and non-disturbing measurement of the phase space of a single muon, reduction of its longitudinal energy spread and confinement within an electrostatic trap. Furthermore muons of lowest energies (below 10 keV) have been injected into the PSC apparatus with their entrance energy being determined by time of flight [3]. This energy information could be used to provide selective triggering of the PSC operation [4].

It is therefore timely to discuss the possibility of applying the PSC technique in its present stage of development to the realization of some of the relevant laser spectroscopical experiments. This paper describes experimental arrangements allowing very precise determinations of the 2s–2p and the hyperfine splittings in the muonic hydrogen atom. Application to the 3d–3p transition is described in a separate report [4] and a contribution to the last atomic conference [5] shows how (with some further PSC) transitions between high n levels of the μp atoms can be induced and detected.

In the scheme considered here the laser experiments take place in a high field solenoid placed downstream of the PSC solenoid and the final laser trigger is provided by the detection of the muon entering the target region. The transfer of the muon from the PSC apparatus to the target makes use of a novel scheme for the extraction of the trapped muons. This conceptually simple method which leads to a muon beam of minimal width is first described. It allows the laser experiments of the last two paragraphs to be realized in a most optimal way.

2. The extraction of the trapped muons

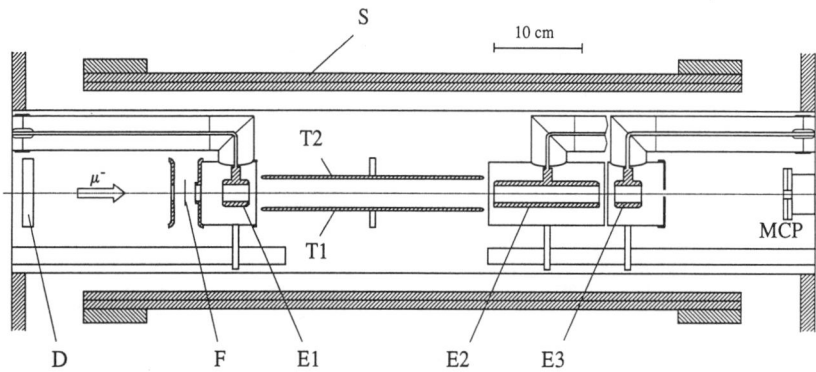

Figure 1: The PSC apparatus. S: solenoid, D: entrance detector (for example PPAC), F: secondary electron emitting foil. E1 – E3: cylindrical electrodes, T1, T2: longitudinal plane electrodes.

The internal components of the PSC apparatus in its presently mounted configuration are shown in Fig. 1. It differs from the initial arrangement [2] by having only three cylindrical electrodes E1, E2 and E3 (trapping taking place between E1 and E2), a different type of entrance detector and the central trapping region containing only two longitudinal parallel plates electrodes of length L. The final positions of the electrodes and detectors can be chosen to match the energy range of the muons that are to be trapped. For highest trapping rates a distance of 50 cm between the detection foil F and E3 can be implemented allowing (within the presently available high voltage pulsing speed) a maximal longitudinal energy $E_\parallel$ of the trapped μ^- of 15 keV. Transverse energy reaches the same order of magnitude. This leads to a high acceptance and results in a trapping rate (at 1 mA proton current in the newly developed PSI πE5 beam line) which exceeds significantly 100 μ^-/s. While such an arrangement can be used for the hyperfine splitting experiment, the 2s–2p measurement requires a lower energy beam with minimal spread in $E_\parallel$. This is more readily obtained by limiting the accepted $E_\parallel$ to less than 8 keV (as achieved in previous tests). Thereby $E_\perp$ can be limited to $E_\perp < 5$ keV and the demonstrated technique of reduction of the energy spread $\Delta E_\parallel$ to less than 1 keV can be used to obtain trapped muons with sharp longitudinal energy (4 keV$< E_\parallel < 5$ keV) and a total energy less than 10 keV.

The basic PSC operations (muon detection, reduction of $\Delta E_\parallel$ and trapping) which are described in detail in [2] together with the low energy injection scheme recently implemented [3] lead to muons trapped one by one between E2 and E3 that spiral around the magnetic field lines (with a spiral radius $\rho = 0.68$ mm for $E_\perp = 5$ keV at B = 5 T) within a distance of r = 1 cm from the solenoid axis. The extraction procedure considered here takes place via a static transverse electric field (1 – 2 kV/cm) that acts vertically on the trapped muons between E1 and E2. This induces a slow horizontal transverse drift velocity ($\vec{v}_D = \vec{E} \times \vec{B}/B^2$) that superposes itself upon the fast longitudinal motion as shown in Fig. 2 for particles with $E_\perp = 0$. The muon makes many round trips before the exit on one side (lower right in Fig. 2) due to reaching a region where the reflection potential V no longer exceeds $E_\parallel$/e.

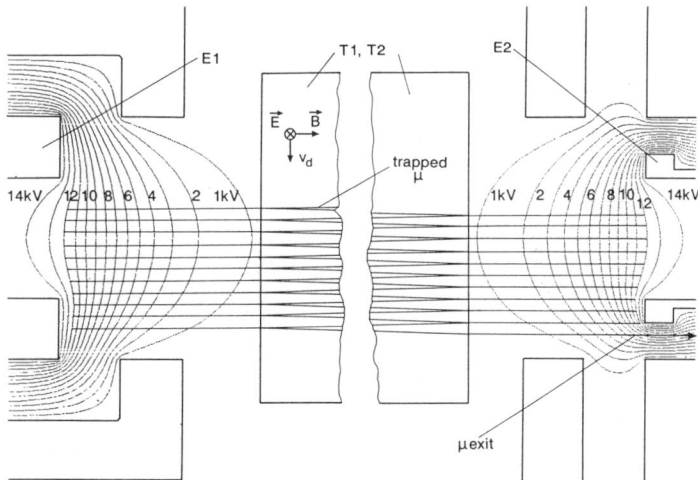

Figure 2: The drift and exit of a trapped muon with 12 keV longitudinal energy, E1, E2: trapping electrodes (with potential lines): T1, T2: plane longitudinal electrodes producing the parallel field $\vec{E}$ between them.

The electrode on the escape side can be shaped in such a way that the outcoming muon beam has a rectangular cross-section b×d with b=2r and d being of the order of $2L\, v_D/v_\parallel$ (for $E_\perp = 0$). For non decaying particles (for example positrons or antiprotons) and a low intensity beam, v_D can be reduced at will and d made negligibly small compared to b (again only if $E_\perp$ is small enough). For the case of muons the lifetime limitation leads to intensity losses for a too small drift velocity. Nevertheless, one obtains a maximal phase space compression factor $f = \tau_\mu/T$ (with τ_μ being the muon lifetime and $T \cong 2L/v_\parallel$ the round trip period). As T can be made as small as 100 ns f can be quite high. The reduction in beam size is a consequence of the violation of the Liouville theorem. The only question is where did it take place? Surely not in the purely static extraction step but clearly during the trapping of the single particles from a continuous beam. This operation transforms the continuous beam longitudinal phase space density $I/v_\parallel \Delta p_\parallel$ (with I being the beam intensity, $\Delta p_\parallel$ the momentum spread and $v_\parallel$ the particle velocity) into a (trapped) longitudinal phase space density of $1/2\, L\Delta p_\parallel$ [6] resulting in $f = 1/IT$ (which reduces to τ/T for particles with lifetime τ). The violation has been made possible by the single particle detection which selected a small portion of longitudinal phase space area surrounding the particle and allowed its movement into the trapped configuration. The job of the extraction procedure is just to return the trapped phase space volume into a continuous beam phase space volume of reduced transverse size. Compared to other extraction procedures [7] the muon comes out in a much wider time distribution but nevertheless the existing position information of the incoming detectors allow the drift time (and thereby also the exit time) to be known within a few hundred ns.

For $T = 100$ ns and an average extraction time of 1 μs over an average drift length of 15 mm one obtains at $E_\perp = 0$ a minimal width $d_{min} = 1.5$ mm. The effect of non zero $E_\perp$ is to expand the total beam width to $d_{min} + 2\rho$ (~ 3 mm at $E_\perp = 5$ keV). The width of the spiral center distribution is not affected by ρ so that the stopping in low density targets can lead to a width of the stopping distribution that approaches d_{min}. As can be seen in the forthcoming applications the resulting rectangularly shaped transverse size of the stopping volume is optimally suited to laser experiments.

3. The 2s–2p transition in the μp–atom

This experiment which, in its basic features, follows the lines of a 1979 SIN (PSI) proposal [8], has recently received a new impetus by the steady increase in the precision obtained for the 1s–2s Lamb shift measurement in atomic hydrogen. It is expected that within a year from now an order of magnitude improvement of the recently published [9] 10^{-5} precision level will be achieved and no fundamental limitation hinders an even higher precision in the forthcoming years. Beyond the 10^{-5} precision level comparison with theory will first require a more precise computation of the QED terms (with work toward the "easily" accessible precision level of 10^{-6} being on the way) and second a more precise knowledge of the proton charge radius r_p. As progress in this direction from electron scattering measurements is expected to be slow and limited the most accessible way to provide the required information is the measurement of the 2s–2p splitting in the μp atom. A minimal precision of only one half of the transition width (about $1.5 \cdot 10^{-4}$ of the transition energy) will already allow a Lamb shift test of QED to the 10^{-6} precision level. With the ten times better precision expected to be achieved in the present experiment the Lamb shift test of QED takes place at a level (10^{-7}) where four loop radiative corrections and hadronic effects need to be computed [10]. It is therefore expected that the combined ep and μp experiments will result in a test of bound state QED of unprecedented precision.

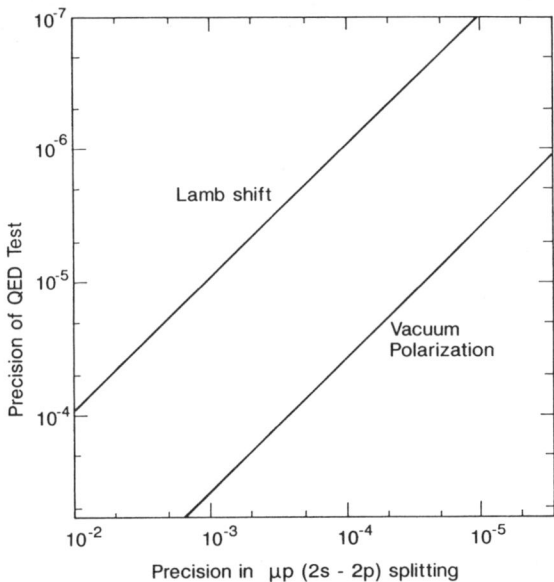

Figure 3: Relationships between the achievable precisions in the muonic 2s–2p splitting and electronic Lambshift or vacuum polarization in hydrogen.

The validity of this QED test scheme has been questioned with the following argument: as the main contribution to the 2s–2p splitting in muonic hydrogen results from vacuum polarization, a precise test of the theoretical computation of this quantity may be required before an absolutely reliable proton radius can be extracted. However, a quick look to Fig. 3 should dissipate any possible doubts. Here, the achievable accuracies of the QED tests are presented for the case of no independent knowledge of the proton radius. The lower line shows the 2.6 % vacuum polarization contribution to the ep shift. It can be observed that at any point the ordinate corresponds to a higher precision than the abscissa. Therefore, in the (expected) situation where the results of both experiments correlate according to the predictions of QED theory, vacuum polarization is automatically tested more precisely than by any other methods. If, on the other hand, the two experimental results are (as can be dreamed about) contradictory, the indication of some new physics (independently as to which part of QED is violated) has sufficiently far reaching implications to validate the proposed scheme.

Before describing the experimental configuration a production scheme for the required tunable 6 μ radiation is presented. It is based on experimental results on Raman shifting of a pulsed dye laser in high pressure H_2 gas. Following recently published work on infra-red third order Stokes radiation production [11] the dye laser wavelength has been switched to 708 nm in order to probe the 6 μ wavelength production region. Preliminary results [12] with a 1 m long H_2 cell led to an optimal operation pressure of 7 atm at which about 100 μJ of stimulated 6 μ radiation (of a quality not worse than twice the gaussian limit) could be obtained when pumped with 42 mJ in a 10 ns long pulse. As a consequence of the sharp threshold effect observed at this intensity level the dye laser to be used in the present application should provide at least the corresponding 4 MW dye laser intensity. This is just at the limit of what the most powerful commercially available eximer-based laser system can achieve, however only at a reduced repetition rate of 20 pulses/s. Such a system is the only one that has a short enough trigger delay time (about 1 μs or less) to allow the triggering to be done by the muon detection. This leads to the overall laser system of Fig. 4 from which about 200 μJ of tunable 6 μ radiation can be expected.

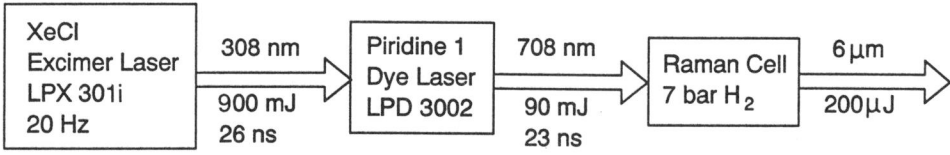

Figure 4: The laser system.

As for the 1979 proposal the experimental method presented here makes allowance for the worse possible situation where the fraction of very long lived (slowed down to below 0.32 eV) μp (2s) formed during the cascade is too low to be used in the experiment. Therefore, an H_2 target pressure of 0.25 torr or less is used in order to insure a 2s decay time sufficiently longer than the cascade time and allows the laser transition to be detected with minimal background from cascade $K\alpha$. The result of a cascade calculation at 0.25 torr is shown in Fig. 5 with the 2s decay time selected according to the computed quenching cross-section of the fast 2s component [14]. With the laser shot about 1 μs after the stopping time the $K\alpha$ following the 2s–2p transition can be detected practically background-free.

The target cell is a 60 cm long low pressure H_2 container placed in a high field solenoid and separated from the vacuum by a 5 μg/cm^2 thin formvar foil (Fig. 6a). The first 20 cm of the cell (Fig. 6b) are used as an entrance detector where a longitudinal electric field produced via cylindrical electrodes accelerates the ionization electrons whose collisions with the H_2 gas result in UV light emission. This far UV light is first converted into the near UV in a thin coating of p-terphenyl deposited on the inner side of the cylindrical glass container. A plastic wavelength shifter scintillator surrounding the cell makes the second conversion step and allows the scintillation light to be collected and detected by photomultipliers placed outside the solenoid. From the results of a previous experimental arrangement [15] it can be deduced that the very efficient light collection system possible here will allow the few keV energy deposition to result in close to 100 % detection efficiency of the entering muon.

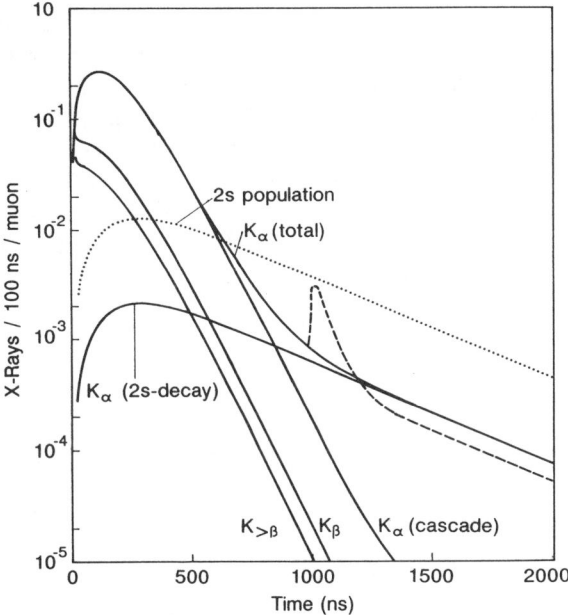

Figure 5: Time distribution of the cascade KX-rays at 0.25 torr (computed according to the model of ref. 13). The broken line gives the expected experimental total $K\alpha$ distribution resulting from a 30% efficient laser transition.

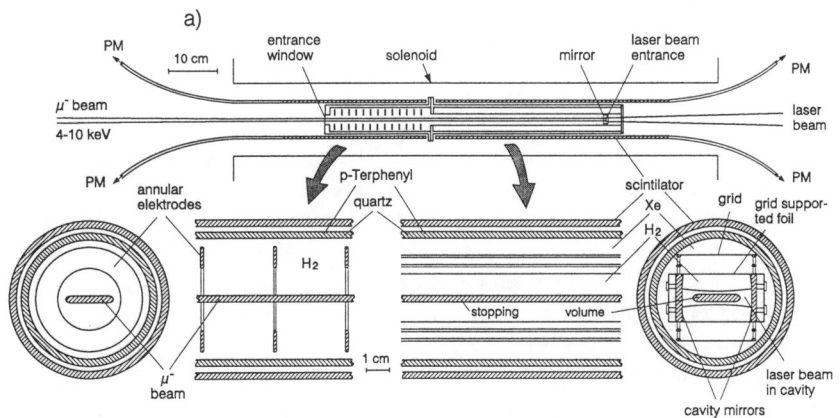

Figure 6: The experimental arrangement for the 2s–2p measurement.

The next 40 cm of the target constitute the effective stopping volume. With their energy (both longitudinal and transverse) reduced appreciably in the entrance foil most of the muons will stop within this volume. As shown in Fig. 6c the 6μ radiation will be confined between two 40 cm long highly reflective ($\sim$ 99.9 %) cylindrical mirrors (slightly curved at both ends for optimal laser light trapping). It is injected into the cavity via a rectangular opening (1 cm × 0.5 mm) in one of the mirrors and undergoes on the average more than 500 reflections. It results in more than 5 mJ/cm^2 radiation intensity on the solenoid axis which is not far from the computed 10 mJ/cm^2 energy required to saturate the $^3S_{1/2} \to {}^5P_{3/2}$ transition. The $K\alpha$ X-rays are detected by 2 driftless [16] Xe–GSPC detectors [17] whose UV scintillation light is twice wavelength shifted [18] and detected in the same way as for the light from the entrance detector. The high solid angle results in about 30 % $K\alpha$ detection efficiency. With 20 triggers/sec the 2s–2p transition is detected with a signal rate at resonance of about 10 events/hour. Within a few days of measurement the splitting is measured with a precision of 10^{-5} which corresponds to a 10^{-7} accuracy limit for the ep Lamb shift.

4. The hyperfine splitting of the μp atom

This oldest μp laser spectroscopy proposal is expected to improve by orders of magnitude the comparison of the best known hyperfine splitting of the ep atom with up to date QED computations and should lead to a better knowledge of the proton structure. It has been classically considered as a difficult experiment [19] but with a specially adapted measurement technique (first described in [1]) and the availability of a PSC muon beam it has returned into the realm of directly accessible experimental enterprises.

Optimal laser action is obtained when the required 6.8 μ radiation is produced with a minimal bandwidth. This can be obtained on the basis of excimer pumped dye lasers by using a system made of a stabilized oscillator and amplifier stages. With the same basic laser chain as in Fig. 3 it should result in a 20 ns pulse with a Fourier limited bandwidth but at a somewhat lower power level ($\sim$ 100 μJ).

The basic measuring scheme is the following. The muon enters the target via a thin entrance foil (whose multilayered composition will be described later), passes a 600 μ thick vacuum gap and stops in a layer of solid protium cooled to below 2.5 °K. Many μp's slow down within the H$_2$ layer and thermalize in the F = 0 state within a short time. They diffuse in the solid H$_2$ and some of them escape back into the vacuum gap and drift slowly toward the entrance foil. In between, the laser is shot into the mid-plane of the vacuum gap thereby exciting some of the μp's into the triplet (F = 1) state. As these μp atoms enter the foil they dissipate their 180 meV excitation energy in a spin-flip reaction and thereby acquire sufficient kinetic energy to find their way through a reactive layer and undergo a transfer reaction in a gold layer. The reactive layer reacts sensitively with the vast majority of slow μp's and hinders them from reaching the gold layer. This leads to a highly selective detection of the hyperfine transition.

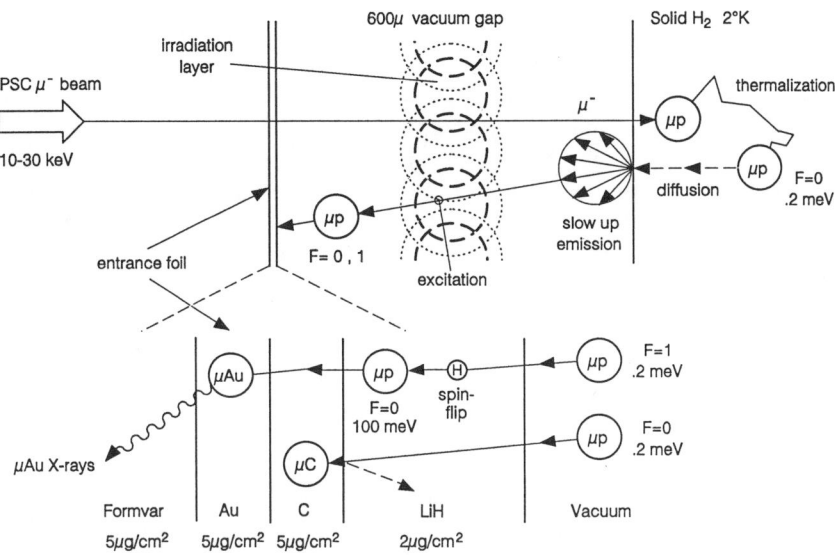

Figure 7: The basic configuration of the hyperfine splitting experiment.

A more detailed account of the main steps is now given (see Fig. 7 for reference).

a) *Formation of a source of ultraslow μp's in vacuum*
   Recent calculations [20] of the energy-loss cross-section of singlet μp's on H_2 show that it increases significantly below 100 meV. This novel situation is optimal for providing efficient thermalization of μp atoms formed in a thin H_2 layer near the surface of the solid. After thermalization, diffusion is still sufficiently fast to result into substantial escape into vacuum before $p\mu p$ formation takes place. These μp's are emitted with the very low average perpendicular velocity of 250 $\mu m/\mu s$.

b) *The laser transition*
   With the target in a high magnetic field (7 T or higher) the PSC muon beam leads to a μp source size of $\sim 3 \times 15$ mm^2. A minute laser multipass cavity (of the same kind as the one described in the previous paragraph) is used. It consists of two 15 mm long and 600 μ wide (the gap width) cylindrical mirrors with a 3 mm distance between them Given the diffraction limitation it allows about 200 reflections with a 100 μ beam radius at the waist and leads to a median "irradiation layer" where the power density averaged over half the gap thickness reaches 500 mJ/cm^2. The linewidth of the transition (to which the laser bandwidth and the Doppler broadening contribute in equal parts) is less than 100 MHz or $2.3 \cdot 10^{-6}$ of the transition frequency. It results with 3 % transition probability [19] for the μp's present within the irradiation

layer during the shot. This takes place about 1.3 μs after the muon stop and the excited μp atoms reach the entrance foil with a 1 μs wide time distribution (at an average 2.5 μs delay time).

c) *Selective detection of μp triplets*

The composition of the multilayer foil is shown in the lower part of Fig. 6 together with the basic interactions taking place. The inner layer of hydrogenic material (LiH for minimal transfer loss) induces the spin-flip reaction of μp triplets whose cross-section at the low energies considered here is as high as $1.5 \ 10^{-17}$ cm^2 [21]. The accelerated μp ($\sim$ 100 meV) pass the next thin carbon foil with minimal transfer loss while the slow μp ($\sim$ 0.2 meV), whose transfer cross-section to C can be computed to be about 13 times higher [22], either transfer or are backscattered. Some backscatter of the fast μp also occur but many of these μp's are rapidly returned by the H_2 layer back towards the foil. This method should allow one to achieve the detection of the signal (fast μp's resulting from the laser transition) with two orders of magnitude more efficiency than that of the background (slow μp's).

d) *The trigger signal*

Silicon counters surrounding the target detect the μp formation (via the 1.9 $K\alpha$ X-ray) with about 40 % efficiency and the gold X-rays are very efficiently detected by a thick NaI (or BGO) assembly placed around the whole target system. The laser trigger is given by a 1.9 keV X-ray signal coming in due time after the PSC signal together with the condition that no NaI pulse appeared in the first few 100 ns following the μp X-ray signal. This very usefull requirement makes use of the high transfer detection efficiency to inhibit laser triggering in the frequent case where a fast μp escapes the H_2 layer and transfers in the entrance foil. Under these conditions, the detection of a transfer at the correct time gives a clean transition signal at an overall efficiency level significantly better than 10^{-4}/trigger. At a rate of 20 triggers/s it leads to a 10^{-7} precision within a few days of measurement.

References

[1] D. Taqqu, Report to the Task Force, Paul Scherrer Institute, 1986.

[2] A. Fuchs, Dissertation, Paul Scherrer Institute, 1992.

[3] P. Wojcekowski, in this volume.

[4] R-91-08 collaboration, unpublished results of May 1992 test run, Paul Scherrer Institute, see also P. Hauser et al., this conference.

[5] D. Taqqu, Internal Report, Paul Scherrer Institute, 1992.

[6] D. Taqqu, Contribution to the 13th Int. Conf. on At. Phys., Munich, 1992.

[7] D. Taqqu, PSI-PR 91-37, Paul Scherrer Institute, 1991.

[8] H. Anderhub et al., Internal proposal SIN, 1979.

[9] M. Weitz, F. Schmidt-Keller and T.W. Hänsch, Phys. Rev. Lett. **68**, 1120 (1991).
[10] J.R. Sapirstein and D.R. Yennie in Quantum Electrodynamics, Advanced Series on Directions in High energy Physics, Vol. **7**, 1990.
[11] G. Sciurba and H.J. Loesch, Opt. Comm. **73**, 489 (1989).
[12] G. Sciurba, private communication.
[13] H. Anderhub et al., Phys. Letters **143B**, 65 (1984).
[14] L. Bracci and G. Fiorentini, Nuovo Cimento **43A**, 9 (1978).
[15] F. Kottmann, Diss. ETH Nr. 7179, 1982.
[16] A. Smith, A. Peacock and P.Z. Kowalski, IEEE Trans. on Nucl. Science **NS-34**, 57 (1987).
[17] See for example for the present application H.P. von Arb et al., Nucl. Instr. Meth. **207**, 429 (1983).
[18] T.K. Edberg et al., Nucl. Instr. & Meth. in Phys. Res. **A316**, 38 (1992).
[19] V.P. Smilga and V.V. Filchenko, Sov. Phys. JETP **58**, 73 (1983).
[20] A. Adamczak and P. Kammel, private communication.
[21] M. Bubak and M.P. Faifman, E4-87-464, J.I.N.R. Dubna, 1987.
[22] L. Bracci and G. Fiorentini, Nuovo Cimento **50A**, 373 (1979).

A Device for Cooling Charged Particle Beams by Moderation and Acceleration

M. MÜHLBAUER, H. DANIEL, F.J. HARTMANN
Physik-Department, E18, TU-München
D-8046 Garching, Germany

Abstract. The design of a device for cooling a beam of very low-energetic charged particles by moderation in matter and simultaneous acceleration in an electrostatic field E has been studied both in closed form and by Monte Carlo calculations. These calculations show that an increase in spectral density of roughly a factor of five can be obtained.

1. The Basic Principle

The device makes use of the phase space compression effect of the stopping power (frictional cooling) [1]

$$S = -\frac{dT}{ds} = -mv\frac{dv}{ds} = -m\frac{dv}{dt} \qquad (1)$$

for a particle with mass m at low velocity v and kinetic energy T ($T \lesssim 4\,\text{keV}$ for muons and $T \lesssim 40\,\text{keV}$ for protons), where s denotes the pathlength, t the time and S increases with T, as shown in fig. 1. The energy loss is compensated for by an electrostatic field E along the beam axis. This electric field defines an equilibrium energy T_∞,

$$S(T_\infty) = q \cdot E, \qquad (2)$$

where the energy loss is equal to the energy gain of a particle with charge q. Under these conditions particles with $T > T_\infty$ will be slowed down and particles with $T < T_\infty$ will be accelerated. The effects of multiple scattering and straggling will simultaneously be strongly reduced.

2. An Analytical Approach

2.1 General Approach: For a wide range of energies below the Bragg maximum the stopping power is well approximated by

$$S = a\sqrt{T} = bv \qquad (a, b = \text{const.}).$$

Neglecting for the moment scattering and straggling we find, with z and r as coordinates along the beam axis and perpendicular to it, respectively,

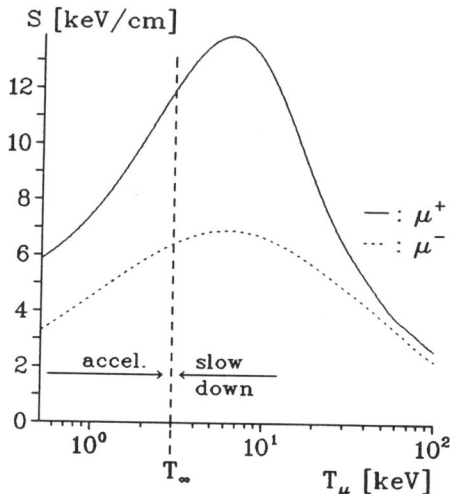

Figure 1: Average stopping power $S(T)$ of H_2 at 40 mbar for μ^+ (scaled proton data from Janni [2]) and μ^- (data from Kottmann et al. [3]).

$$\frac{dv_z}{dt} = \frac{1}{m}\left(S(T_\infty) - \frac{v_z}{v}S(T)\right) = \frac{b(v_\infty - v_z)}{m}, \quad (3)$$

$$\frac{dv_r}{dt} = -\frac{1}{m}\frac{v_r}{v}S(T) = -\frac{bv_r}{m} \quad (4)$$

for a particle moving nearly parallel to the beam axis under the influence of both S and E. We obtain:

$$v_z - v_\infty = (v_{z_0} - v_\infty)\exp\left(-\frac{b(t - t_0)}{m}\right), \quad (5)$$

$$v_r = v_{r_0}\exp\left(-\frac{b(t - t_0)}{m}\right), \quad (6)$$

where v_∞ is the equilibrium velocity and the subscript 0 denotes values at $t = t_0$. Both a longitudinal motion difference $v_z - v_\infty$ and a transverse motion v_r die out *exponentially* with t, the time constant being independent of the particle velocity v. Hence the beam divergence decreases.

2.2 Multiple Scattering: Of course multiple scattering and straggling, which have been neglected so far, may become very important and therefore need a quantitative treatment. From the Gaussian approximation we find for the increase of the mean square $\langle\Theta_s^2\rangle$ of the scattering angle Θ_s per pathlength s

$$\frac{d\langle\Theta_s^2\rangle}{ds} = \left(\frac{10\,\text{MeV}}{T}\right)^2 \cdot L_{\text{rad}}^{-1}, \quad (7)$$

Table 1: Effects of multiple scattering and straggling on angular and energy distribution

T_∞ [keV]	Multiple Scattering			Straggling	Total
	$\langle\Theta_\infty\rangle$ [°]	$\sqrt{\langle\Delta\Theta_\infty^2\rangle}$ [°]	$\sqrt{\langle\Delta T_\infty^2\rangle}$ [keV]	$\sqrt{\langle\Delta T_\infty^2\rangle}$ [keV]	$\sqrt{\langle\Delta T_\infty^2\rangle}$ [keV]
3.0	19	5.1	≈ 0.5	≈ 0.3	≈ 0.6
2.5	21	5.7	≈ 0.4	≈ 0.3	≈ 0.5
2.0	24	6.3	≈ 0.4	≈ 0.2	≈ 0.4

where $L_{\rm rad}$ denotes the radiation length. On the other hand energy loss and electrostatic field lead to a decrease of the beam divergence. For a particle moving with constant velocity at a small angle Θ to the beam axis we obtain

$$\frac{d\Theta^2}{ds} = 2\Theta\frac{d\Theta}{ds} = -\frac{S\Theta^2}{T}. \tag{8}$$

In the case of equilibrium $T = T_\infty$ the sum of eqs. (7) and (8) must vanish for $\Theta = \Theta_s = \Theta_\infty$ and we obtain a Gaussian angular distribution with a mean angle

$$\langle\Theta_\infty\rangle = \frac{\sqrt{\pi\langle\Theta_\infty^2\rangle}}{2} = \frac{\sqrt{\pi}}{2}\cdot\frac{10\,{\rm MeV}}{\sqrt{T_\infty S(T_\infty)L_{\rm rad}}}. \tag{9}$$

An angle $\langle\Theta_\infty\rangle > 0$ results in an increase of the mean pathlength s and therefore of the average energy loss:

$$S(T_\infty) \to \frac{S(T_\infty)}{\cos\langle\Theta_\infty\rangle} \tag{10}$$

That is, we get a shift to a lower equilibrium energy and also an increase of the width of the corresponding energy distribution. Typical values for the effects of the multiple scattering on the angular and energy distributions are shown in table 1 and fig. 2.

2.3 Straggling: The second effect neglected so far, straggling, can be treated in a similar way. In Gaussian approximation the increase of the mean square of the energy fluctuation $\langle\Delta T^2\rangle$ per pathlength s is given by

$$\frac{d\langle\Delta T^2\rangle}{ds} = \frac{m_e}{m}\cdot T\cdot S(T) \tag{11}$$

for low energies $v^2 < 3Zv_B^2$, where Z is the atomic number of the scatterer, m_e the electron mass and v_B the Bohr velocity [4]. Under the influence of energy loss and electrostatic field we find for a particle at an energy $T = T_\infty + \Delta T$ according to eq. (3)

$$\frac{d\Delta T^2}{ds} = 2\Delta T\frac{d\Delta T}{ds} = 2\Delta T(S_0 - S(T)). \tag{12}$$

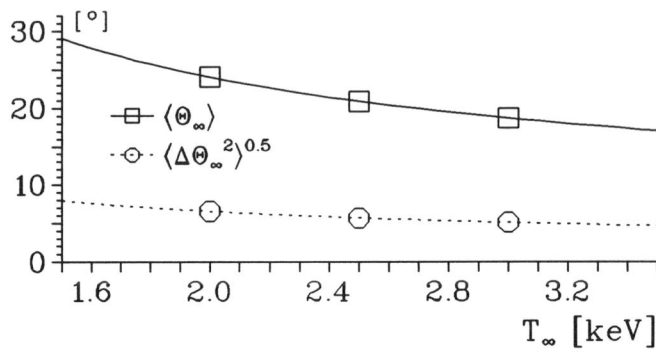
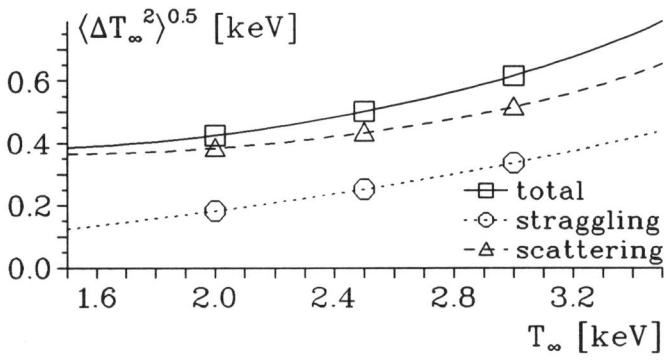

Figure 2: Effects of multiple scattering and straggling on angular and energy distribution

The sum of eqs. (11) and (12) must also vanish in the case of equilibrium, yielding a width of the energy distribution

$$\sqrt{\langle \Delta T_\infty^2 \rangle}(S_0 - S(T)) = \frac{m_e}{2m} \cdot T_\infty \cdot S_0. \quad (13)$$

This equation can be solved numerically. Typical values for $\langle \Delta T_\infty^2 \rangle$ are given in table 1 and fig. 2.

3. The Monte Carlo Calculations

3.1 The Method: In order to obtain results in an independent way we performed detailed Monte-Carlo studies for a frictional cooling device. We simulated the motion of negative muons through an acceleration tube filled with hydrogen gas at low pressure ($p(H_2) = 10 - 100$ mbar). The applied electrostatic field is assumed to be homogeneous. The track of the muon is computed step by step, from one collision with a hydrogen atom to the next one. The scattering angle Θ and the energy loss of the muon are taken at random for each collision between the muon and a hydrogen atom.

Scattering is treated as Rutherford scattering using a screened potential

$$\phi(r) = \begin{cases} (-2a_B/r + 1) \cdot E_B & \text{for } r \leq 2a_B \\ 0 & \text{for } r > 2a_B \end{cases}, \quad (14)$$

where $a_B = \hbar^2/m_e e^2$ and $E_B = e^2/2a_B = 13.6\,\text{eV}$ are Bohr radius and Bohr energy, respectively [3]. With this assumptions we obtain a total scattering cross section $\sigma = 4\pi a_B^2$ and therefore a mean free pathlength between two collisions $\langle s \rangle = 1/(n\sigma) \approx 0.12\,\text{cm} \cdot \text{mbar}/p(\text{H}_2)$. This model gives an excellent approximation for the rare wide-angle scattering events and was found sufficiently accurate for small-angle scattering. The average energy loss per collision $\Delta T = \langle s \rangle S(T)$ is calculated from the $S(T)$ values shown in fig. 1, and the energy straggling is derived according to Besenbacher et al. [4] and Bohr [5].

In a second series of simulations we replaced the homogeneous H_2-moderator by a discrete moderator with many thin carbon foils ($5\,\mu\text{g}/\text{cm}^2$). For the description of multiple scattering we used the Gaussian approximation and added an extra tail to the distribution found to get an appropriate number of wide-angle scattering events. The average energy loss was derived from the proton data given by Janni [2] and the muon data measured by Baumann et al. [6].

3.2 Results: The Results for a μ^- beam moving through 20 cm of hydrogen at 40 mbar ($T_\infty = 3.0\,\text{keV}$) or a comparable area thickness of carbon ($T_\infty = 4.0\,\text{keV}$) are shown in figs. 3 and 4. The equilibrium angular distribution is reached after a few centimetres. In case of hydrogen we get an distribution with $\langle \Theta_\infty \rangle \approx 20°$ for the outgoing beam corresponding to eq. (9). For the carbon moderator the distribution is widened due to the higher atomic number. Nevertheless we obtain an increase up to a factor of five in spectral density for both setups (fig. 4). The equilibrium energy is achieved after $s = 10 - 15\,\text{cm}$

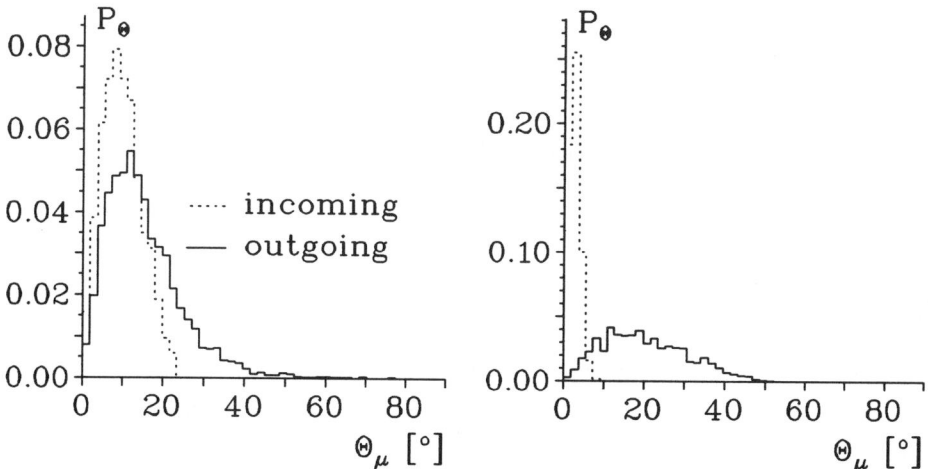

Figure 3: Angular distribution of the incoming and outgoing muons (left: H_2; right: C)

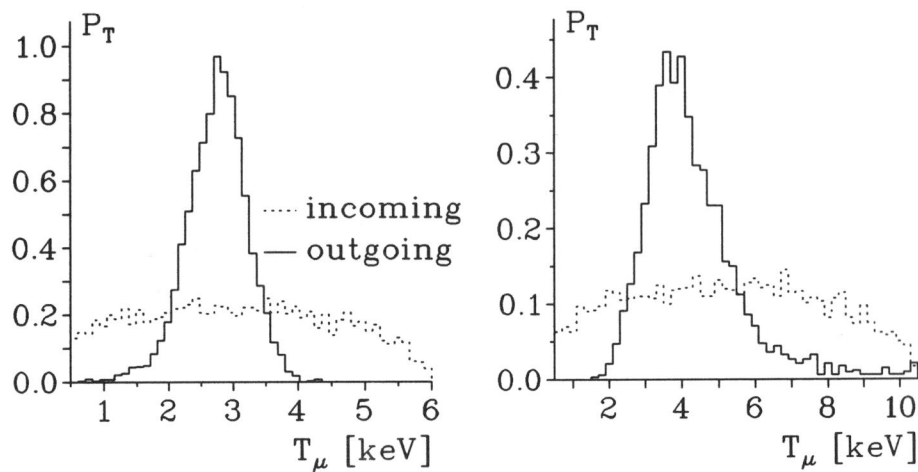

Figure 4: Energy distribution of the incoming and outgoing muons (left: H_2; right: C)

and we obtain $\sqrt{\langle \Delta T_\infty^2 \rangle} \approx 0.5\,\text{keV}$ for the hydrogen moderator. Only a few percent of the muons are lost.

4. Summary

Closed form calculations and Monte Carlo simulations indicate that a low-energy charged particle beam can be compressed in energy by frictional forces. We found an increase in spectral density for both hydrogen and carbon moderator. We obtained small beam divergence. Large-angle scattering gives only a small deterioration.

Acknowledgements

This work was supported by the German BMFT.

References

[1] H. Daniel, Muon Catalyzed Fusion **4** (1988) 425

[2] J.F. Janni, At. Data Nucl. Data Tables **27** (1982) 147

[3] F. Kottmann, private communication

[4] F. Besenbacher et al., Nucl. Instrum. Methods **168** (1980) 1

[5] N. Bohr, Philos. Mag. **30** (1915) 581

[6] P. Baumann et al., PSI Nucl. Part. Phys. Newsl. (1991) 59

Progress in Soft X-Ray Detection: The Case of Exotic Hydrogen

Jean-Pierre EGGER, Didier CHATELLARD and Eric JEANNET
Institut de Physique de l'Université, Breguet 1,
CH-2000 Neuchâtel, Switzerland

Abstract. Modern X-ray detection techniques with CCDs and their implications in the field of exotic hydrogen are discussed. Examples of recent achievements are presented and possible future developments are mentioned.

1. Introduction

An X-ray detector must have the best possible spatial and energy resolution, high quantum efficiency and a low particle background rate. Its use in both, the dispersive (position) mode in combination with a Bragg crystal spectrometer or in energy (non-dispersive) mode must be possible. In dispersive mode the position resolution is most important and the energy resolution is only used as a cut for background suppression. However the overall efficiency (including solid angle, crystal reflectivity etc.) is low. In the other mode the energy resolution and the background rejection capabilities become most important.

Detectors for low energy X-rays include gas filled proportional and scintillation counters and solid state devices such as conventional Si(Li) and Ge diodes, microchannel plates, silicon drift detectors and CCDs (charge coupled devices). A detailed comparison of these detectors is presented for example in [1] for the energy range from 1 keV to 10 keV. As can be seen, CCDs have the best combined position and energy resolution except for new developments like microcalorimeters [2] or superconducting tunnel junctions [3] which both still are severely size-limited.

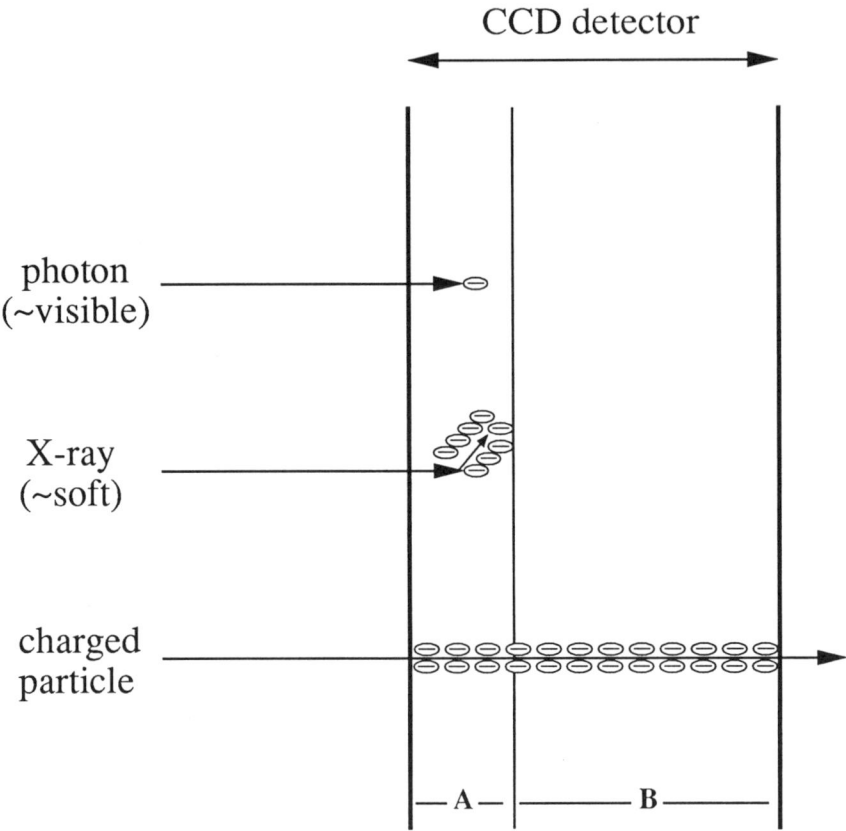

Figure 1: Photon and particle detection in a CCD. The number of electron-hole pairs created is one per photon for visible light, $\approx$ 270 per keV for X-rays and approximately 80 per micron of Si for minimum ionizing charged particles. A corresponds to the depleted zone and B to the substrate.

2. Operation and performance of CCDs as X-ray detectors

The CCDs [4] are operated in vacuum and cooled down to approximately 160 K in order to limit dark current and therefore allow for several hours of exposure time. They operate in a similar way to conventional silicon solid state detectors in that the incoming X-rays, following absorption by photoelectric effect, are converted to electron-hole pairs where each pair requires 3.68 eV for its creation. A schematic view of the detection of visible photons, X-rays and charged particles is given in fig.1.

The CCD position resolution is simply given by the pixel size; in our case 22μm $\times$ 22μm.

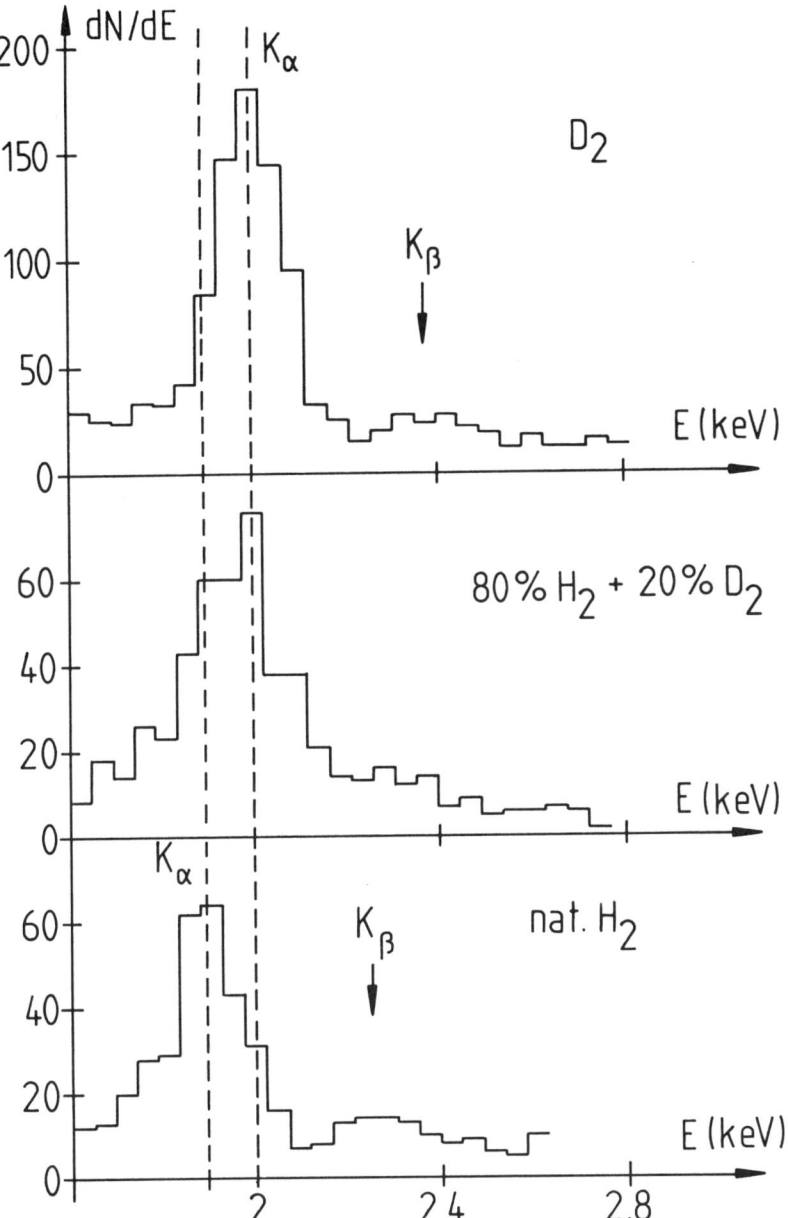

Figure 2: Raw muonic X-ray spectra from natural H_2, HD and D_2 liquid targets detected by CCDs at a distance of 16cm from the center of the target. In addition to the target windows of 100μm Be a Kapton window of 12.5μm separated the target insulation vacuum from the CCD vacuum. No shielding was used. One channel corresponds to 46.6 eV. The energy resolution is approximately 150 eV FWHM. The K_α and K_β

peaks for muonic hydrogen and deuterium are well separated (the K_α positions at 1.90 keV for pμ and 2.00 keV for dμ are marked with dotted lines). X-ray yields for higher K transitions are very small at liquid densities. Signal to background ratio is ≈ 10. Even without a fit one can see directly by comparing the K_α position in the 3 plots that there is a sizeable excited state μ transfer from pμ to dμ since the HD K_α line shape is already dominated by the dμ contribution although there is only 20% deuterium present.

The energy resolution of a CCD is given by

$$\Delta E\ FWHM(eV) = 2.355 \times 3.68 \left(N^2 + \frac{FE}{3.68}\right)^{1/2}$$

where N is the rms transfer and readout noise of the CCD, F the Fano factor and E the X-ray energy. The conversion energy for an electron-hole pair in Si is 3.68 eV. Lowest possible readout noise can be obtained by reducing output transistor noise and by using the *correlated double sampling* technique to eliminate reset noise [5]. However this requires slow-scan operation (reading time of 1 pixel is $\geq 50\mu s$) and therefore readout is slow. From the formula above, the best possible energy resolution with Si CCDs can be estimated by considering N^2 very small. The result is 70 eV FWHM at 2 keV and 140 eV FWHM at 8 keV.

Examples of muonic hydrogen X-ray energy spectra are presented in fig.2; antiprotonic hydrogen is shown in fig.3 and pionic hydrogen in figs. 4 and 5. The energy resolution is 150 eV FWHM at 2 keV in fig.2, 140 eV FWHM at 1.73 keV in fig.3 and 150 eV FWHM at 2.43 keV in fig.4. Work is in progress to further improve these energy resolutions by identifying all possible noise sources and minimizing them.

The CCD background rejection capabilities are very powerful and work as follows: Events are collected into the CCD for a finite time and then read out. This collection time must not be too long, to avoid a double hit on a single pixel. In most of the cases, the energy of an X-ray is deposited into one single pixel. Sometimes, however, the energy is split between two pixels ($\approx$ 15-35 % of the cases, depending on energy and CCD depletion depth). Background events like charged particles, neutrons and gammas, on the other hand, are usually split into several pixels. There are three reasons for this difference. Firstly the background particles above tend to disturb the surroundings more on impact than low-energy X-rays and secondly the background events enter the detector from all directions, therefore having a higher probability to hit several pixels. The third reason can be understood by examining fig.1. When a suitable voltage is applied on the substrate (B), the electrons of the created pairs in B will migrate and some of them will eventually reach neighboring pixels whereas the X-ray electrons, confined in region A, deposit their charge in one pixel. It was then

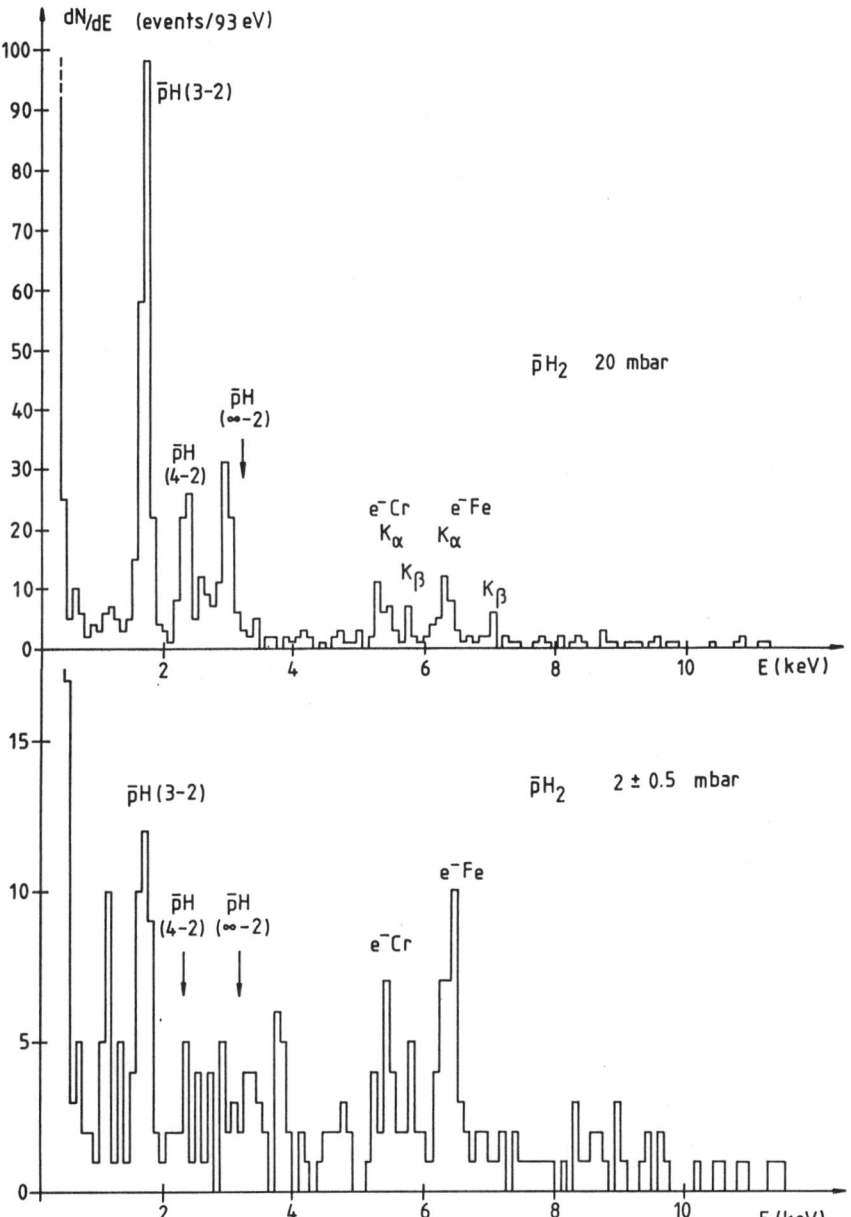

Figure 3: Antiprotonic hydrogen spectra taken with CCDs inside the cyclotron trap at LEAR for hydrogen target pressures of 2 and 20 mbar. This short measurement was carried out during the P118 test experiment [6] for development of a research program on ultra low energy antiprotons at CERN. A 72 MeV/c antiproton beam was used in slow extraction mode. One spill was necessary for each spectrum. The p̄H Balmer series X-rays are clearly visible. Signal/background at 20 mbar is ≈ 20. Fe and Cr electronic X-rays are due to the stainless steel grid necessary to protect the CCDs from the active injection device (kicker) and pulsed extraction electrodes used in this test.

A careful examination of the 2 mbar spectrum in the 8-10 keV energy region shows that a clean measurement of the $\bar{p}$H K_α line could be carried out with a small internal bremsstrahlung contribution as the only background.

---

possible to adjust the voltages for maximizing one pixel events with an X-ray source and maximizing multiple pixel events with other sources. A good event is then characterized by an isolated pixel with the proper energy surrounded by eight pixels which must be part of the noise peak. A simple lower limit on the surrounding pixels was used for fig.2. In fig. 3, 4 and 5 more sophisticated criteria were introduced [7]. This resulted in the loss of some events, but background was reduced to almost zero. Indeed in fig.5 which is a position spectrum in the focal plane of a crystal spectrometer [8] an additional energy cut was performed and background was lowered to 1 event per cm^2 and per 120 h in the πE3 area at PSI with beam on. The background rejection capabilities are further illustrated in fig.6. The computer generated images of CCD surfaces without background rejection are shown on the left and after rejection on the right.

The theoretical CCD efficiency is given in fig.7. However X-ray loss due to windows, target liquid or gas etc. is not included. The sharp variation of the efficiency at 1.84 keV corresponds to the Si absorption edge. The effect of this edge can be seen in the hydrogen spectrum in fig.2 (asymmetry of the K_α peak). A good way to measure the CCD efficiency including X-ray losses is to detect X-rays from antiprotonic nitrogen at low gas pressure where the circular transitions dominate with a yield of practically 100 % [9]. An example (although with little statistics) is presented in fig.8.

3. Examples of exotic hydrogen physics

3.1 Spectroscopy of muonic hydrogen. The predictions of quantum mechanics can be verified to a very high degree of accuracy in muonic hydrogen in view of its simplicity and because some higher order corrections are not negligible like in ordinary hydrogen.

Lets take the 3D-3P transition as an example. When the Stark mixing is not taken into account, the transition probability between levels with the same principal quantum number is very small. However this probability can be increased by shining infrared light with the corresponding frequency on the μ^-p system to stimulate the transition. An increase in 3D-3P transitions will increase the 3P-1S (K_β) versus the 2P-1S (K_α) yield. The frequency of the stimulated transition can then be deduced from the observation of the K_β/K_α intensity ratio. The X-ray energies are 2.25 keV and 1.90 keV. This type of experiment was proposed at PSI [10]. First tests should begin in early 1993.

3.2 Muon catalyzed fusion (μCF). The μCF process is started by injecting a free muon into a mixture of hydrogen isotopes where it is stopped. Following the initial

Coulomb capture, the muon cascades to lower levels via Stark mixing, Auger, inelastic and radiative processes. This muonic cascade, even in the simplest muonic atoms (muonic hydrogen pμ, deuterium dμ and tritium tμ) is not sufficiently understood quantitatively.

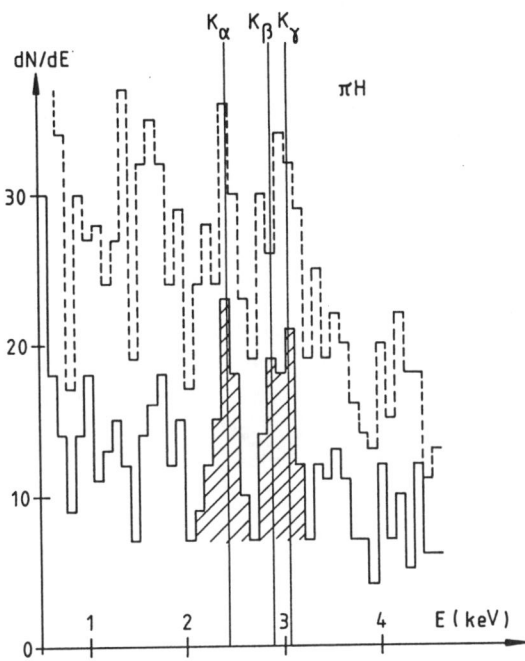

Figure 4: Pionic hydrogen X-ray energy spectrum at a gas pressure of 15 bar taken with CCDs inside the cyclotron trap on the πE3 channel at PSI. Incoming pions of 80 MeV/c were used. Measuring time was 30 min. The CCDs were emptied every 30 s. Comparison of a simple background cut (broken line) with a more sophisticated treatment (full line [7]).

In particular the probability q_{1S} to reach the ground state of a given muonic hydrogen isotope has never been directly measured. A precise knowledge of q_{1S} gives information about muon transfer in excited atomic states and defines therefore the initial conditions for the complex cycle of muon-induced processes. Fig.2 presents a first direct measurement of muonic X-ray spectra from natural H_2, HD and D_2 liquid targets measured with CCDs during a test run on the μE4 channel at PSI [11]. Even without a fit, the comparison of the K_α position in the three plots shows a sizeable μ transfer from pμ to dμ since the HD K_α line shape is cleary dominated by the dμ contribution although there is only 20% deuterium present.

Systematic measurements will start in early 1993.

3.3 Antiprotonic hydrogen. Since the antiproton is a strongly interacting particle the energy of the 1S level in antiprotonic hydrogen will be shifted (ϵ) and broadened (Γ) due to the strong interaction. In order to determine these parameters, accurate energy measurements are needed. Furthermore, calculations by Borie and Leon [12] show that K and L X-ray intensities are very low except at the lowest hydrogen gas pressures. Therefore the maximizing of stopping density and the reduction of background is of prime importance. An interesting way of obtaining a high stopping density of antiprotons in a low density gas target is to use a *cyclotron trap* [13]. The principle of the trap is to wind up the range curve of the incoming charged particle beam in a weakly focussing magnetic field of the order of 2.5 T produced by a superconducting split coil magnet. After radial injection the particles spiral towards the center while loosing energy in a few thin degraders and finally the target gas. Obviously the trap works best for particles with long life times (antiprotons, muons). Fig.3 presents antiprotonic hydrogen spectra taken with CCDs inside this cyclotron trap for hydrogen target pressures of 2 and 20 mbar. The quality of the data obtained from this test shows that the strong interaction parameters ϵ_{1S} and Γ_{1S} could be determined with significantly better accuracy. Therefore a precise comparison with the different theoretical predictions could be made. A detailed experimental and theoretical review of antiprotonic hydrogen up to 1989 can be found in [14].

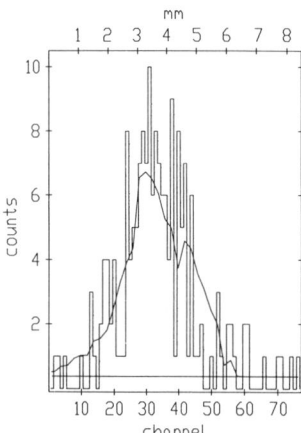

Figure 5: Pionic hydrogen 3P-1S (K_β) X-ray position spectrum in the focal plane of a Si crystal spectrometer [15]. One channel corresponds to 5 pixels (110 μm). Background suppression was carried out according to [7]. Signal/background ratio is $\approx$ 15. The line shape was taken from a fit to the argon $K_{\alpha 1}$ calibration spectrum.

3.4 Pionic hydrogen. The measurement of the strong interaction shift ϵ_{1S} and broadening Γ_{1S} is a suitable way to study the pion-nucleon strong interaction at zero energy since these quantities can be directly related to the S-wave scattering lengths a_1 (isospin 1/2) and a_3 (isospin 3/2). The scattering lengths can also be obtained indirectly from dispersion relations using phase shifts from low energy elastic πp scattering and charge exchange measurements and extrapolating to zero energy [16]. This procedure is indirect since extensive theoretical treatments of the experimental data are necessary. In addition, results from different scattering experiments lead to solutions for the low energy parameters which are mutually inconsistent [17]. This is not surprising since, because of their tendency to decay rapidly, low energy pions are extremely difficult objects to handle in an experiment.

In a first step at PSI ϵ_{1S} was measured with an accuracy of $\approx 5\%$ [8]. The result is ϵ_{1S} = -7.12 $\pm$ 0.32 eV (attractive) yielding the scattering length combination $1/3(2a_1 + a_3)$ = 0.086 $\pm$ 0.004 m_π^{-1}. Work is in progress to improve the setup for an ϵ_{1S} measurement at the 1% and Γ_{1S} at the 10% level. Run time for this experiment is scheduled on the new πE5 beam line this summer with the cyclotron trap, a new high resolution crystal spectrometer and new large size CCDs [18].

An important application of the πN scattering parameters at zero energy is the investigation of the low energy structure of QCD through the calculation of the so-called pion-nucleon σ-term. A summary of the present situation may be found in [19].

3.5 Kaonic and hyperonic hydrogen. Early measurements of ϵ_{1S} and Γ_{1S} in kaonic hydrogen [20] were plagued by small statistics, large backgrounds and Stark effect since a liquid hydrogen target had to be used in order to have a sufficient stopping density. An improved experiment could be carried out at the future KAON (Vancouver) or at a Φ-factory like DAΦNE (Frascati) with CCDs. However even with a precise measurement of ϵ_{1S} and Γ_{1S}, some problems arise when the s-wave scattering length is deduced. From the analysis of low energy K$^-$p scattering, the K$^-$p scattering length is known to be positive. However this is due to the presence of the baryonic resonance $\Lambda(1405)$ which strongly couples to the isospin zero channel of the K$^-$p system [21].

Figure 6: Computer generated images of CCD surfaces irradiated with different particles. These figures present the background suppression capabilities of CCDs placed in the focal plane of a Si crystal spectrometer. The images on the left correspond to the raw spectra whereas the result of the background suppression is displayed on the right. The 3 cases are (a) a strong Ar K_α X-ray calibration source (2.95 keV) obtained by fluorescence of ^{55}Mn X-rays in Ar gas plus little background, (b) some pionic hydrogen X-rays (3P-1S, 2.88 keV) and background and (c) a few pionic hydrogen X-rays in a severe background (πE3 channel at PSI, 200 μA primary proton current, slits wide open, 2.5 m from channel exit).

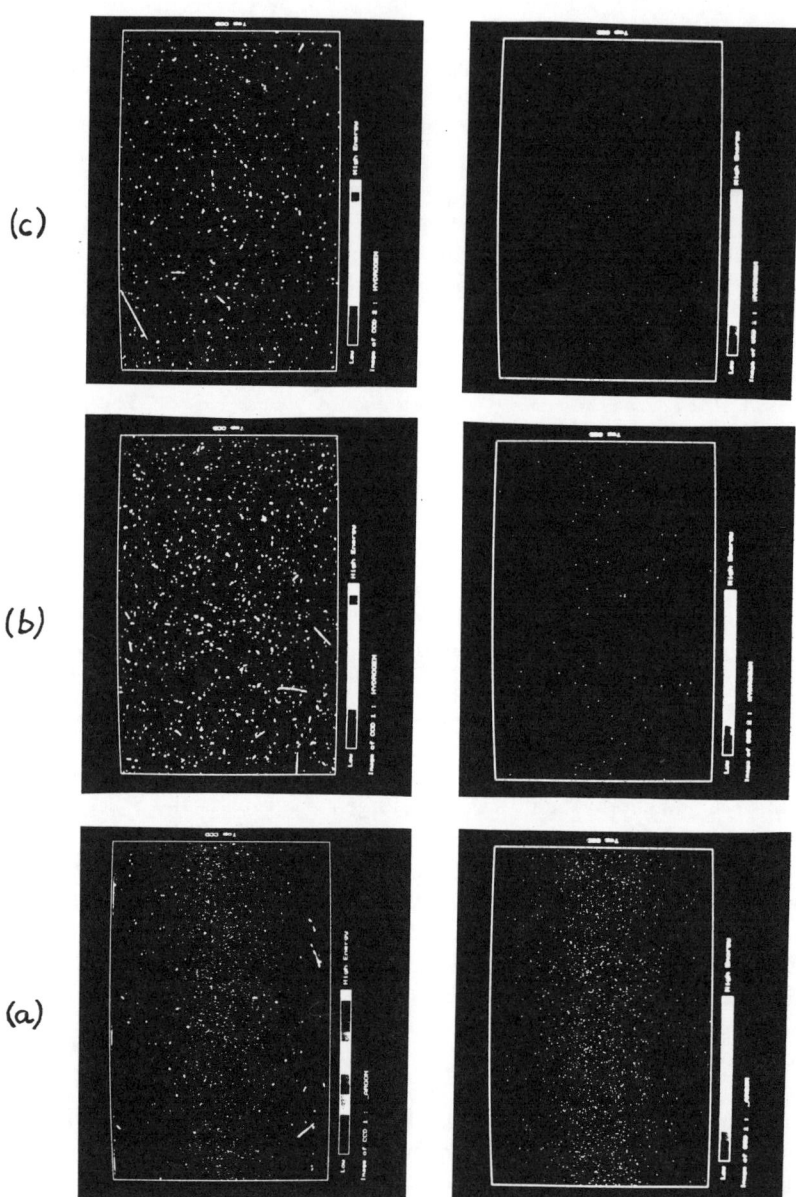

The energy of the sigmaonic hydrogen 2P-1S transition is of the order of 10 keV. This transition could be measured simultaneously with the kaonic hydrogen 2P-1S transition since de K^- can be stopped and the Σ^- produced and stopped in the same hydrogen target ($K^-p \to \Sigma^-\pi^+$ with a very low energy Σ^- appears in $\approx 44\%$ of the cases). For the Σ^-p atom the $\Sigma(1385)$ resonance which is also below the K^-p threshold should not, because of its isospin, complicate the situation in the way the $\Lambda(1405)$ does in the K^-p system.

Xionic and omegaonic hydrogen X-rays might also be obtained at KAON or other places (COSY, Jülich, for example) with a setup similar to the so-called *Gatchina configuration* [22].

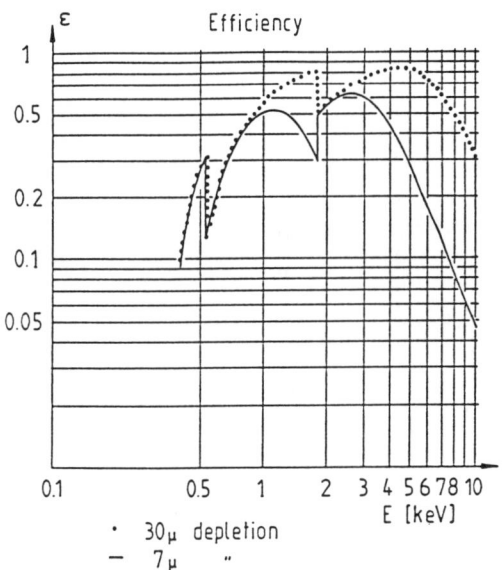

Figure 7: Efficiency as a function of X-ray energy for CCDs with 7 μm and 30 μm depletion layer.

4. Conclusions and outlook. CCDs have a unique potential as X-ray detectors for energies between 1 keV and 10 keV since no other X-ray detector can match their overall performance at this time. However, further significant improvements in detector quality will be hard to achieve without *major* programs of development. As examples, lets mention a few areas where such programs might prove useful:

- In order to overcome the *natural* limit for the energy resolution of 70 - 140 eV FWHM, the new developments on microcalorimeters and superconducting tunnel junctions are very promising but the serious problem of size limitation has to be solved before these detectors are of practical use.

- A more constant detector efficiency as a function of X-ray energy can be obtained with Ge CCDs but there might be other problems with Ge.

- CCD readout is slow if excellent pixel transfer efficiency is needed, as is in the case of good energy resolution. CCD readout could be speeded-up by incorporating an array instead of a single on-chip charge detection amplifier plus output electronics.

Tests have shown that CCDs can work in strong magnetic fields and directly in the isolation vacuum of a cold target or in an accelerator or beamline vacuum.

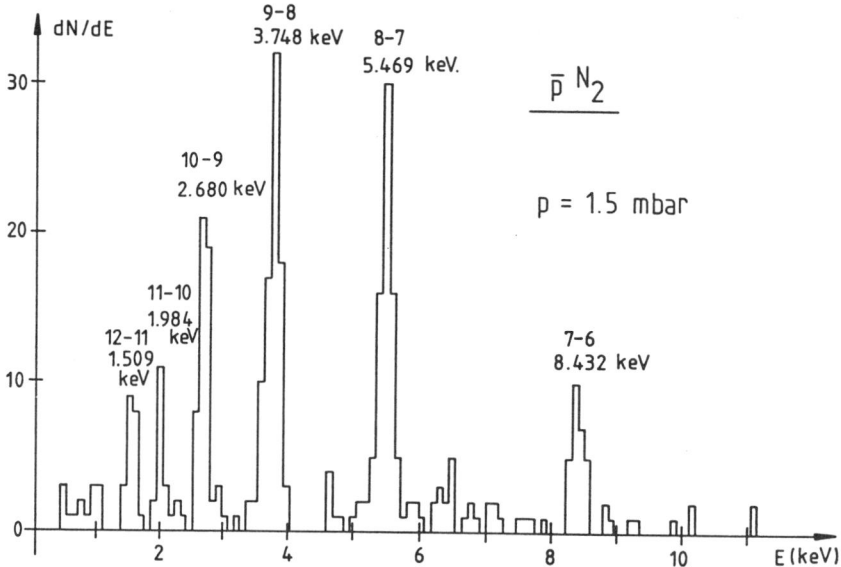

Figure 8: CCD measurement of antiprotonic nitrogen X-rays for a target pressure of 1.5 mbar. At this low pressure the yield for circular transitions (n → n-1, l max.) is practically 100%. The different peaks give therefore an experimental efficiency curve including X-ray losses in windows etc.

Many other applications for X-ray CCDs can be imagined. They include:

- X-ray astronomy. The bulk of the detectable X-ray flux from outside the solar system and the emission and absorption lines of the most abundant elements lie in the energy region from 0.1 to 10 keV, nicely covering the CCD detection range. Problems arise in space with cooling and irradiation damage. Tests at PSI by exposing CCDs to different particle beams could help to understand the irradiation damage problem. Future X-ray space programs include the Russian mission SPECTRUM X-G and the European effort XMM.

- Detection of single low energy electrons and positrons. Single electrons and positrons in the energy range up to 10 keV should interact in a CCD like X-rays of the same energy, provided the surface layer of the CCDs (electrodes etc.) is < 1 μm. In experiments like the muonium to antimuonium conversion at PSI, a unique identification of positrons should be possible. However CCD readout speed and triggering have to be improved.

- Biology and solid state physics. With the advent of the new synchrotron light sources at high intensity storage rings like ESRF at Grenoble, many new experiments requiring high position and energy resolution for X-ray detection become feasible. These include protein crystallography and autoradiography in biology [23], soft X-ray spectroscopy and materials research. CCDs are promising detectors in this field, especially if faster readouts can be built.

It should be noted however that CCDs of current design cannot be used in the energy region above 15 keV which includes most *medical* X-rays for fundamental reasons (Si becomes transparent to these X-rays).

Acknowledgements. We would like to thank all our colleagues that collaborated with us, especially our friends at IMP-ETHZ, PSI and Vienna. We also would like to thank the organizers of this conference, Lukas Schaller and Claude Petitjean. This work was partially funded by the Swiss National Science Foundation.

References

[1] J.L. Culhane, Nucl. Inst. and Meth. **A310**(1991)1

[2] D. McCammon et al., *Proc. Low Temp. Detectors for Neutrinos and Dark Matter*, 1990, Gif-sur-Yvette, in press

[3] D. Twerenbold, Physica **C168**(1990)381 and references therein

[4] EEV (English Electric Valve), Waterhouse Lane, Chelmsford, Essex, CM1 2QU, England

[5] D. Varidel, J.-P. Bourquin, D. Bovet, G. Fiorucci and D. Schenker, Nucl. Inst. and Meth. **A292**(1990)147

[6] E. Aschenauer, D. Cauz, P. DeCecco, J. Eades, K. Elsener, G. Gorini, D. Horvath, I. Krafcsik, V. Lagomarsino, G. Manuzio, R. Poggiani, L.M. Simons, G. Testera, G. Torelli and F. Waldner, *The P118T test experiment at LEAR (CERN)*

[7] D. Sigg, Diplomarbeit, **ETHZ-IMP**(1990)

[8] W. Beer, M. Bogdan, P.F.A. Goudsmit, H.J. Leisi, A.J. Rusi El Hassani, D. Sigg, St. Thomann, W. Volken, D. Bovet, E. Bovet, D. Chatellard, J.-P. Egger, G. Fiorucci, K. Gabathuler and L.M. Simons, Phys. Lett. **B261**(1991)16

[9] R. Bacher, P. Blüm, D. Gotta, K. Heitlinger, M. Schneider, J. Missimer, L.M. Simons and K. Elsener, Phys. Rev. **A38**(1988)4395

[10] F. Kottmann, C. Petitjean, L.M. Simons, D. Taqqu, D. Chatellard, M. Denoréaz, J.-P. Egger, E. Jeannet, F. Ciocci, G. Dattoli, A. Doria, G.P. Gallerano, L. Giannessi, G. Giubileo, G. Messina, A. Renieri, A. Vignati, F. Della Valle, E. Milotti, C. Rizzo, A. Vacchi and E. Zavattini, PSI proposal R-92-06.1

[11] P. Ackerbauer, P. Baumann, E.D. Bovet, W.H. Breunlich, T. Case, D. Chatellard, K.M. Crowe, K. Daniel, J.-P. Egger, F.J. Hartmann, E. Jeannet, M. Jeitler, P. Kammel, B. Lauss, K. Lou, V. Markushin, J. Marton, C. Petitjean, W. Prymas, W. Schott, R.H. Sherman, T. von Egidy, P. Wojciechovski and J. Zmeskal, PSI proposal R-81-05 and PSI Nuclear and Particle Physics Newsletter, 1991, in press

[12] E. Borie and M. Leon, Phys. Rev. **A21**(1980)1460

[13] L.M. Simons, Phys. Scripta **T22**(1988)90

[14] C.J. Batty, Rep. Prog. Phys. **52**(1989)1165

[15] W. Beer, M. Bogdan, J.F. Gilot, P.F.A. Goudsmit, H.J. Leisi, A.J. Rusi El Hassani, D. Sigg, St. Thomann, W. Volken, D. Bovet, E. Bovet, D. Chatellard, J.-P. Egger, G. Fiorucci, K. Gabathuler and L.M. Simons, Nucl. Inst. and Meth. **A311**(1992)240

[16] O. Dumbrajs et al., Nucl. Phys. **B216**(1983)277 and references therein

[17] J. Gasser, H. Leutwyler and M.E. Sainio, Phys. Lett. **B253**(1991)252 and references therein

[18] E.C. Aschenauer, A. Badertscher, W. Beer, D. Chatellard, M. Denoréaz, J.-P. Egger, K. Gabathuler, P.F.A. Goudsmit, E. Jeannet, H.J. Leisi, E. Matsinos, J. Missimer, H.Ch. Schröder, D. Sigg, L.M. Simons and Z.G. Zhao, PSI Nuclear and Particle Physics Newsletter, 1991, in press

[19] J. Gasser, H. Leutwyler, M.P. Locher and M.E. Sainio, Phys. Lett. **B213**(1988)85 and ref. [17]

[20] J.D. Davies et al., Phys. Lett. **B83**(1979)55;
M. Izycki et al.,Z. Physik **A297**(1980)11;
P.M. Bird et al., Nucl. Phys. **A404**(1983)482

[21] F. Scheck, *Leptons, Hadrons and Nuclei*, North-Holland, Amsterdam(1983)page 79

[22] V.I. Marushenko et al., JETP Lett. **23**(1976)72

[23] A.R. Faruqi, Nucl. Inst. and Meth. **A310**(1991)14

Measurement of the Stopping Power for μ^- and μ^+ at Energies between 3 keV and 50 keV

P. WOJCIECHOWSKI, P. BAUMANN, H. DANIEL, F.J. HARTMANN,
M. MÜHLBAUER, W. SCHOTT
Physik-Department, E18
Technische Universität München
D-8046 Garching, Germany

A. FUCHS, P. HAUSER, K. LOU, C. PETITJEAN, D. TAQQU
Paul Scherrer Institut
CH-5232 Villigen, Switzerland

F. KOTTMANN
Institut für Hochenergiephysik
ETH Zürich
CH-8093 Zürich, Switzerland

Abstract:

An experiment has been performed in order to measure the stopping power of carbon for μ^- and μ^+ at energies between 3 keV and 50 keV. The energy loss of negative and positive muons was determined by measuring their time of flight before and after the target. For these measurements a parallel plate avalanche counter and a microchannel plate were used. The Bragg peak has been observed for the first time for muons. The Barkas effect was measured down to the Bohr velocity.

1. Introduction

Measuring the stopping power S for low energy muons is interesting for a number of reasons:

1. one can study the Barkas effect [1,2], which is the difference in S for negative and positive particles;

2. the electronic structure of the target can be probed;

3. the dependence of μ^+e^- and $\mu^+e^-e^-$ formation on muon energy may be determined;

4. last but not least, the feasibility of new phase space compression techniques based on frictional cooling of charged particle beams can be studied [3,4].

The Barkas effect was observed already 35 years ago by comparing track lengths of π^- and π^+ [1] and Σ^- and Σ^+ [2], respectively, in emulsions. More measurements of this effect with π^- and π^+ in emulsions were done in 1969 [5]. The first measurement with μ^- and μ^+ was a counter experiment on Al and Cu down to velocities $v \approx 5\alpha c$ (with α the fine-structure constant) [6]. The stopping power differences were deduced from the respective spectral flux densities of the muons. The Barkas effect for protons and their antiparticles, finally, was determined by comparing stopping power data from an experiment at LEAR down to $v \approx 3\alpha c$ [7] with proton data presented in [8] and [9]. The aim of the present experiment was to measure the Barkas effect for muons near the Bragg maximum down to velocities as low as $v \approx \alpha c$ by comparing the stopping power of μ^- and μ^+ measured with the same experimental setup.

2. Principle of the measurement

In summer 1991 a test experiment was performed in the πE1 area of the Paul Scherrer Institute, Villigen (Switzerland), in order to measure the stopping power of carbon for slow negative and positive muons. For these measurements part of the existing device for phase space compression (PSC) of muons was used [10]. The setup is shown in Figure 1.

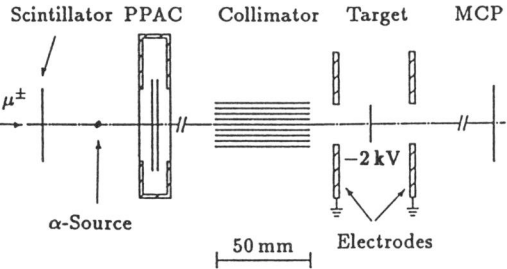

Figure 1: Schematic view of the experimental arrangement. The Wien filter, the quadrupoles in front of the setup and the solenoid are not shown. Distance between parallel plate avalanche counter (PPAC) and target: 198 mm, between target and microchannel plate detector (MCP): 721 mm.

The aim was to measure the energy loss ΔT of the muons in the target. This is the difference between the kinetic energies T_1 before and T_2 behind the target. T_1 and T_2 were determined with the help of the time-of-flight (TOF) technique. The first start signal came from a parallel plate avalanche counter (PPAC). The application of such a transmission counter for muons was new. Therefore a scintillation counter of thickness 100 μm was used in order to study and control the behavior of the PPAC. The target, a thin carbon foil, was the end of the first and the beginning of the second TOF section. At the end of the second section there was a microchannel plate detector (MCP). The muons lost part of their energy in the target and secondary electrons were knocked out of the target foil, which was kept at a voltage of −2 kV. This gave the electrons enough energy to reach the MCP further downstream in a much shorter time than the muons. With the help of the signal from the secondary electrons it was possible to determine the time when the muon hit the target. This provided the stop signal for the first TOF measurement and simultaneously the start signal for the second one. The muon detection in the MCP gave the stop signal for the second TOF measurement.

The use of a superconducting solenoid surrounding the setup was essential as its magnetic field of 3 T made the diverging muons spiral around the field lines and directed them onto target and MCP, respectively. Depending on the multiple scattering angle the particles gyrated in the magnetic field. This gave rise to an uncertainty in the determination of the energy from TOF. In order to cut off muons having a large radius of gyration a collimator was placed in front of the target. It consisted of a stack of parallel plastic foils (thickness 0.1 mm) with a distance of 3 mm between the foils and a length of 50 mm in beam direction.

A weak alpha source was mounted in front of the PPAC to provide a time calibration during the whole run. At the low entrance momentum ($p_\mu = 13.8 \text{ MeV}/c$), the electron contamination in the beam was quite large. Therefore a Wien filter was placed in front of the setup to clean the beam.

3. First results

The energy loss in carbon was measured for two different target thicknesses, 11 μg/cm² and 5 μg/cm². In Figures 2 to 4 some spectra and the results from the measurements with the 11 μg/cm² target *without* collimator are shown on the left hand side in each case, whereas the spectra and results from the 5 μg/cm² target obtained *with* collimator are presented on the right hand side. Figure 2 shows typical energy loss spectra taken with μ^+. The corresponding spectra for μ^- can be found in Figure 3. The cutoff of the spectra in b) corresponds to the limit $\Delta T \leq T_1$.

The energy loss ΔT vs. the energy T_1 of the muon is presented in Figure 4. The data points for μ^+ lie well above those for μ^-. They indicate a strong increase in the Barkas effect with decrease of energy down to the Bragg maximum. The results shown in this figure are from the raw data (not corrected for beam divergence) and hence only preliminary. As the collimator was not used during the measurements of the 11 μg/cm² target, it was not possible to distinguish the μ^+ events from background in the lowest energy region (see Figure 4,a).

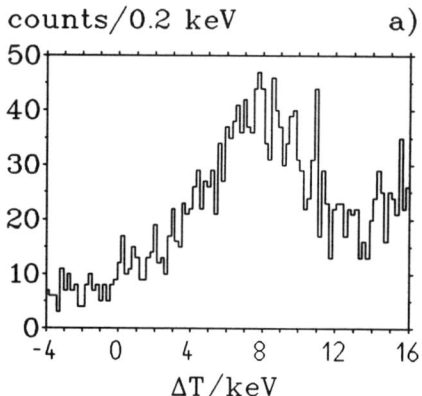
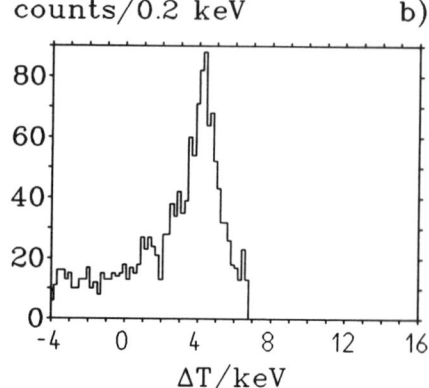

Figure 2: Spectra of μ^+ energy losses in carbon. a) $11\,\mu g/cm^2$ foil measured without collimator. Kinetic energies T_1 in front of the target between $21\,keV$ and $24\,keV$. b) $5\,\mu g/cm^2$ foil measured with collimator. T_1 between $6\,keV$ and $9\,keV$.

The results for μ^+ presented here may be compared to available proton data [11]. The same stopping power for protons S_p and positive muons S_{μ^+} is expected, as long as the particles have the same velocity. For that reason we recalculated the proton data with the help of the scaling law

$$S_{\mu^+}(T_1) = S_p(T_1 m_p/m_\mu) \qquad (1)$$

where T_1 is the kinetic energy of the incoming μ^+. The nonvanishing thickness of our target was taken into account. The result for the $5\,\mu g/cm^2$ target is shown in

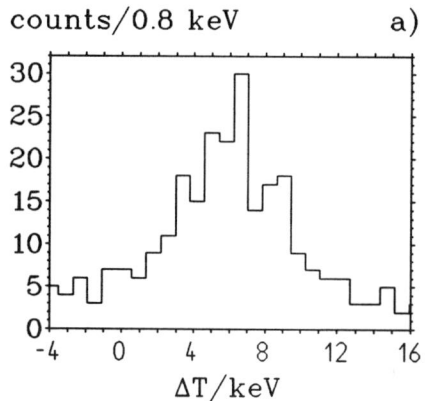
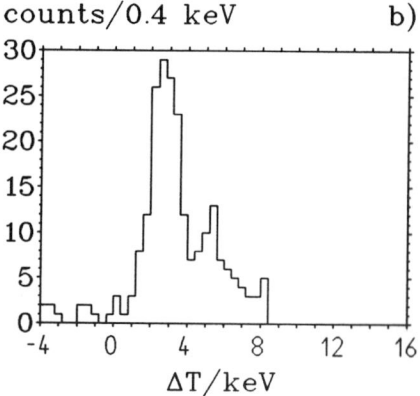

Figure 3: Spectra of μ^- energy losses in carbon. a) $11\,\mu g/cm^2$ foil measured without collimator. Kinetic energies T_1 in front of the target between $21\,keV$ and $24\,keV$. b) $5\,\mu g/cm^2$ foil measured with collimator. T_1 between $6\,keV$ and $9\,keV$.

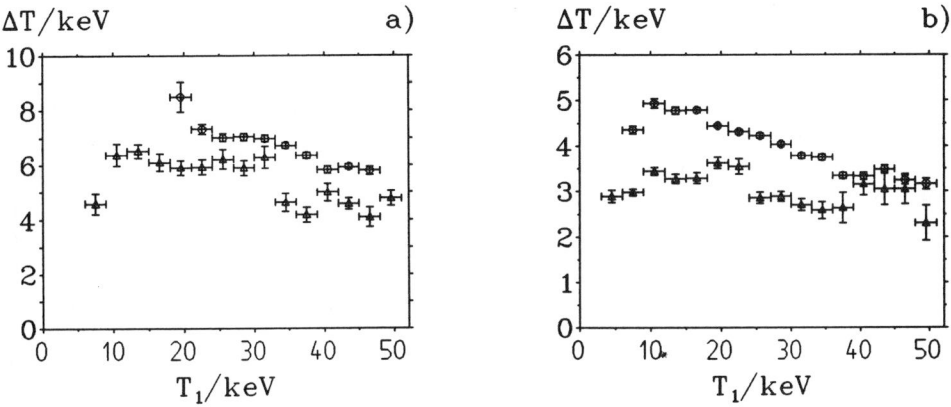

Figure 4: Energy loss ΔT for negative (triangles) and positive (circles) muons in carbon vs. kinetic energy. Only the statistical errors are shown for ΔT. The bars in T_1 direction indicate the range of T_1 values for the corresponding data point. a) 11 μg/cm² foil measured without collimator. b) 5 μg/cm² foil measured with collimator.

Figure 5.

Comparison of Figures 4,b and 5 shows that the shape of the proton stopping power curve has been reproduced for μ^+ down to energies below the Bragg maximum. The higher ΔT measured can be attributed to multiple scattering effects and to the preliminary calibration procedure used. Detailed data analysis will improve the measured absolute stopping power values appreciably but will not affect the obtained ratios μ^+ to μ^- stopping power significantly.

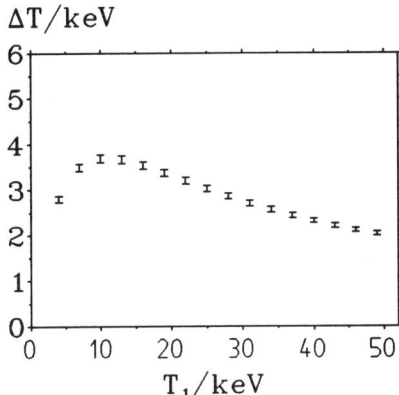

Figure 5: Expected energy loss for μ^+ in the 5 μg/cm² target derived from proton data [11]. See text for details.

4. Summary

The first preliminary results of an experiment which was performed in order to measure the stopping power of carbon for μ^- and μ^+ at energies between 3 and 50 keV were presented. The energy loss of the particles was determined by measuring their time of flight in front and behind a target. For these measurements a parallel plate avalanche counter and a microchannel plate have been used. Furthermore we made use of the fact that secondary electrons were emitted when a muon hit the target. The Bragg peak has been observed, with the maximum located at roughly 12 keV incident muon kinetic energy. There the stopping power for μ^+ is higher than that for μ^- by roughly a factor of 1.4. Complete data analysis together with future more precise measurements should allow us to obtain accurate stopping power values down to about 5 keV.

Acknowledgements

This work is supported by the German BMFT.

References

[1] W.H. Barkas, W. Birnbaum, and F.M. Smith, Phys. Rev. **101**, 778 (1956).

[2] W.H. Barkas, N.J. Dyer, and H.H. Heckman, Phys. Rev. Lett. **11**, 26 (1963).

[3] H. Daniel, Muon Catalyzed Fusion 4, 425 (1989).

[4] H. Daniel, F.J. Hartmann, and M. Mühlbauer, PSI Nuclear and Particle Physics Newsletter 1991, Annex I, Annual Report 1991, p. 61.

[5] H.H. Heckman, and P.J. Lindstrom, Phys. Rev. Lett. **22**, 871 (1969).

[6] W. Wilhelm, H. Daniel, and F.J. Hartmann, Phys. Lett. **98B**, 33 (1981).

[7] R. Medenwaldt, S.P. Møller, E. Uggerhøj, T. Worm, P. Hvelplund, H. Knudsen, K. Elsener, and , E. Morenzoni, Nucl. Instr. Methods **B58**, 1 (1991).

[8] H.H. Andersen, and J.F. Ziegler, Hydrogen Stopping Powers and Ranges in All Elements (Pergamon, New York, 1977).

[9] P. Mertens, and P. Bauer, Nucl. Instr. Methods **B33**, 133 (1988).

[10] D. Taqqu, Nucl. Instr. Methods **A247**, 288 (1986).

[11] J.F. Janni, At. Data Nucl. Data Tables **27**, 147 (1982).

Public Lecture

Un Modo Insolito di Studiare le Proprietà Nucleari, Atomiche e Chimiche

G. TORELLI
Dipartimento di Fisica dell'Università di Pisa e Sezione INFN di Pisa
Via Livornese 582/a
56010 - S. Piero a Grado (Pisa)

Abstract. Any attempt to explain physics, and generally science, to not involved people is strictly connected to a deep understanding of the language and of the general culture of the audience. This talk has been given in Italian for an audience of Italian language and culture; it is a nonsense, I believe, to translate this talk in English.

La pace e la bellezza di questo luogo portano a quel distacco ed a quella serenità di pensiero necessari per cercare di presentare un tema scientifico a persone non impegnate professionalmente in questo tipo di attività; questo è appunto il compito che mi è stato affidato stasera. Gli organizzatori di questa conferenza non potevano trascurare questo tema, visto che la fondazione che ci ospita prende nome appunto da "Stefano Franscini", fondatore, ben 155 anni fa, della "Società Ticinese di Istruzione Pubblica" ed ha tra i suoi compiti statutari anche quello di contribuire alla diffusione della cultura al di fuori di un ambito puramente specialistico. Questo compito che il Prof. Lukas Schaller ed il Dr. Claude Petitjean mi hanno affidato non è un compito facile ed io ringrazio molto entrambi per la stima che mi hanno dimostrato affidandomelo, stima che credo dovuta ai lunghi anni di conoscenza e comune lavoro; farò del mio meglio e spero di non deluderli troppo.

Parlare di fisica in italiano mi fa molto piacere e potrei farlo in piena tranquillità, se non dovessi parlare in *buon italiano*, comprensibile cioè a chi non conosce quel gergo professionale, abituale tra fisici italiani quando parlano di lavoro, che comprende parole tecniche in inglese mal pronunciato e parole italiane con un significato profondamente diverso da quello comune. Ben più difficile è descrivere in modo chiaro quello che facciamo, comunicarvene il fascino e convincervi della sua utilità; la difficoltà non nasce tanto dalla struttura logica della fisica, quanto dalla mancanza di un linguaggio sintetico comunemente noto.

Questa mancanza mi costringerà a ripetere spesso precise definizioni di concetti e quantità appesantendo il discorso, o rendendolo poco chiaro, quando mi dimenticherò di farlo, come mi succederà più di una volta. La necessità di un linguaggio così specializzato non è colpa nostra, dei fisici intendo, ma è un risultato della evoluzione del mondo che ci ha costretti e limitati ad acquisire conoscenze e capacità estremamente specializzate per poter portare al livello attuale la conoscenza, la cultura, il sapere della società nel suo complesso. Gli uomini colti di qualche secolo fa erano persone aggiornate sull'evoluzione dell'intera cultura nel mondo occidentale, ed erano attivi in un campo specifico, spesso assai ampio; nel secolo scorso un uomo colto era in grado di seguire lo sviluppo di grandi aree di interesse: letteratura, arti figurative, scienze esatte, scienze naturali etc. Attualmente è molto difficile e faticoso mantenersi aggiornati sullo sviluppo della fisica nel suo complesso, e trovare il tempo per partecipare ancora attivamente alla ricerca; c'è qualcuno che è in grado di farlo, naturalmente, ma ciò richiede non comune intelligenza e capacità di apprendimento e di sintesi. Abitualmente ognuno di noi è ben informato su quanto succede nei suoi campi di ricerca, e conosce le linee di sviluppo delle grandi teorie generali.

Ma forse è il momento di smettere di dire banalità e cominciare un discorso un po' più serio. A questo punto devo scegliere se abbandonarmi al piacere di strabiliarvi raccontando fatti e situazioni curiosi e meravigliosi in modo un po' magico e quindi poco comprensibile, o cedere al "furor didacticus" cercando di guidarvi passo passo a comprendere il dettaglio del nostro lavoro ed annoiandovi terribilmente dopo cinque minuti. Spero di rimanere lontano da entrambi gli estremi e mi limiterò a cercare di descrivere il mondo in cui avvengono i fenomeni che noi studiamo in modo che vi sia possibile rappresentarvelo mentalmente e non solo confrontare cifre e numeri. Per raggiungere rapidamente questo scopo non posso portarvi ad una rappresentazione del mondo microscopico partendo dalla vostra esperienza quotidiana del mondo macroscopico, come suggerirebbe la maieutica di Socrate; sono piuttosto costretto a fornirvi alcune informazioni a priori, in cui vi prego di credere, e su questa base descrivere il micromondo e le sue relazioni con il macromondo. Vi annoierò un po', perchè voi già conoscete probabilmente questi dati e mi limiterò quindi al minimo indispensabile.

Feynman, nella introduzione al suo testo di "Fisica Generale" [1], dice che " se l'umanità dovesse scomparire e potesse trasmettere una sola informazione alle civiltà future questa informazione dovrebbe essere: la materia è fatta di atomi"; questa appunto è la prima cosa che voglio ricordarvi, le altre sono:

- gli atomi sono composti da un nucleo, pesante, e da elettroni, molto più leggeri, che girano intorno al nucleo;

- i nuclei sono fatti di protoni e neutroni, particelle dello stesso peso;

- gli elettroni hanno carica elettrica negativa, i protoni carica elettrica positiva, i neutroni non hanno carica elettrica;

- normalmente, ossia quando un atomo è in condizioni stabili, il numero di elettroni che girano intorno al nucleo è uguale al numero di protoni nel nucleo, in modo che l'atomo nel suo complesso non ha carica elettrica [è neutro];

- elettroni, protoni e neutroni sono particelle elementari; ce ne sono altre di cui parleremo quando e se necessario;
- le particelle esercitano delle forze l'una sull'altra, queste forze o interazioni sono classificate sulla base della loro intensità e dei fenomeni cui sono legate:
  1. interazione forte, che tiene insieme i nuclei;
  2. interazione elettromagnetica, che tiene insieme gli atomi;
  3. interazione debole, di cui parleremo poi.

Ed ora arriviamo ai μ [2]; i μ (o muoni) sono particelle elementari, sono i fratelli maggiori degli elettroni, hanno volume trascurabile come gli elettroni e carica negativa o positiva, ma pesano molto di più. Usando come unità il peso del protone, l'elettrone pesa 2000 volte meno [mezzo grammo contro un chilo] ed il μ invece pesa solo 10 volte meno [ben 100 grammi]. I μ sono particelle instabili, la loro instabilità è appunto dovuta alle interazioni deboli, ed appena prodotti cominciano a suddividersi in altre particelle più leggere.

$$\mu \to e + \nu + \nu \tag{1}$$

Questo processo si chiama decadimento; il μ decade in un elettrone ed in due particelle neutre senza massa chiamate neutrini di cui non ci occuperemo ulteriormente; l'elettrone ha una energia piuttosto alta. La probabilità di decadimento per unità di tempo è costante e ciò produce una diminuzione esponenziale della probabilità che il μ sopravviva. In un decadimento si chiama vita media il tempo necessario per ridurre a 1/2.71 la probabilità di sopravvivenza della particella; la vita media del μ è $\tau = 2,20$ μs, poco più di due milionesimi di secondo. Si può misurare il tempo che passa tra l'arrivo del μ nell'apparato sperimentale ed il suo decadimento, fare un grafico in cui si divide l'asse dei tempi in tanti intervalli successivi Δt e riportare per ogni intervallo il numero di volte in cui un μ vi è decaduto; si ottiene in tal modo una curva esponenziale la cui pendenza è proporzionale alla vita media; la Fig. 1 mostra appunto un esempio dei risultati dell'ultima misura della vita media del μ fatta a Saclay qualche anno fa [3].

Tutte queste particelle elementari sono naturalmente invisibili direttamente, ma sappiamo costruire molti tipi di apparecchi che forniscono un segnale elettrico quando sono attraversati da una particella carica con una energia abbastanza grande; questi apparecchi si chiamano rivelatori o, in gergo, contatori perchè permettono di contare le particelle; si possono rivelare direttamente solo le particelle cariche, ma la radiazione elettromagnetica come i raggi X trasmette molto facilmente la sua energia agli elettroni del materiale che forma i rivelatori e quindi si possono rivelare abbastanza facilmente anche i raggi X. Non si trovano in natura μ che attendano di essere usati perchè i μ decadono appena prodotti;

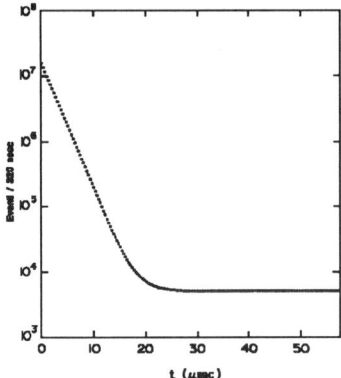

Figure 1: Distribuzione temporale dei decadimenti di μ negativi fermati in idrogeno liquido; da [3].

occorre quindi produrre appositamente i μ usando acceleratori di energia abbastanza alta; tutto il nostro lavoro si svolge quindi in grandi laboratori dotati di acceleratori adatti; uno di tali laboratori si trova a Villigen, vicino a Baden [Paul Scherrer Institut - Villigen, Argovia CH].

Arriviamo adesso al mondo macroscopico, quello che conosciamo direttamente con i nostri sensi. Il mondo macroscopico è un mondo elettromagnetico, costituito da atomi, legati da forze elettromagnetiche, che a loro volta si riuniscono in molecole, ancora per forze elettromagnetiche, ed in corpi amorfi o in cristalli, ancora per forze elettromagnetiche; per fare un paradosso la fisica dello stato solido, la chimica, la biologia etc... non sono altro che lo studio delle molteplici possibilità di combinazione delle forze elettromagnetiche tra un numero grandissimo di atomi di circa 90 tipi diversi, i 90 elementi del sistema periodico.

Ma come appare questo mondo macroscopico, questo pezzetto di gomma o questo lapis, descritto dal punto di vista di una particella elementare, del nostro μ, per esempio; ossia assumendo come unità di misura i pesi e le dimensioni e le forze di una particella elementare?

Cominciamo dal nucleo e riferiamoci a quattro elementi: Idrogeno, Carbonio, Ferro e Piombo; il più leggero e diffuso nell'universo, la base del mondo organico, il più solido meccanicamente ed il più pesante tra gli elementi comuni e diffusi. Le loro caratteristiche sono riportate nella Tabella 1; Z è il numero di protoni nel nucleo e quindi anche il numero di elettroni nella nuvola elettronica, A è il numero totale di protoni e neutroni nel nucleo, ed r_n è il raggio del nucleo, una particella non può andare più vicina al nucleo senza romperlo o perturbarlo gravemente. Passando dall'Idrogeno, il cui nucleo è formato da un solo protone, al Piombo, con ben 207 protoni e neutroni [nucleoni], il raggio nucleare cambia solo di un fattore 6. La colonna successiva riporta invece il raggio atomico r_a, ossia il raggio della nuvola elettronica o la distanza dal nucleo a cui ruota l'elettrone più lontano; il rapporto tra Piombo ed Idrogeno è in questo caso minore di 2.5. Guardandola dal di dentro la materia è molto più uniforme di quanto non sembri; ed è anche molto vuota perchè il raggio del nucleo è piccolissimo rispetto al raggio dell'atomo, da ventimila volte più piccolo per il Piombo a cinquantamila volte più piccolo per l'Idrogeno. Questo è il grande vuoto di cui è fatta la materia, anche la più solida e densa, quando la si guardi dal punto di vista di una particella. Una particella può muoversi nella nuvola elettronica senza disturbarla troppo. Quindi il volume occupato, nel senso che non è possibile ad una particella entrarci dentro senza romperlo, è costituito dal solo volume del nucleo. Non è facile darvi una rappresentazione visiva di una simile situazione usando oggetti della nostra esperienza quotidiana; si può paragonare il nucleo ad una sfera di cento metri di raggio posta al centro della terra mentre il volume dell'intera terra è il volume vuoto in cui si muovono pallini da caccia che rappresentano gli elettroni. Un'altra particella, pallino da caccia se è un elettrone o mongolfiera se è un protone, può muoversi in questo vuoto con probabilità molto bassa di urtare direttamente qualcosa. Sentirà, beninteso, le forze che le altre particelle esercitano su di lei ed eserciterà forze a sua volta, ma potrà attraversare l'atomo senza processi drammatici.

Table 1: Numero atomico Z, peso atomico A, raggio nucleare r_n, raggio atomico r_a e distanza interatomica d per quattro elementi: Idrogeno H, Carbonio C, Ferro Fe e Piombo Pb.

	Z	A	r_n	r_a	d
H	1	1	1.4	0.8	26.6
C	6	12	2.5	0.9	1.8
Fe	26	56	5.8	1.7	2.3
Pb	82	207	8.3	1.8	3.1
			$\times 10^{-15}$ m	$\times 10^{-10}$ m	$\times 10^{-10}$ m

Vediamo adesso le posizioni relative degli atomi; la colonna successiva riporta la distanza media tra atomi in Idrogeno gassoso, Carbonio sotto forma di diamante, Ferro e Piombo metallici; in tutti i solidi gli atomi distano in media tra loro circa una o due volte il loro diametro, e sono tenuti in queste posizioni da forze molto deboli; nei gas la distanza media aumenta di un fattore dieci o venti e gli atomi non sono fermi o quasi, ma si muovono rimbalzando tra di loro e sulle pareti. Lo spazio tra atomo ed atomo è altro spazio libero in cui può muoversi una particella.

Un atomo, contrariamente ad una particella, non può invadere lo spazio interno di un altro atomo, perchè ha grosso modo le stesse dimensioni ed è tenuto insieme dalle stesse forze, ma può muoversi liberamente tra gli atomi di un solido, perturbando le forze che li legano, e può scavarsi anche una piccola nicchia stabile.

Possiamo visualizzare tutto questo pensando a batuffoli di cotone: ne posso fare un piccolo mucchio ed ognuno occupa il suo volume e non penetra dentro gli altri, ma posso prendere una punta sottile e muoverla come voglio dentro quella massa, spostando qualche filo di cotone senza distruggere niente.

Riassumendo, abbiamo tanti atomi che formano la materia, palline da tennis ad una diecina di centimetri l'una dall'altra, tenute in posizione da tante piccolissime molle molto deboli; un'altra pallina può sempre entrare in questa struttura deformando le molle; ogni pallina ha al suo centro un nucleo con tanti elettroni intorno; una particella può muoversi in tutto questo spazio.

Ho creduto necessario darvi questa immagine realistica di come è fatto il mondo in cui i nostri μ si muovono per convincervi che l'idea di un proiettile sparato contro un muro non è una buona rappresentazione di una particella che penetra in un corpo, solido o gassoso che sia.

L'immagine adatta a comprendere cosa succede ad un μ che entra ad alta velocità in un solido o in un gas è quella di una pallina lanciata in un gran vuoto con tante palline cariche pesanti [nuclei] e tantissimi puntini carichi leggeri [elettroni]. Fino a quando la sua velocità è piu grande della velocità degli elettroni, il μ si comporta come una palla da tennis che investe tante palline da ping-pong, le palline schizzano via da tutte le parti e la palla da tennis rallenta la sua corsa. Quando la velocità del μ diventa confrontabile con quella degli elettroni le cose diventano un po' più complicate perchè cominciano a diventare importanti anche per il μ le forze elettriche tra cariche che tengono insieme gli elettroni ed i nuclei atomici e gli atomi tra di loro. Noi siamo interessati ai μ negativi, quanto detto fino ad ora non dipende dal segno della carica elettrica e va bene per μ positivi e negativi, ma d'ora in poi il discorso descrive specificamente quanto succede al μ negativo che è in tutto identico ad un elettrone, a parte la massa e la possibilità di decadere. Nelle condizioni che abbiamo descritto, quando ha raggiunto una velocità abbastanza ridotta il μ negativo può sostituirsi ad un elettrone e formare atomi muonici.

Ed ora comincia il nostro gioco; il μ comincia a sentire le forze interatomiche e va dove queste forze lo guidano, quindi, osservando quello che il μ fa, noi possiamo conoscere la struttura interatomica. In altri termini noi abbiamo introdotto in una struttura in equilibrio una particella simile agli elettroni, ma riconoscibile, che perturba il sistema, ma non lo distrugge; seguendo questa particella possiamo conoscere meglio il sistema; come al solito dobbiamo perturbare un poco il sistema, e vedere come reagisce, per capire come è fatto.

Ma come si può misurare, rivelare quello che il μ fa, in queste condizioni? In realtà noi siamo completamente ciechi dal momento in cui il μ entra nel nostro apparato ed è rivelato dal contatore posto tra il tubo sotto vuoto che ci porta i μ ed il materiale, solido o gassoso, in cui il μ perde energia e si ferma. Non è un tempo lungo, qualche miliardesimo di secondo, ma in questo tempo avviene un fondamentale cambiamento di condizioni: si passa da un μ libero ad un μ legato in un atomo. Sfortunatamente non è possibile rivelare niente di quello che accade da quando il μ arriva a quando è saldamente legato in un

Un Modo Insolito di Studiare le Proprietà Nucleari, Atomiche e Chimiche 359

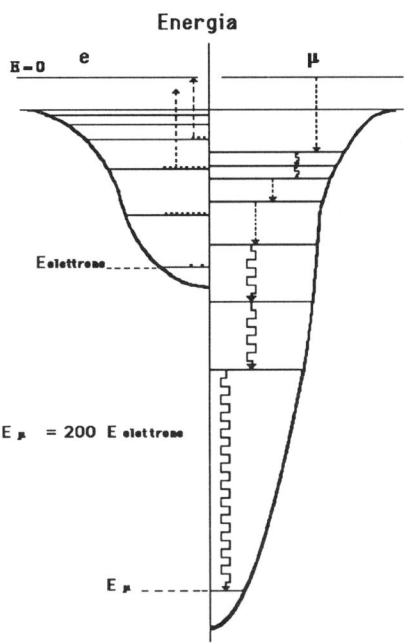

Figure 2: Rappresentazione schematica non in scala della buca di potenziale per un elettrone e per un μ in un atomo; le linee rappresentano i livelli di energia stabili; le frecce rappresentano transizioni [passaggio da uno stato ad un'altro] con emissione di un elettrone [frecce dritte] o di un raggio X [frecce ondulate].

atomo; possiamo solo preparare una situazione sperimentale quanto più definita possibile e poi rivelare dove è andato a legarsi il μ, e si deve dedurre cosa è successo in questo intervallo di tempo dai dati misurati prima e dopo.

Vedremo tra un minuto come rivelare a quale atomo si è legato il μ; per seguire un filo logico conviene ora vedere come avviene lo scambio tra il μ e l'elettrone. La cattura atomica diventa molto probabile quando l'elettrone atomico ed il μ libero si muovono all'incirca con la stessa velocità; l'elettrone è in una buca di potenziale, ha una energia totale negativa, perchè è legato, il μ invece ha una energia positiva perchè è libero [Fig. 2]; potete proprio pensare ad un pavimento con una buca su cui rotola una pallina, se la pallina rotola troppo forte cade nella buca e ne riesce, se va abbastanza piano invece cade nella buca e ci resta.

La cattura Auger, così si chiama questo processo, consiste appunto in uno scambio di ruoli tra l'elettrone ed il μ; il μ si ritrova legato nella posizione in cui era legato l'elettrone, e l'elettrone si ritrova libero con l'energia del μ.

Il μ si è sostituito all'elettrone alla stessa distanza dal nucleo e, diciamo così, ruotando intorno al nucleo con la stessa velocità dell'elettrone; ma il μ è molto più pesante e queste condizioni non sono più condizioni stabili per il μ mentre lo erano per l'elettrone. Possiamo immaginare che le possibili posizioni stabili di un elettrone intorno ad un nucleo formino come una scala che scende verso il nucleo [la scala verticale è proporzionale all'energia], gli elettroni riempiono tutti i gradini partendo dal più basso e questa configurazione è stabile. Per il μ la situazione è analoga, ma la scala del μ scende molto più vicino al nucleo perchè il μ è molto più pesante dell'elettrone; il μ si trasferisce inizialmente sul suo gradino più vicino in energia al gradino occupato dall'elettrone con cui si scambia. Questo gradino, nella scala del μ, è piuttosto alto e la scala è tutta libera, così il μ comincia a scendere la scala e usa l'energia che così guadagna per espellere altri elettroni dall'atomo o per emettere raggi X. Questo processo di diseccitazione si chiama cascata elettromagnetica.

Come abbiamo visto, questi raggi X possono essere rivelati e sono appunto loro che ci informano sull'avventura del μ; si può misurare con precisione l'energia di questi raggi X e distinguere da che atomo vengono ed a quale transizione, a quale gradino, corrispondono. Non che sia facile, ma ci si riesce [Fig. 3]; e tutto quello che sappiamo sul processo di cattura e su altri processi che poi vedremo è dedotto dalla misura di questi raggi X, della loro energia e del ritardo con cui sono emessi rispetto all'arrivo del μ nell'apparato sperimentale.

Abbiamo descritto la cattura atomica del μ da parte di un atomo isolato, come gli atomi dei gas nobili; abitualmente gli atomi sono legati in molecole ed allora la cattura è un po' più complicata. Prendiamo una molecola biatomica per esempio, ossia una molecola formata da due atomi; il μ si sostituisce ad un elettrone della molecola, cioè ad un elettrone legato ad entrambi gli atomi, e comincia a scendere una scala analoga a quella che abbiamo visto; ad un certo punto la scala si divide in due rami [Fig. 4] ed il μ deve scegliere dove andare. La probabilità di andare a destra o a sinistra dipende dalla struttura della molecola, e misurare questa probabilità ci permette di capire un po' di più come è fatta questa molecola.

Guardiamo per esempio cosa succede per una molecola semplice come quella degli ossidi formati dai vari elementi del sistema periodico legati con l'Ossigeno. Questo istogramma [Fig. 5] riporta appunto la probabilità che il μ si leghi all'altro elemento in funzione del suo numero atomico Z, ossia del numero degli elettroni, e mostra che questa probabilità varia periodicamente con la posizione dell'elemento nel sistema periodico [4]. Confrontare misure di questo tipo con le informazioni che si conoscono dalla Chimica permette di capire non solo il processo di cattura dei μ, ma anche quello di cattura degli elettroni e la distribuzione degli elettroni nella molecola.

Prendiamo adesso un caso completamente diverso e supponiamo di voler sapere cosa c'è dentro un vaso senza aprirlo o romperlo, o di voler conoscere la presenza o meno di un

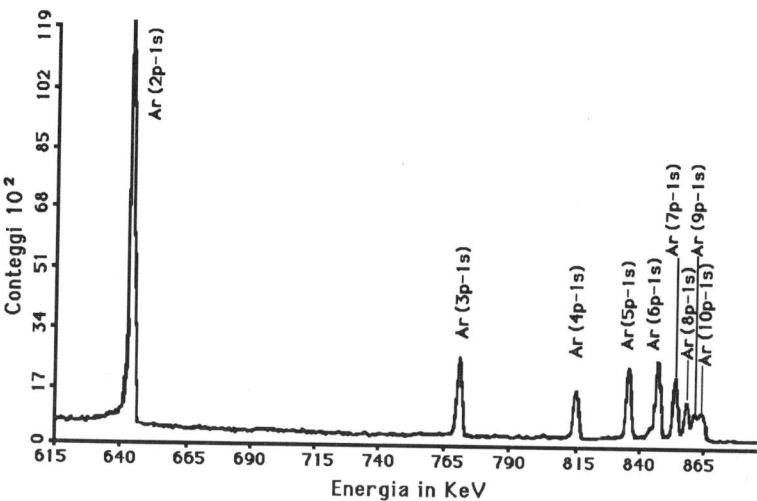

Figure 3: Un esempio della distribuzione in energia dei raggi X prodotti da μ catturati in Argon; si possono individuare le transizioni tra i diversi livelli.

certo elemento in un frammento che non si vuole distruggere, nemmeno in parte, con una analisi chimica (il caso è di un certo interesse in archeologia per esempio); i μ ce ne forniscono il mezzo, basta far fermare nel materiale in questione dei μ ed analizzare lo spettro di energia dei raggi X emessi; se ne può dedurre quali sono gli elementi presenti ed il loro peso relativo.

Abbiamo parlato finora di fare chimica usando i μ; ma i μ possono anche essere usati per studiare altri aspetti dell'atomo in cui sono legati. Prendiamo i gradini della scala che abbiamo gia visto, ogni gradino è suddiviso in sottogradini, e l'altezza dei sottogradini dipende dalla distanza del μ dal nucleo e da molti altri fattori, che hanno importanza via via minore e che noi sappiamo o crediamo di saper calcolare. Una misura accurata dei sottogradini permette quindi di controllare se e quanto sia corretto il nostro modo di descrivere la natura.

Quando il μ è abbastanza vicino al nucleo, sui gradini bassi della scala, anche le interazioni deboli con i protoni e neutroni, di solito trascurabili, diventano importanti ed aprono al μ un'altra via per scomparire: il μ può sempre decadere con la stessa probabilità, ma può anche essere catturato da un protone [cattura nucleare].

$$\mu + p \to n + \nu \tag{2}$$

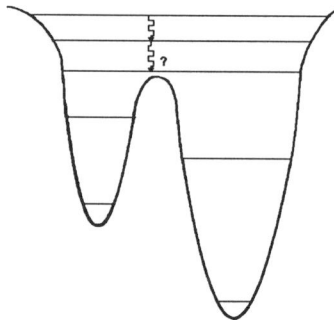

Figure 4: Rappresentazione schematica non in scala della buca di potenziale per un μ in una molecola biatomica; le linee rappresentano i livelli di energia stabili; le frecce rappresentano transizioni [passaggio da uno stato ad un'altro] con emissione di un elettrone [frecce dritte] o di un raggio X [frecce ondulate].

Questa nuova via per andarsene diminuisce la vita media dei μ. Se studiamo la probabilità di questo processo in nuclei molto semplici: Idrogeno, Deuterio, Elio, otteniamo informazioni su come funziona la interazione debole che è responsabile di questa cattura.

I due ultimi tipi di misure che vi ho descritti sono stati e sono ancor oggi di grande importanza per controllare la validità dei principi generali che sono alla base della grande teoria - il modello standard - che descrive in modo unitario le interazioni elettromagnetiche e deboli delle particelle elementari; anche i valori numerici di alcune costanti fondamentali del modello standard sono dedotti da misure su atomi muonici o sui decadimenti del μ.

Quando misuriamo la vita media dei μ in nuclei pesanti, ed osserviamo quello che succede al nucleo dopo la cattura nucleare ricaviamo informazioni sulla struttura di questi nuclei complicati; usiamo quindi gli atomi muonici per studiare la fisica nucleare.

Ed affrontiamo ora l'ultimo e sorprendente aspetto della fisica degli atomi muonici: cosa succede quando un μ si lega ad un atomo con un solo elettrone, cioè Idrogeno, Deuterio o Trizio. Per semplicità parlerò sempre di Idrogeno, e quindi di protoni, a meno che non sia necessario fare distinzione tra i tre. Per capire perchè il caso di un atomo con un solo elettrone sia cosi differente da quello degli atomi con più elettroni dobbiamo tornare indietro a quando abbiamo parlato della cattura atomica. Il μ, appena catturato, è in cima alla sua scala e comincia a scendere i gradini. Se l'atomo iniziale aveva più elettroni, il μ scende espellendo dall'atomo gli elettroni residui; l'atomo diventa quindi elettricamente carico, respinge gli altri atomi e non ha più rapporti con il mondo circostante. Se l'atomo iniziale ha un solo elettrone, il μ si sostituisce a questo elettrone e non ha altri elettroni a disposizione da espellere per scendere la sua scala. Questo dovrebbe rendere la cascata più lenta perchè espellere elettroni è più facile e veloce che emettere raggi X. Ma questo

Un Modo Insolito di Studiare le Proprietà Nucleari, Atomiche e Chimiche 363

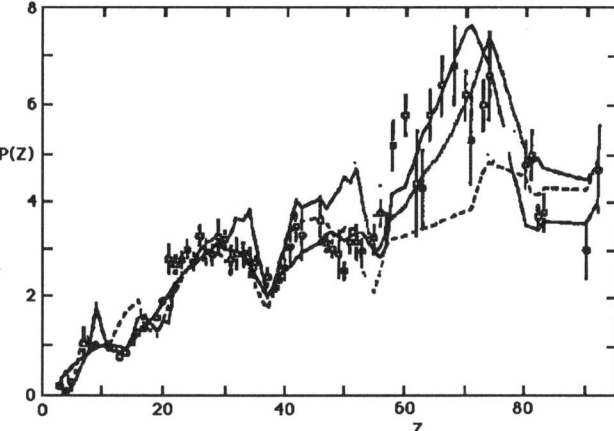

Figure 5: Probabilità di cattura di un μ da parte dell'elemento X, misurato come rapporto tra il numero di catture in X ed il numero di catture in Ossigeno, per ossidi X_nO_m, in funzione del numero atomico Z dell'elemento X [4].

speciale atomo muonico è molto piccolo, in confronto ad un normale atomo di Idrogeno, perchè il μ sta 200 volte più vicino al nucleo dell'elettrone corrispondente, e per di più è neutro, quindi non respinge più gli altri atomi, anzi entra dentro la nuvola elettronica degli altri atomi dove trova di nuovo elettroni da espellere per scendere in fretta la sua scala. Questo succede e la cascata elettromagnetica torna ad essere veloce se ci sono tanti atomi in giro in modo che l'atomo muonico possa passare buona parte del suo tempo dentro altri atomi; il tempo della cascata quindi dipende dalla pressione del gas; queste cose infatti si possono notare solo in gas a bassa pressione, perche se la pressione è alta o siamo in solidi ci sono così tanti atomi a disposizione che il tempo di cascata è sempre brevissimo.

Il μ impiega in ogni caso un tempo breve rispetto alla sua vita media per arrivare in fondo alla scala e la probabilità di cattura nucleare è bassa per l'Idrogeno ed i suoi isotopi Deuterio e Trizio, quindi l'atomo muonico pμ, dμ o tμ vive ancora a lungo e può cominciare una sua avventura andando a spasso dentro la nuvola elettronica degli altri atomi. Se il pμ incontra un atomo diverso, il μ abbandona il suo protone e si lega al nuovo atomo con un meccanismo abbastanza simile a quello della cattura atomica [Fig. 6], questo processo si chiama trasferimento. Il μ trasferito si viene a trovare ad un gradino ancora alto della nuova scala, comincia a scenderla nel solito modo ed emette raggi X che ci permettono di vedere a che atomo il μ si è trasferito e quando.

Se ci sono pochi atomi diversi dall'Idrogeno nell'ambiente in cui il pμ di trova, pochi vuol

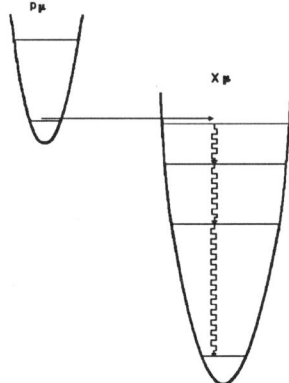

Figure 6: Rappresentazione schematica non in scala della buca di potenziale per un μ nell'atomo pμ ed in un altro atomo Xμ e del processo di trasferimento; le linee rappresentano i livelli di energia stabili; le frecce ondulate rappresentano transizioni [passaggio da uno stato ad un'altro] con emissione di un raggio X.

dire molto meno di uno per milione di atomi di Idrogeno, il trasferimento è poco probabile ed il μ può formare una molecola pμp vale a dire che il μ si lega a due protoni e li tiene molto vicini l'uno all'altro, formando un blocchetto compatto con una sola carica elettrica ed il peso di due protoni, un oggetto non diverso da un nucleo di Deuterio dal punto di vista chimico. Questo pseudo-Deuterio cattura quindi un elettrone e forma una specie di atomo, può poi legarsi ad un altro atomo di Idrogeno e formare una pseudo-molecola (pμp)p con due elettroni che tengono insieme il protone semplice p e lo pseudo-Deuterio (pμp); etc . Tutti questi processi sono divertenti, molto utili per capire i meccanismi della dinamica chimica e per capire cosa succede al μ, ma l'interesse per lo studio degli atomi muonici era stato provocato, sul finire degli anni quaranta, dalla speranza di una utilizzazione pratica di tali atomi, speranza che i risultati sperimentali e teorici ottenuti fino a circa quindici anni fa sembravano cancellare completamente.

Questa speranza si basa su di una idea molto semplice suggerita indipendentemente da Frank [5] e da Sakharov [6] nel 47-48, vi posso appunto mostrare copia della copertina del manoscritto di Sakharov di cui Semion Gershstein mi ha fatto dono [Fig. 7].

L'idea presentata è questa: il μ vince la repulsione elettrostatica e costringe due nuclei di Idrogeno o suoi isotopi a stare molto vicini per un tempo relativamente lungo; la distanza è così piccola ed il tempo così lungo che l'interazione forte [forza nucleare] può produrre la fusione dei due nuclei in un nucleo più pesante, lasciando libero il μ e producendo energia. Tutte le combinazioni interessanti sono presentate nella Tabella 2, ma parleremo

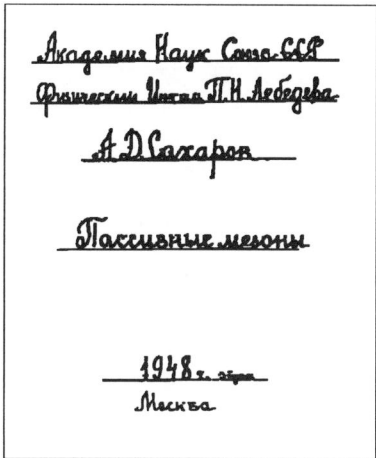

Figure 7: Copertina del manoscritto presentato da A.D. Sakharov all'Accademia delle Scienze dell'URSS; Il titolo è appunto "Il mesone passivo".

del processo più semplice pp per evitare inessenziali complicazioni.

La prospettiva è affascinante perchè questo tipo di fusione nucleare ha bisogno di μ, ma non è capace di produrli, pur producendo l'energia; il processo quindi è controllabile e non esplosivo: basta smettere di mandare μ e tutto si ferma. Produrre μ costa molto in energia, direttamente ed indirettamente, e per avere un bilancio energetico positivo occorre che ogni μ, nella sua breve vita, produca molti processi di fusione, dell'ordine di un migliaio almeno. Per avere un alto numero di fusioni nella vita media del μ occorre che il tempo di ciclo [Fig. 8] sia una piccola frazione della vita media, ossia occorre che il μ si fermi, venga catturato, formi una molecola, produca la fusione e torni libero in pochi miliardesimi

Table 2: Reazioni di fusione nucleare catalizzate da μ e relativa energia liberata.

$p + p + \mu$	$d + e + \mu$	2.2 MeV
$p + d + \mu$	$He^3 + \gamma + \mu$	5.4 MeV
$p + d + \mu$	$He^3 + \mu$	5.4 MeV
$p + t + \mu$	$He^4 + \gamma + \mu$	20.0 MeV
$d + d + \mu$	$t + p + \mu$	4.0 MeV
$d + d + \mu$	$He^3 + n + \gamma + \mu$	3.2 MeV
$d + t + \mu$	$He^4 + n + \mu$	17.6 MeV
$t + t + \mu$	$He^4 + n + n + \mu$	10.0 MeV

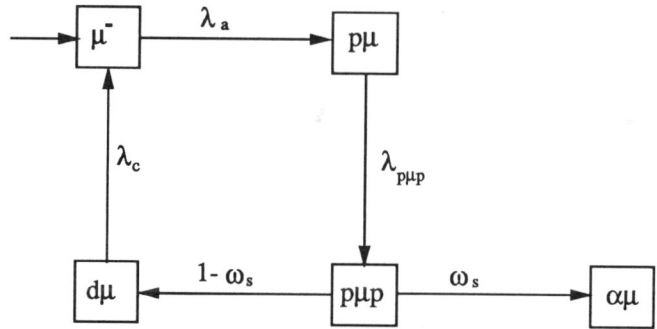

Figure 8: Ciclo ideale semplificato di fusione nucleare catalizzata da μ.

di secondo.

Sperimentalmente fusioni nucleari prodotte da questo meccanismo sono state viste dieci anni dopo che l'idea era stata pubblicata [7], ma i valori dei tempi di reazione misurati erano tanto lunghi che alla fine degli anni settanta si era quasi persa la speranza di raggiungere un bilancio energetico positivo. Tutte le misure erano state fatte o in gas a temperatura ambiente o in Idrogeno liquido ed i modelli teorici prevedevano processi adiabatici, graduali, che non dipendevano molto dalle energie relative delle varie particelle. In quel periodo un gruppo di teorici del JINR di Dubna [8] cominciò ad elaborare un modello dettagliato del processo di formazione molecolare che era il passo più lento dell'intero ciclo. I risultati dei loro calcoli sono stati inattesi e sorprendenti: la probabilità di formazione molecolare dipende molto dall'energia relativa perchè al normale processo adiabatico si aggiunge, nelle condizioni opportune, un processo di formazione risonante. Quindi il tempo necessario per un ciclo di fusione può diminuire di un fattore 100 o 1000, e non è più assurdo sperare di arrivare ad una fusione controllata per questa via. Come è ovvio questi risultati teorici hanno suscitato nuovo lavoro sperimentale per verificarli e per trovare le migliori condizioni di lavoro etc etc. ; non siamo ancora in fondo, ma questo metodo per ottenere la fusione controllata attualmente sembra molto promettente per future applicazioni pratiche.

Questo risvegliarsi di interesse per la fisica degli atomi muonici e dei μ a bassa energia, dopo un periodo di stanca in cui sembrava che non ci fosse più nulla da apprendere al riguardo, è un fenomeno che ho avuto modo di vedere più volte nei miei quarant'anni di lavoro, che sono molti per me, ma niente nella storia della scienza: sembra di non sapere più cosa cercare, rimangono da definire solo dettagli, e poi un nuovo modo di guardare le cose, una inaspettata connessione col mondo circostante, scientifico e non, rimettono tutto in moto con nuovo slancio.

Io credo che una simile dinamica sia fisiologica per un settore vitale, forse i settori senza crisi non hanno crisi perche non sono ancora riusciti ad arrivare in fondo al primo ciclo? Non so, in ogni caso per nostra fortuna adesso questo nostro specifico settore degli atomi muonici è in pieno sviluppo, ed ancor più tutto il grande campo della fisica fondamentale che è tornata a riunire in una unica disciplica lo studio dell'infinitamente piccolo e dell'infinitamente grande, dai componenti fondamentali della materia alla formazione ed evoluzione dell'universo, e cerca di descrivere il tutto con la stessa teoria.

Spero di essere restato nei limiti di una conversazione amichevole ragionevolmente breve, di non avervi annoiato troppo e di avervi trasmesso un po' del mio entusiasmo, ringrazio di nuovo Schaller e Petitjean e garantisco loro che ho fatto del mio meglio.

Bibliografia

[1] *The Feynman lectures on Physics*, R.P. Feynman, R.B. Leighton, M. Sands, 1-2, Addison-Wesley Publishing Company, Inc., Reading, Massachussets, Palo Alto (London).
[2] M. Conversi, E. Pancini and M. Piccioni, Phys. Rev. **68** (1945) 232.
[3] G. Bardin and al., Nucl. Phys. **A352** (1981) 365.
[4] F. J. Hartmann in *E.M. Cascade and Chenistry of Exotic Atoms, Proc. of the Fifth Course of the Int. School of Physics of Exotic Atoms, Erice, Italy, May 14-20, 1989*, edited by L. M. Simons, D. Horvath and G. Torelli, pg. 23.
[5] F.C. Frank, Nature **160** (1947),525.
[6] A.D. Sakharov, Report of the Physics Institute, Academy of Sciences (1948).
[7] L.W. Alvarez and al., Phys.Rev. **105** (1957)1127.
[8] E. A. Wesman, J.E.T.P. Lett. **5** (1967) 91;
S.S. Gershtein et L.I. Ponomariev, Phys. Lett. B72 (1977) 80.

Aprile 1992

LIST of CONFERENCE PARTICIPANTS

Name	Address	Country
1. P.Baeriswyl	Univ. de Fribourg, CH-1700 Fribourg	Switzerland
2. D.Bakalov	Univ. de Trieste, I-23127 Trieste	Italia/Bulgaria
3. W.Bertl	PSI-West, CH-5232 Villigen	Switzerland
4. M.Camani	Dip. del terr., CH-6500 Bellinzona	Switzerland
5. T.A.Case	Univ. of California, Berkeley, CA 94720	USA
6. P.David	Univ.Bonn, D-5300 Bonn	Germany
7. Ms. P.de Cecco	PSI-West, CH-5232 Villigen	Switzerland/USA
8. J.Deutsch	Univ. de Louvain, B-1348 Louvain	Belgium
9. J.P.Egger	Univ. de Neuchâtel, CH-2000 Neuchâtel	Switzerland
10. S.S.Gershtein	IHEP Serpukhov, 142 284 Moscow	Russia
11. M.Harston	Math. Inst., Oxford OX1 3LB	England
12. F.J.Hartmann	TU München, D-8046 Garching	Germany
13. M.D.Hasinoff	Univ. of British Columbia, Vancouver	Canada
14. P.Hauser	PSI-West, CH-5232 Villigen	Switzerland
15. R.Jacot-Guillarmod	Univ. de Fribourg, CH-1700 Fribourg	Switzerland
16. M.Jeitler	Inst. ME-Physik, OeAW, A-1090 Wien	Austria
17. K.P.Jungmann	Univ. Heidelberg, D-6900 Heidelberg	Germany
18. P.Kammel	Inst. ME-Physik, OeAW, A-1090 Wien	Austria
19. F.Kottmann	IHP, ETHZ, PSI, CH-5232 Villigen	Switzerland
20. A.V.Kravtsov	LNPI, 188 350 Gatchina, St.Petersburg	Russia
21. K.Lou	PSI-West, CH-5232 Villigen	Switzerland
22. F.Maas	Univ. Heidelberg, D-6900 Heidelberg	Germany
23. V.E.Markushin	Kurchatov Institute, 123 182 Moscow	Russia
24. G.Marshall	TRIUMF, Vancouver B.C. V6T 2A3	Canada
25. M.Meyberg	PSI-West, CH-5232 Villigen	Switzerland
26. R.Milotti	Univ. de Trieste, I-23127 Trieste	Italia
27. E.Morenzoni	PSI-West, CH-5232 Villigen	Switzerland
28. M.Mühlbauer	TU München, D-8046 Garching	Germany
29. Ms. F.Mulhauser	Univ. de Fribourg, CH-1700 Fribourg	Switzerland
30. C.Petitjean	PSI-West, CH-5232 Villigen	Switzerland
31. C.Piller	Univ. de Fribourg, CH-1700 Fribourg	Switzerland
32. L.I.Ponomarev	Kurchatov Institute, 123 182 Moscow	Russia
33. C.Rizzo	Univ. de Trieste, I-23127 Trieste	Italia
34. R.Rosenfelder	PSI-West, CH-5232 Villigen	Switzerland
35. L.A.Schaller	Univ. de Fribourg, CH-1700 Fribourg	Switzerland
36. L.Schellenberg	Univ. de Fribourg, CH-1700 Fribourg	Switzerland
37. H.Schneuwly	Univ. de Fribourg, CH-1700 Fribourg	Switzerland
38. G.Semenchuk	LNPI, 188 350 Gatchina, St.Petersburg,	Russia
39. R.T.Siegel	William and Mary, Williamsburg, VA	USA
40. L.M.Simons	PSI-West, CH-5232 Villigen	Switzerland
41. P.Souder	Syracuse Univ., Syracuse, NY 13244	USA

Name	Address	Country
42. D.Taqqu	PSI-West, CH-5232 Villigen	Switzerland
43. G.Torelli	Univ. de Pisa, I-56010 Pisa	Italia
44. A.Vacchi	Univ. de Trieste, I-23127 Trieste	Italia
45. D.Viel	PSI-West, CH-5232 Villigen	Switzerland/USA
46. H.K.Walter	PSI-West, CH-5232 Villigen	Switzerland
47. P.Wojciechowski	TU München, D-8046 Garching	Germany
48. E.Zavattini	Univ. de Trieste, I-23127 Trieste	Italia
49 J.Zmeskal	Inst. ME-Physik, OeAW, A-1090 Wien	Austria

Monte Verità

Centro Stefano Franscini
Ascona – ETH Zürich

Edited by
Konrad Osterwalder, ETH Zürich, Switzerland

The conference center "Stefano Franscini", situated on the Monte Verità above Ascona, was opened in summer 1989, and is under the auspices of the Eidgenössische Technische Hochschule Zürich. International research groups, invited by lecturers of Swiss universities, are given the opportunity to hold seminars on aspects of their respective fields. The meetings cover a great variety of topics and are carefully screened by an international advisory board. Some of the most interesting reports are now being published in our new interdisciplinary series **Monte Verità** and provide excellent accounts of the actual state of the respective research areas. Each volume is unique in that it places particular emphasis on making the connections between the individual contributions explicit and their long-term importance clear.

W. Czaja
Institut de Physique Appliquée,
EPFL, Lausanne, Switzerland **(Ed)**

Synchrotron Radiation: Selected Experiments in Condensed Matter Physics

1991. 188 pages. Hardcover
sFr. 58.– / DM 68.–
ISBN 3-7643-2594-1

In this volume, 10 contributions describe selected experiments in condensed matter physics with synchrotron radiation, the subject matter at an international workshop held at the Centro Stefano Franscini in Ascona, Switzerland, in July 1990. The experiments concerned magnetic properties, electrical properties of clusters, liquid metals and magnetic semiconductors, interface problems (Schottky barriers), etc. They demonstrate the enormous impact and challenge which lies in the use of a synchrotron light source in many areas of solid state physics. In another paper, applications of the new light source are discussed in relation with the determination of crystal structures. Finally, information on the properties and experimental possibilities of the new synchrotron in Trieste are given. This installation is one of the three new third generation synchrotrons now under construction and promises once again to enlarge the number of possible applications.

Please order through your bookseller or write to:

Birkhäuser Verlag AG
P.O. Box 133
CH-4010 Basel / Switzerland
FAX: ++41 / 61 / 271 76 66

For orders originating in the USA of Canada:

Birkhäuser
44 Hartz Way
Secaucus, NJ 07096-2491 / USA

Birkhäuser Verlag AG
Basel · Boston · Berlin

Prices are subject to change without notice. 12/92

Monte Verità

Centro Stefano Franscini
Ascona – ETH Zürich

Edited by
Konrad Osterwalder, ETH Zürich, Switzerland

The conference center "Stefano Franscini", situated on the Monte Verità above Ascona, was opened in summer 1989, and is under the auspices of the Eidgenössische Technische Hochschule Zürich. International research groups, invited by lecturers of Swiss universities, are given the opportunity to hold seminars on aspects of their respective fields. The meetings cover a great variety of topics and are carefully screened by an international advisory board. Some of the most interesting reports are now being published in our new interdisciplinary series **Monte Verità** and provide excellent accounts of the actual state of the respective research areas. Each volume is unique in that it places particular emphasis on making the connections between the individual contributions explicit and their long-term importance clear.

N. Setter / E.L. Colla
EPFL
Lausanne, Switzerland **(Eds)**

Ferroelectric Ceramics
Tutorial reviews of materials, theory, processing, and applications

1992. 392 pages. Hardcover
sFr. 98.– / DM 118.–
ISBN 3-7643-2838-X

One of the fascinating aspects of the field of ferroelectric ceramics is its interdisciplinary nature. It was the purpose of the summer school on ferroelectric ceramics at the Centro Stefano Franscini in Ascona, Switzerland, in September 1991 to help to build bridges between people from the different disciplines and to draw for them, in the form of tutorial lectures, some of the different facets of ferroelectrics. The book is a written version of this summer school and represents a cross section of basic topics of current interest. Materials are presented (L.E. Cross) from the point of view of the user. The important topic of ferroelectric domains and domain walls is addressed (J. Fousek and H. Schmid). In the part devoted to theory, three subjects of current interest are covered: phase transition in thin films (D.R. Tilley), weak ferroelectrics and dielectric losses (A.K. Tagantsev). The fast growing field of ferroelectric thin films is presented through reviews of theory, processing of films (K. Sreenivas) and thin film devices (J.F. Scott). Multilayer ceramic processing is believed to be a key technology for ferroelectric components (A. Bell). Degradation and reliability problems are discussed (R. Waser). Various recent applications are reviewed (W. Wersing), and a separate article addresses the important area of multylayer piezoceramics (S. Takahashi). Finally, the use of ferroelectric ceramics as smart materials is presented (R.E. Newnham).

Please order through your bookseller or write to:

Birkhäuser Verlag AG
P.O. Box 133
CH-4010 Basel / Switzerland
FAX: ++41 / 61 / 271 76 66

For orders originating in the USA of Canada:

Birkhäuser
44 Hartz Way
Secaucus, NJ 07096-2491 / USA

Birkhäuser

Birkhäuser Verlag AG
Basel · Boston · Berlin

Prices are subject to change without notice. 12/92